UNDERSTANDING
DNA
The Molecule & How It Works

Third Edition

From reviews of earlier editions

A systematic and comprehensive analysis of the structure of DNA from a wonderfully fresh perspective. The book is a systematic effort to understand this fascinating molecule from the inside out, building from the first, and simplest, principles .... I recommend it very highly.
Trends in Genetics

We see DNA structures so often that it is often taken for granted that the molecule should not be anything but an aesthetically appealing, spiraling helix. But why should it assume such a nice structure? The book offers an absolutely delightful answer to this and other similarly mischievous questions. 'Understanding DNA' is a great book that will surely prove to be a valuable teaching tool.
The Biochemist

Among the strengths of the book are the clarity of the explanations of some quite difficult concepts and the novel way in which certain ideas are treated, perhaps causing the reader to think again about certain aspects of DNA structure. I enjoyed reading this book and would encourage colleagues working in the general area of DNA research to read it.
Heredity

Stylish ... beautifully crafted, with a logical step-by-step approach to the subject. A book from which the advanced undergraduate will benefit, and which will also generate a refreshing perspective for experts.
Nature

Authoritative and lucid.
Aaron Klug

UNDERSTANDING DNA

The Molecule & How It Works

Third Edition

by

Chris R. Calladine
Department of Engineering
University of Cambridge, Cambridge, UK

Horace R. Drew
CSIRO Division of Molecular Science
Sydney Laboratory, Australia

Ben F. Luisi
Department of Biochemistry
University of Cambridge, Cambridge, UK

Andrew A. Travers
Medical Research Council Laboratory of Molecular Biology
Cambridge, UK

ELSEVIER
ACADEMIC
PRESS

Amsterdam   Boston   Heidelberg   London   New York   Oxford   Paris
San Diego   San Francisco   Singapore   Sydney   Tokyo

First Edition 1992
Second Edition 1997
Third Edition 2004

Elsevier Academic Press
525 B Street, Suite 1900, San Diego, California 92101-4495, USA
http://www.elsevier.com

Elsevier Academic Press
84 Theobald's Road, London WC1X 8RR, UK
http://www.elsevier.com

British Library Cataloguing in Publication Data
A catalogue record for this book is available from the British
Library

Library of Congress Cataloging-in-Publication Data
A catalog record for this title is available from the Library of
Congress

ISBN 0-12-155089-3

Typeset by Charon Tec Pvt Ltd, Chennai, India
Printed and bound in Italy

04 05 06 07 08 9 8 7 6 5 4 3 2 1

The cover picture shows a complex between a protein called 'HMG-D' from fly chromosomes, and a particular sequence of DNA to which it binds strongly.

The two strands of double-helical DNA are shown in white and yellow respectively, while the protein is shown with less detail in red.

The strongly curved and untwisted structure of DNA in this complex illustrates our modern understanding of the molecule's biological action, in terms of its three-dimensional structure, which may be recognized and bound specifically by a regulatory protein.

Thus the DNA structure itself contains important information, in addition to the well-known one-dimensional Genetic Code written in the sequence of bases A, T, C and G.

Contents

About the authors

Chris Calladine is Emeritus Professor of Structural Mechanics at the University of Cambridge. In addition to researching many aspects of structural engineering, he has applied the methods of structural mechanics to the study of bacterial flagella, DNA and proteins.

Horace Drew solved several of the first DNA crystal X-ray structures with Richard Dickerson at Caltech, and subsequently spent 5 years researching DNA and chromosome structures with Aaron Klug at the MRC Laboratory of Molecular Biology in Cambridge, England. He now lives in Australia and is a Principal Research Scientist at CSIRO Molecular Science, Sydney Laboratory.

Ben Luisi studied hemoglobin structure with Max Perutz in Cambridge, and protein–DNA interactions with Paul Sigler at Yale University. He is a Wellcome Trust Senior Fellow in the Department of Biochemistry, University of Cambridge.

Andrew Travers is a staff scientist at the MRC Laboratory of Molecular Biology in Cambridge, England. He has studied transcriptional control in bacteria and flies, the wrapping of DNA in nucleosomes, and the role of HMG proteins in cells.

Preface

We also now appreciate that molecular biology is not a trivial aspect of biological systems. It is at the heart of the matter. Almost all aspects of life are engineered at the molecular level, and without understanding molecules we can only have a very sketchy understanding of life itself. All approaches at a higher level are suspect until confirmed at the molecular level.

Francis Crick, *What Mad Pursuit*, 1988

This is a book about DNA, the most central substance in the workings of all life on Earth. It is a book about the way in which DNA works at a molecular level. We have used the title *Understanding DNA…* because our subject has now reached the stage where many aspects of it are well enough understood for us to be able to give a clear and uncluttered presentation of the main ideas. But we shall not disguise the fact that there is still a great deal which is not known or understood.

The book can be read at two different levels. First, it can be taken as an easy-to-read textbook for undergraduate or graduate students of chemistry and biology at university. Second, it may be read by ordinary people who have no prior knowledge of biochemistry, but who want to understand something of the fundamental processes of life. The sort of people we have in mind here are those who have learned something about DNA from popular magazines, newspapers, and TV programs. They know, for example, that DNA contains the 'genes' of classical genetics – those units of inheritance which pass on characteristics such as red hair or a long nose from parent to child, or even crippling diseases such as sickle-cell anemia or thalassemia. They probably also know that DNA is a long molecule, like a computer tape – the tape which tells our bodies how to grow and how to digest food and (perhaps) how to behave. And they may even know, if they are into quiz games and the like, that the initials 'DNA' stand for 'Deoxyribo-Nucleic Acid,' a certain kind of acid found in the cell nucleus, which was first identified over 100 years ago. People like this, who are curious to know more, will be able to learn a lot from this book about how DNA performs its tasks in our bodies at a molecular level.

This third edition of the book comprises 11 chapters. Chapter 1 is a general introduction to molecular biology: it is aimed at the non-specialist reader, and so it may be passed over by a student who already knows some biology. Chapters 2, 3 and 4 give some lessons about various aspects of the molecular structure of DNA, such as why it is helical, and how it can bend around proteins; this is basic material, which is nevertheless not yet available in other textbooks. Chapters 5 and 6 discuss the three-dimensional structure of DNA at a higher level. These chapters include some mathematics and geometry that may be unfamiliar to non-specialists and biology students; but we take care to present the key ideas by means of clear diagrams wherever possible. Chapter 7 gives an overview of the organization of chromosomes, which are large particles that contain both protein and DNA: there the DNA wraps about the protein into several different levels of structure. Chapter 8 discusses the mechanism of 'direct reading' of DNA sequences by proteins: this is an area that has expanded greatly since the first edition appeared in 1992. Chapter 9 explains the various experimental techniques which scientists use to study DNA. Chapter 10 describes the way in which DNA techniques are increasingly being used in medicine; while Chapter 11, which is new to this edition, summarizes the fast-growing area of cytosine methylation and DNA epigenetics. We end with a Postscript on what we have left out, followed by three Appendices on matters too detailed for the main part of the book. At the end of most chapters we give a bibliography of works to which we have referred in the text; and we provide some further reading for the student, and also some pointers to web-based resources. We have substantially updated the reference lists for this third edition. We have also supplied a few exercises at the end of most chapters.

We have made many changes from the two earlier editions apart from those mentioned above. Thus, we have updated the text and figures where necessary, particularly in Chapters 7 to 10 and the Appendices: the two new members of our author team, B.F.L. and A.A.T., have played a major part in these revisions. As in the earlier editions, we have tried to write in plain English, with minimum use of jargon which might confuse the reader.

Understanding DNA: the molecule & how it works should be suitable as a small text to accompany the very large, general textbooks which are now used widely in university biochemistry courses; or else it may be employed as a main text for a course specializing in DNA structure, provided the students have a background in biology and are willing to pursue more detailed readings in the scientific literature, as suggested. Or, of course, it may just be read as a book.

Many friends and colleagues have helped us greatly in various ways in the preparation of this book. We are grateful to Nick Cozzarelli, Mustafa El Hassan, Malcolm Ferguson-Smith, John Finch, Robert Henderson, Ron Hill, Chris Hunter, Maxine McCall, Garth Nicholson, Dinshaw Patel, Tim Richmond, Masashi Suzuki, David Tremethick, Takeshi Urayama and Sue Whytock for providing photographs and diagrams etc. which we have used; to Dick Dickerson for giving us data on X-ray structures; to Aaron Klug, John Melki, Kiyoshi Nagai, Daniela Rhodes, Deidre Scadden, Chris Smith, Jean Thomas and Michael Waring for commenting freely on various drafts of the manuscript of the previous and the present edition; and to the late Julian Wells for encouraging us in the first place to write a book on DNA. We are grateful to Japan Graphers and Kyoritsu Shuppan Co., Ltd (publishers of the Japanese version of our first edition) for the chapter opening icons, and to Elliott Stollar for help with the cover picture. Tessa Picknett has been a constant source of editorial advice and encouragement. The work of Caryn Wilkinson in revising and adding to the manuscript disk, and of Dennis Halls in updating and making more diagrams, has been beyond praise. Lastly we thank our respective wives, Mary, Maxine, Sandra and Carrie for their help of many kinds over the years; and we dedicate this new edition to them with gratitude.

| C.R.C. | H.R.D. | B.F.L. | A.A.T. |
| Cambridge | Sydney | Cambridge | Cambridge |

CHAPTER 1

An Introduction to Molecular Biology for Non-Scientists

One day two of us were having lunch together at a Cambridge College. We got into a conversation with one of our neighbors at the table, who was a senior historian. After a while he asked us what we did, and we explained that we were scientists, working with the very tiny molecules of biology. Then he said, 'I don't see how you do it'.

'Do what?'

'Work all the time with things that you can't see'.

You see, even people of great intelligence and learning, who spend their lives gathering evidence and pondering it deeply, nevertheless think in ways very different from those of modern biologists; it is hard for them to imagine what atoms and molecules look like. How hard will it be, then, for the beginning student to do the same?

For this reason it is necessary for us to start this book by explaining carefully about the *sizes* of things. A single DNA molecule is too small to be seen by eye. But if you have a big clump of lots of DNA molecules together, then the substance becomes visible, and appears as a clean, white, stringy, and viscous mass, somewhat like molasses sugar. Yet, you *can* see single DNA molecules by using special equipment involving X-rays, or an electron microscope, or an atomic force microscope; and we shall show some pictures of individual DNA molecules later in the chapter.

But first, in order to gain an intuitive feeling for the microscopic world, let us compare the size of DNA to the sizes of things in general, and especially to the sizes of other things that are too small to be seen by the naked eye. Figure 1.1 shows a scale of typical sizes. It is a logarithmic scale, and each division represents a factor of 10. The scale (on the left) covers 10 orders of magnitude from 1 m down to 0.000 000 000 1 or 10^{-10} m. Near the top we have the largest objects

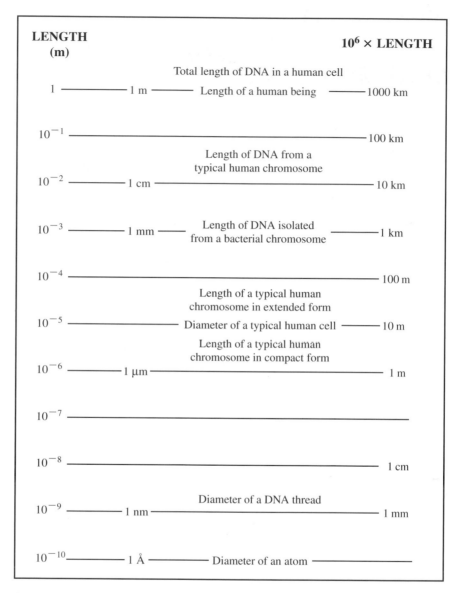

Figure 1.1 The relative lengths of things on a microscopic scale.

that we shall be thinking about: human beings are roughly 1 m long, as an order of magnitude. At the bottom of the scale are the smallest objects that we shall be concerned with: atoms, which are typically of diameter 10^{-10} m (or 1 Ångstrom unit, Å). Exactly halfway between these two extremes, on the present kind of scale, we have the diameter of a typical human cell at about 10^{-5} m or 10 μm.

Some things are larger than a cell on our scale, while others are smaller. It is perhaps surprising that the length of DNA isolated in

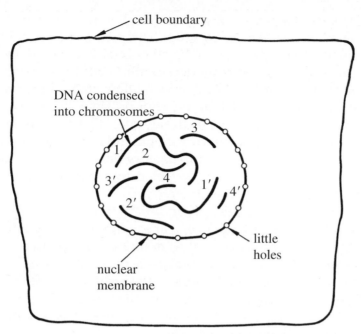

cell boundary

DNA condensed
into chromosomes

little
holes

nuclear
membrane

Figure 1.2 Schematic picture of a typical cell from higher organisms. The chromosomes, which contain a mixture of protein and DNA, come in homologous pairs. This picture shows a hypothetical cell with four such pairs that can be distinguished by their sizes. They are labeled 1,1'; 2,2'; etc.

pure, thread-like form from a single human chromosome[1] (3 cm, or 3×10^{-2} m) or from a single bacterium (1 mm, or 10^{-3} m) can be so much longer than the cell from which it came (10^{-5} m). But this illustrates a very important point: the DNA is compacted in length by a factor of as much as 10 000 when it is embedded in a living cell. DNA is a very narrow thread of diameter of just 2×10^{-9} m, and although the DNA from a single human cell has a total length of 2 m, it could conceivably be compacted into a tight ball, like a ball of string, of diameter 2×10^{-6} m. But Nature chooses to pack the DNA into a form somewhat less dense than this, at a length of about 3×10^{-6} m for a single, compact chromosome. After all, if DNA was packed too tightly into a cell, then the information along its length, known as 'genes'[1] (to be discussed below), would probably become inaccessible.

Our bodies are made from billions of individual cells, and DNA is the control center of each and every cell. This DNA is something like a computer tape that stores many programs for a large computer to run. It is present in each cell in the form of a number of chromosomes. Chromosomes are arranged in pairs (see Fig. 1.2), and the

1 See note on p.15

two members of a pair are nearly identical or 'homologous'[1] copies of one another, just like the back-up disks on a computer: one copy saves the functional program if the other copy becomes defective.

The mass of DNA is surrounded in most cells by a strong membrane with tiny, selective holes, which allow some things to go in and out, but keep others either inside or outside. Important chemical molecules go in and out of these holes, like memos from the main office of a factory to its workshops; and indeed the individual cell is in many ways like an entire factory on a very tiny scale. The space in the cell which is not occupied by DNA and the various sorts of machinery is filled with water.

On the right of Fig. 1.1 is a second scale of lengths which would apply if we were to enlarge every linear dimension by a factor of $1\,000\,000$ or 10^6. It can be useful sometimes to make an imaginary enlargement of this sort. When we do so, the relative sizes and proportions of objects remain the same, of course. Note that the length of DNA from a typical chromosome on this expanded scale is about 30 km, while its diameter is just 2 mm. Very few objects in the physical world are so long and so narrow.

Now that we have gained a general idea of the relative sizes of things, and of the importance of DNA in the control of a living cell, let us see what DNA and chromosomes really look like.

Human chromosomes become compact and squat when cells are about to divide, and they can be seen easily by the use of a light microscope. The human chromosomes shown in Fig. 1.3 have been sorted and arranged by size into pairs. Each chromosome, as in Fig. 1.2, has duplicated itself in preparation for cell division, and has then reduced its length 10-fold, so that the duplicate copies can separate from one another without tangling as the cell divides. One half of each X-shaped duplicate chromosome will go to each new cell.

Many more chromosomes are shown here than in the simple, schematic drawing of Fig. 1.2. Human cells contain 46 chromosomes in 22 homologous pairs (numbered 1 to 22) plus the non-homologous X and Y chromosomes that determine sex. All animals and plants have chromosomes that look like these, but in different numbers: for example, a fruit fly has eight chromosomes, in three homologous pairs plus X and Y. Spreads of chromosomes, such as those shown in Fig. 1.3, are very useful for medical purposes; for example, to check the health of an unborn child while still in the womb.

At other stages in the life of a cell, far from cell division, the chromosomes are generally more extended and less condensed, and so they cannot be seen by use of a light microscope. That is why we had to draw the single chromosomes of Fig. 1.2 in schematic form, because no true pictures of such chromosomes exist. Nevertheless,

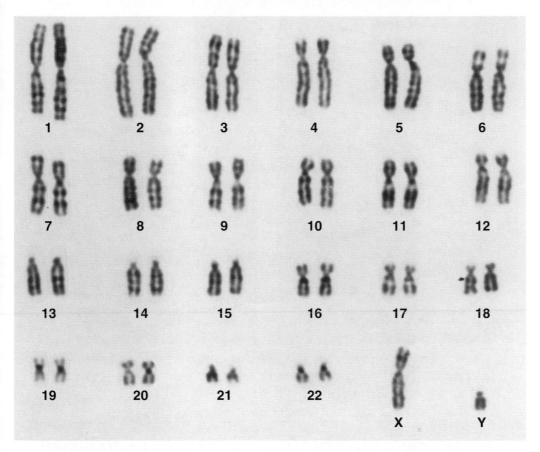

Figure 1.3 Photographs of human chromosomes in duplicate, as isolated just before cell division (at metaphase[1]) and then sorted by length into pairs. Each number identifies the two chromosomes of a homologous pair. X and Y are non-homologous chromosomes that determine a person's sex as female (XX) or male (XY). The two duplicate copies of any individual chromosome form an 'X' shape because they have not yet separated entirely. Scale: chromosomes 3 are approximately 10^{-5} m or $10\,\mu$m long. Courtesy of Malcolm Ferguson-Smith.

in a few particular tissues of the fruit fly, these single-copy, extended chromosomes happen to duplicate themselves about 1000 times over, without becoming compact. They eventually contain 1000 identical DNA molecules laid side-by-side, in parallel register. These 'poly-tene' chromosomes are just like the ones shown in Fig. 1.2, but they are much wider. Due to their greatly increased size, these mon-strous, amplified fly chromosomes can be seen clearly by the use of a light microscope, as shown in Fig. 1.4 at different magnifications.

In Fig. 1.4(a) the extended, DNA-containing chromosomes can be seen along with the tight membrane in which they are wrapped; the entire assembly is known as the 'nucleus'.[1] Here the chromosomes

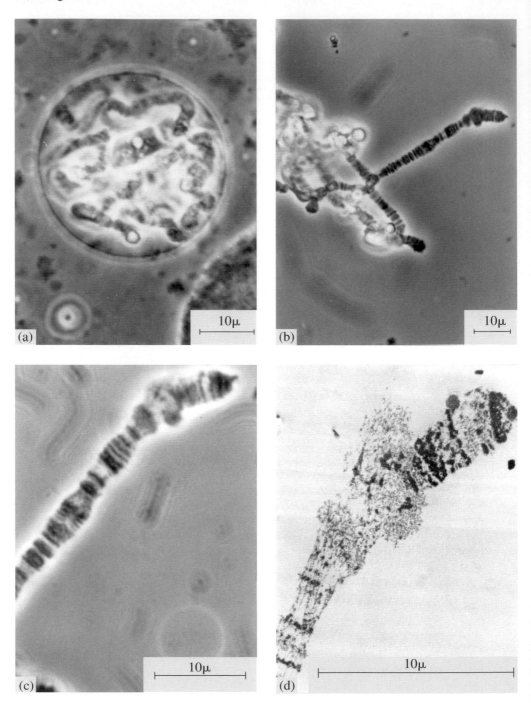

Figure 1.4 Polytene chromosomes from the salivary gland of a fruit fly, as seen in the electron microscope. Length scales: 100 000 Å or 10 μm. Courtesy of Ron Hill and Margaret Mott.

protein spool   two loops

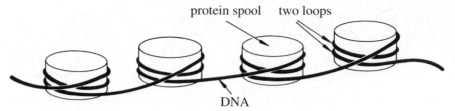

DNA

Figure 1.5 A DNA thread making two coils (left-handed) around each of a series of protein spools, as might be seen in the chromosomes of Figs 1.2, 1.3 and 1.4 if the magnification were higher. Each DNA–protein spool is about 100 Å (or 10^{-8} m) across.

are like little worms or eels surrounded by water and trapped inside a bubble. When we break open the bubble and look more closely, as in Fig. 1.4(b), (c) and (d), we can see that each chromosome is divided along its length into many clear striations of dark and light, which are known as 'bands' and 'interbands'. The bands are dense clumps of protein[1] plus DNA, whereas the interbands are sparse regions of low density. No one knows why the variations in dark and light are so sharp, and indeed so reproducible from one fly to the next. Presumably, these divisions are marked off by certain patterns in the molecules along the DNA thread. But it makes good sense in biological terms to divide a chromosome along its length: after all, computer tapes store information in the same sort of way as discrete files, each with a beginning and an end, along the length of a tape.

At a finer level, more is known. Each thread of the band and interband material in fruit-fly chromosomes (or indeed in human chromosomes) is composed of a well-defined mixture of protein and DNA. This protein is of a special kind that makes 'spools', and the DNA wraps twice around each spool, as shown schematically in Fig. 1.5, into a series of double loops. The wrapping of DNA twice about each spool reduces its overall length by a factor of about 6.

When we remove the protein spools, we are left with a very long, string-like DNA molecule. The DNA from the longest individual human chromosome, if it were enlarged by a factor of 10^6, so that it became the width of ordinary kite string, would extend for about 100 km. Imagine sitting in a train traveling from Cambridge to London, or from Los Angeles to San Diego, and looking out of the window for the whole trip at a single DNA molecule and watching the genes go by!

The basic form of this immensely long DNA fiber is shown in Fig. 1.6. It consists of two strands which coil around each other to make a 'double helix'. The term 'helix' is just another word for 'screw' or 'spiral'. The sense of wrapping of these two strands is

Figure 1.6 DNA double helix showing base-pairs as short vertical lines. Now the DNA thread of Fig. 1.5 is shown in its true form, as a double spiral containing two chains of DNA. The diameter of the spiral or helix is about 20 Å.

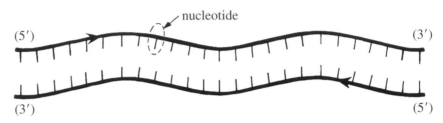

Figure 1.7 The two strands of DNA separated, showing a nucleotide. Each nucleotide is about 6 Å wide.

usually clockwise as you go forward, or right handed – the same sense as an ordinary corkscrew.

If we uncoil the two strands, as shown in Fig. 1.7, then each strand may be seen to consist of a series of units called 'nucleotides'. These are linked to one another with a certain 'directionality', known technically as '5-prime to 3-prime', in a head-to-tail sense that we shall explain below. The two strands run in opposite directions, as shown by the labels 5′ and 3′, and by the arrows.

Each nucleotide is made of about 20 atoms, such as carbon, nitrogen, and oxygen. These atoms can again be grouped into smaller parts which are connected in a particular way. The three parts of a nucleotide are its sugar, phosphate, and base; in the diagram of Fig. 1.8, they are labeled S, P, and B, respectively. For present purposes, we may draw the sugar as a five-sided ring, because the atoms in a sugar join to form such a ring. We have taken the liberty of drawing the ring in the form of an arrow, in order to indicate its directionality; and for the same reason we have put the letters P, S, and B, so that they read in this same direction from left to right. If we took out the central nucleotide in Fig. 1.8, rotated it through 180° about a vertical axis and tried to re-connect it, we should find that it would not link up.

We can draw the base as a rectangle because the atoms there join to form a flat, rigid shape. There are usually four different kinds of nucleotides in DNA, which share the same sugars and phosphates but have different bases attached to the chain. These four different

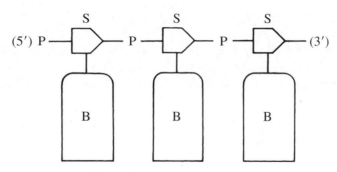

Figure 1.8 Three nucleotides, showing their sugar (S), phosphate (P), and base (B) components.

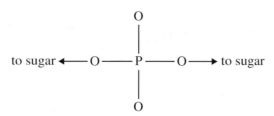

Figure 1.9 The phosphate group as part of a sugar–phosphate chain. Each atom is 1–2 Å wide; atom types are phosphorus (P) and oxygen (O).

bases form pairwise interactions that join the two strands of DNA together weakly (see Chapter 2).

The three parts of a nucleotide can also be studied in isolation, as collections of just a few atoms. For example, a phosphate contains one phosphorus and four oxygen atoms, as shown in Fig. 1.9. The atoms themselves are made of protons, neutrons, and electrons. They are connected together by the sharing of electrons; and the various kinds of chemical bond that connect atoms to each other are all aspects of this. We could go on to discuss atoms within the realm of subatomic physics, but there would be no point in this, because our knowledge of living things begins with chemistry (or biochemistry) and extends into biology. So we can stop our description of DNA with the relatively simple picture of atoms shown in Fig. 1.9, without losing anything.

If you have not studied biology before, you may be puzzled that we have put so much emphasis on the *cell*. Would it not be more sensible to start with the parts of the body, such as limbs, eyes, and lungs? The answer to this is that all of these various organs and tissues, etc. are built up from cells by essentially the same process.

You are probably familiar with the way in which a house is built. The various components – bricks, cement, tiles, timber planks, window frames, etc. – are first delivered to the building site and then assembled in accordance with the architect's plan or 'blueprint'.

This plan is a sheet of paper on which are drawings of the finished house taken from several points of view. There may well be only one copy of this plan, and it may be kept in the pocket of the foreman builder and taken out and consulted from time to time, as required. Such a scheme of construction is typical for non-living things like houses, motor cars, and gadgets.

In contrast, the scheme of construction for living things – whether they are plants or animals – is different from this in almost every way. Thus, construction of a human begins with a single cell – an egg from the mother which has been fertilized by a sperm from the father. This composite cell contains DNA from both the mother and the father; such DNA contains the complete genetic information for construction of a human being. Growth occurs by the process of cell division: each cell divides into two new cells, and these cells in turn divide, and so on. Just before any cell divides, it duplicates all of its DNA, so that every new cell contains a complete set of DNA, which again contains all the genes of the organism. (We saw some pictures of duplicated chromosomes in Fig. 1.3.)

Only a small fraction of all the genes present on this DNA are activated in any given type of cell. Thus, cells which develop into an eye use only the genes which program for the growth of eye-cells. How cells 'know' which kind of organ they belong to is a large and only partly understood area of research, *developmental biology*, which we shall not go into here.

We mention all of this because it may seem, to a non-scientist, to be enormously wasteful for Nature to provide a complete set of DNA in every one of the billions of cells of every animal or plant: would it not be more straightforward and efficient just to have a single copy of the design information, just as in the construction of a house? However, a little thought indicates that the scheme for providing every cell with a complete set of DNA is, in fact, an extremely simple way of providing this necessary information, in all places where it is required – even though, of course, the scheme requires a vast amount of repetitive copying and duplicating of DNA. The machinery for duplicating DNA is accurate enough for the entire scheme to work well, and the double-helical structure of DNA is extremely convenient, as we shall see in Chapter 2, for the purposes of duplication.

We are now ready to clarify a further point: how does DNA carry the information necessary to run the activities of a cell as a factory, as its control-center or main office? The most basic way in which DNA runs the activities of a cell is to specify the composition and structure of protein molecules. Proteins come in a wide range of shapes and sizes, and play a wide range of roles in the life of a cell. Some proteins are strong and rigid, and form the building-components for muscles,

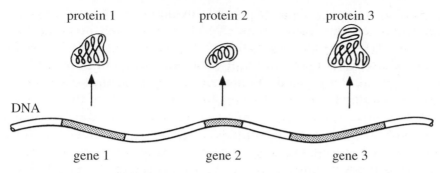

Figure 1.10 Genes in the DNA code for proteins. The DNA thread shown here represents the double helix depicted in Fig. 1.6, but here it is about 100 times longer. Both genes and proteins can vary in size, in direct proportion to one another. Each protein molecule is shown with a schematic chain-like structure.

tendons, and finger-nails. Other proteins, or 'enzymes',[1] catalyze a large number of chemical reactions, such as digestion of food or the synthesis of hormones. Other proteins carry oxygen in the blood, while still others form the protein spools around which DNA wraps in a chromosome. Where do all these proteins come from? How does a cell know which proteins to make?

It turns out that the DNA runs a very definite 'program' every time the cell needs to make a given protein molecule. The program tells the cell exactly what kind of protein to make, and approximately how much of it. This program is the well-known 'gene' of classical genetics; and each DNA molecule contains many genes along its length, as shown in Fig. 1.10. Sitting in our train traveling from Cambridge to London, we could see thousands of them as we looked out of the window watching the DNA go by!

The proteins, although very different from one another in their physical and chemical properties, share a common scheme of construction: they are all long-chain molecules, consisting of a single, unbranched chain that is made from the end-to-end joining of many small units known as 'amino acids'. Proteins are made up from amino acids, just as DNA is made up from nucleotides. For example, the sweetener in Diet Coca-Cola is a very tiny protein made by man from just two amino acids; while hemoglobin in the blood is a much larger protein, made from 600 amino acids. Altogether there are 20 different kinds of amino acid that make proteins; and this wide variety of building blocks allows for the construction of very many different proteins, with their enormous range of physical properties.

A program or gene in the DNA tells the cell in what order to assemble these amino acids and for what length of protein chain. The order of amino acids in a protein is set by the order of nucleotides in the corresponding piece of DNA. Because there are 20 possible amino acids, and only four possible nucleotides, the cell must use more than one nucleotide in the DNA to specify each amino acid in a protein. The universal rule is that *three nucleotides* specify *one* amino acid. The length of protein chain varies with the length of the DNA 'program', and is typically from about 100 to 1000 amino acids.

The cellular machinery that enables a certain set of three nucleotides in the DNA to be associated with only one of the 20 possible amino acids is extremely complex. Likewise, the conversion of the magnetized stripes on a regular computer tape into printed characters on a page of computer output must be rather complicated. But we do not have to worry about this kind of machinery in order to explain the simple code or cipher by which DNA makes proteins. Usually, this cipher is presented in the form of a table called 'The Genetic Code' (Table 1.1). This same code is used for specifying

Table 1.1 The Genetic Code

1st base	2nd base				3rd base
	T	C	A	G	
T	Phe	Ser	Tyr	Cys	T
	Phe	Ser	Tyr	Cys	C
	Leu	Ser	STOP	STOP	A
	Leu	Ser	STOP	Trp	G
C	Leu	Pro	His	Arg	T
	Leu	Pro	His	Arg	C
	Leu	Pro	Gln	Arg	A
	Leu	Pro	Gln	Arg	G
A	Ile	Thr	Asn	Ser	T
	Ile	Thr	Asn	Ser	C
	Ile	Thr	Lys	Arg	A
	Met	Thr	Lys	Arg	G
G	Val	Ala	Asp	Gly	T
	Val	Ala	Asp	Gly	C
	Val	Ala	Glu	Gly	A
	Val	Ala	Glu	Gly	G

Any series of three bases (or nucleotides) in the DNA prescribes for an amino acid in the protein chain, or gives a 'stop transcribing' signal. The bases are always read from left to right. The chain usually starts with ATG or methionine (Met). Abbreviations used: A, adenine; G, guanine; C, cytosine; T, thymine (or U, uracil in RNA). Ala, alanine; Arg, arginine; Asn, asparagine; Asp, aspartic acid; Cys, cysteine; Gln, glutamine; Glu, glutamic acid; Gly, glycine; His, histidine; Ile, isoleucine; Leu, leucine; Lys, lysine; Met, methionine; Phe, phenylalanine; Pro, proline; Ser, serine; Thr, threonine; Trp, tryptophan; Tyr, tyrosine; Val, valine.

proteins in all living things, whether they are bacteria, plants, or animals, with only a few minor exceptions.

The four different nucleotides in DNA are called adenine, guanine, cytosine, and thymine, or simply A, G, C and T. The 20 possible amino acids in a protein have names such as 'methionine', usually abbreviated to 'met' or 'M'. As shown in Table 1.1, each possible set of three nucleotides in the DNA specifies one amino acid. For example, 'TTT' specifies phenylalanine, while 'AAA' calls for lysine, and 'GCT' gives alanine. In each case, the letters are read from left to right. There are $4 \times 4 \times 4 = 64$ combinations of DNA triplet, and often two of these triplets, such as 'AAG' and 'AAA', specify the same amino acid. Thus, a series of nucleotides, such as 'TTTAAAAAGGCT', specifies a portion of protein with amino-acid sequence 'Phe–Lys–Lys–Ala', as shown in Fig. 1.11. Certain triplets along the length of DNA also do the special work of telling the protein chain where to start and stop.

Finally, the completed protein chain usually folds by itself into a precisely determined shape, which depends on the exact arrangement of the amino acids in its chain. A correct three-dimensional shape is essential for the physical or chemical activity of a protein.

You should not think that all DNA does is to make proteins like a piece of computer tape. In fact, only a small fraction of the long DNA molecule in a chromosome – about 1 percent in humans – contains programs to make specific proteins. The vast majority of DNA in our bodies does things that we do not presently understand. There is plenty of room here for people to make new discoveries.

We said above that we can safely ignore the complex machinery which the cell uses in order to make proteins. This is true for the most part, but we do need to know about a few features of the process. We described above how the DNA-containing chromosomes are surrounded by a membrane in the nucleus of every cell

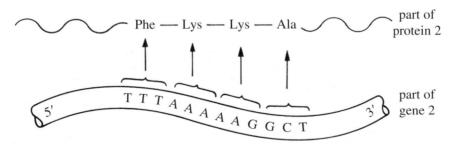

Figure 1.11 Part of the gene codes for part of the protein chain. Each set of three nucleotides (or bases) in the DNA specifies one amino acid, according to the scheme of Table 1.1.

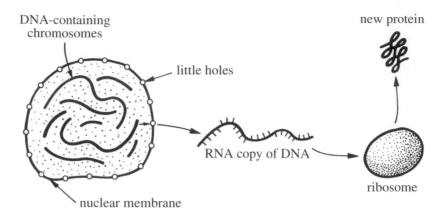

Figure 1.12 From DNA to RNA to protein. The DNA is copied into 'messenger-RNA', which then travels outside the nucleus to the ribosome, where it specifies the assembly of some particular protein, according to the series of nucleotides in its chain as in Table 1.1. For simplicity, we have not mentioned here that many genes in higher organisms contain substantial amounts of non-coding DNA, or 'introns', that are cut out at the level of RNA before the RNA copy of DNA leaves the nucleus.

(except in bacteria[1] – and bacteria-like single-cell microbes called archaea[1], where there are no nuclear membranes). It has been discovered that proteins are made in the cellular space *outside* of this membrane. What happens is that the DNA of a particular gene first makes a copy of itself that can pass through the small holes in the membrane; then this copy goes off to the protein-making machinery, which is called a 'ribosome'. This process is shown schematically in Fig. 1.12. These copies of the DNA program are made from a slightly different kind of molecule called 'RNA'. At the level of the picture shown in Fig. 1.8, RNA has an extra oxygen atom on the sugar ring as compared with DNA. A second difference is that RNA lacks a carbon and three associated hydrogen atoms on the thymine (T) base, and so this base is renamed 'uracil' (U). The RNA copy itself is called 'messenger-RNA' because it carries instructions on how to make a particular protein from the chromosome to the ribosome.

Interestingly enough, messenger-RNA makes up only a small fraction of the total RNA in any cell, about 5 percent of the total; and there is also much more RNA than DNA altogether. A lot of RNA is copied from DNA but never used to make protein. These abundant RNA molecules are different from messenger-RNA, and are mainly of two types: 'transfer-RNA' and 'ribosomal-RNA'. Transfer-RNA carries single amino acids from elsewhere in the cell to the ribosome; there the amino acids, while still bound to the transfer-RNA, can pair up with specific triplets of nucleotides on the messenger-RNA chain, and so make a protein in accordance with the cipher of Table 1.1. In other words, transfer-RNA performs the function of the vertical arrows shown in Fig. 1.11, once the DNA has been copied to messenger-RNA. We explain more about this in Chapter 2. Ribosomal-RNA makes up the bulk of the ribosome in association with a few proteins; it helps to align a series of transfer-RNA molecules along a chain of messenger-RNA, so that a series of amino acids can be joined chemically to form a long protein. It seems remarkable that so many steps in the synthesis of protein involve an RNA intermediate. This observation has led many people to speculate that life on Earth began with RNA, rather than with DNA, as the substance of genes.

Now we have presented all of the essential background information that is required for you to understand how DNA works in biology. Other books could be written on RNA or protein, but here we focus exclusively on the role of DNA. We shall work from small to large, starting with the basic chemistry of DNA and ending with the role of DNA in medicine and the new field of 'epigenetics'. Despite much recent progress, the great revelations in biology will no doubt belong to the new twenty-first century; we cannot pretend that our present knowledge is any more than a rough and incomplete foundation on which others can build.

Note

1. The linguistic or historical derivations of words marked thus are given in Appendix 1.

Further Reading

Alberts, B.M., Johnson, A., Lewis, J., Raff, M.C., Roberts, K., and Walter, P. (2002) *Molecular Biology of the Cell* (4th Edn). Garland, New York. A good general reference volume.

Chargaff, E. (1963) *Essays on Nucleic Acids*. Elsevier, New York. An important historical document describing the state of knowledge about DNA in the late 1950s.

Crick, F.H.C. (1963) The recent excitement in the coding problem. *Progress in Nucleic Acid Research* **1**, 163–217. A good scientific review of how the Genetic Code was discovered.

Crick, F.H.C. (1988) *What Mad Pursuit*. Weidenfeld & Nicolson, London. An anecdotal history of the discovery of the structure of DNA and of the Genetic Code. Also, the source of the quotation used in the Preface.

Judson, H.F. (1979) *The Eighth Day of Creation*. Simon & Schuster, New York. A very thorough history of how molecular biology became a separate branch of science.

McCarty, M. (1994) A retrospective look at the discovery of the genetic role of DNA. *The FASEB Journal* **8**, 889–90. A first-hand account of how it was discovered that DNA is the substance of genes.

Yusopova, G.Zh., Yusupov, M.M., Cate, J.H., and Noller, H.F. (2001) Path of messenger RNA through the ribosome. *Cell* **106**, 233–41. A high-resolution picture by X-ray methods of messenger-RNA bound within a protein-synthesis factory or ribosome.

Zhimulev, I.F. (1996) Morphology and structure of polytene chromosomes. *Advances in Genetics* **34**, 1–490. A comprehensive summary of information about giant 'polytene' chromosomes of flies.

Exercises

1.1   Every human cell, in the non-dividing state, contains a total of about 6×10^9 base-pairs of DNA. The diameter of a typical human cell is $10 \, \mu m = 10^5 \, \text{Å}$.

a   By treating the DNA as a cylinder of length $3.3 \, \text{Å}$ per base-pair, calculate the total length of DNA in any cell. Compare this to the diameter of the cell.

b   By treating the DNA as a cylinder of radius $10 \, \text{Å}$, calculate the total volume of DNA in any cell. Compare this to the total volume of the cell, if it is assumed to be spherical. (Volume of cylinder $= \pi r^2 l$, where r is the radius and l the length; volume of sphere $= \frac{4}{3}\pi R^3$, where R is the radius.)

c   Consider a typical chromosome, which contains 1/46 of the total DNA of the cell, on average. Find the total length of the DNA in this chromosome, and then the diameter of a solid sphere into which that volume of DNA could be compacted, in principle. Compare the diameter of this compact sphere with the mean length of an actual, metaphase chromosome, given that the DNA length–compaction ratio for such a chromosome is about 10 000.

1.2a   Use Table 1.1 to give the sequence of amino acids in a protein chain which is coded by the following base sequence:

GCCAAGCAACTCATTCAAGGT
1 2 3

Start reading at base 1.

b   Now repeat the process by beginning to read at base 2; and then again by beginning to read at base 3. (Observe that the amino-acid sequence of a protein chain depends critically on which 'reading frame' is used in the DNA.)

1.3a An extra base G is now inserted between bases C and T at positions 10 and 11 of the DNA sequence given in Exercise 1.2. Starting at base 1, read out the new sequence of amino acids.

 b The sequence given in Exercise 1.2 is now altered instead by the deletion of C at position 10. Starting at base 1, read out the sequence of amino acids. (These are both known as 'frame-shift' mutations.)

1.4a Translate the following DNA base sequence into a sequence of amino acids for a protein molecule:

ACGCTATGTCACATGGTACCTAACGTAT

On this occasion, do not begin reading at position 1, but rather in accordance with the true, biologically used start-scheme described in Table 1.1; and do not read beyond a STOP triplet.

 b Search the base sequence given below for potential STOP triplets. What amino-acid sequence will be assembled, starting as in a?

GCTCATGGTCATTCGTAACAGTTAGGCCATGACCG

CHAPTER 2

Why a Helix?

It is crucially important, not only in biology but in all fields of science, to understand the inner workings of Nature as well as its external form. For example, consider the funny-looking object shown from two perspectives in Fig. 2.1. It has five different colors, but what *is* it? How would you go about making such an object? A poor scientist would study the object superficially, and give a name to each particular feature, such as the intersection of five edges; he or she would be deeply concerned with measuring the angles between edges on the surface of the object, in order to look for some sort of pattern. But a more perceptive scientist would stop and think: what is the internal structure of the thing?

Now look at the object again, but this time focus on only one color, say yellow: behold, the yellow parts form a cube in space, tilted somewhat on its side. Each of the other four colors also describes a cube. In fact, this toy is called 'The Compound of Five Cubes'.

Spotting the five distinct cubes – it was helpful of the makers to use five different colors – is the first clue to understanding the internal structure of the object. This clue leads us to think about matters of geometrical symmetry; and if we were to follow this line of mathematical thinking, we would eventually understand why there are five cubes rather than, say, six or four. Indeed, we might then be in a position to make other, kindred objects by using the same underlying, structural principles.

This example illustrates an important point in science. By perceiving the internal structure of an object, you learn much about Nature; but by studying only its external form, especially in great detail and with high concern for nomenclature, you learn relatively little that is worthwhile.

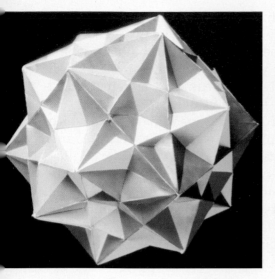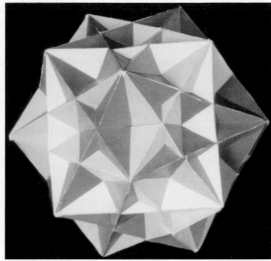

Figure 2.1 A puzzling object in two views. (From a cut-out book by E. Jenkins and M. Bear, (1985). Tarquin Publications, Diss, UK.)

Here we are going to describe the internal structure of DNA. The internal structure is the key to understanding how DNA works. In writing this book, we have tried to make everything as easy as spotting the structure of the five-fold cube. We may sometimes have to do a little mathematics, because mathematics tightens up the various relationships that we shall discover. There are a lot of things to learn about DNA, but we have tried to select the most important and general points. First, we shall learn why DNA makes a *double helix*, as we saw in Chapter 1, instead of some other shape; and then we shall learn how the double helix is held together at its core.

You should recall from Chapter 1 that DNA is made of three things: phosphates, sugars, and bases, and that these components are linked together chemically in a particular way. Now the phosphates are very soluble in water. If farmers use too much phosphate on their land as fertilizer, some of the phosphate runs off into the nearby ditches and rivers, causing algae to grow wildly and so killing the fish. Phosphates are very soluble in water.

Sugars are also soluble in water. There would be no point in putting sugar into coffee or tea if it would not dissolve: if it just sat at the bottom of the cup, it would not flavor the liquid. So sugars, as well as phosphates, dissolve in water.

But what about the bases? People have no intuitive feel for the four DNA (or RNA) bases: guanine, adenine, cytosine, and thymine (or uracil), because they do not recognize them in everyday life. It is easy to find out whether bases dissolve in water; just put some into a test tube, add water, and watch. You can buy adenine and uracil from any chemical company (they cost little) and use about 50 mg of each. When you do this experiment you find that neither adenine nor uracil dissolves in ordinary water. Yet further simple experiments show that adenine dissolves in weak acid and that uracil dissolves in ammonia, an alkali – but not vice versa.

Now we mentioned in Chapter 1 that the space in our cells which is not occupied by important components such as DNA, RNA, and enzymes, is filled with water. This water is not at all acidic or alkaline. To use a technical term, it is at 'neutral pH'. It follows, therefore, that adenine and uracil (and indeed all of the bases listed above) will be practically insoluble in the aqueous environment of our bodies. Although we are not familiar with bases A, G, C, T, and U in everyday life, we are very familiar with other substances that will not dissolve in water, such as grease and oil. These are all 'water hating' or 'hydrophobic' substances.

The insolubility of bases in water does not really pose a problem for the cell, because these bases do become soluble in water once they are attached to a sugar and a phosphate to form a 'nucleotide', which is the building block of DNA or RNA (see Fig. 1.8). But this insolubility does place strong constraints on the overall conformation of any large DNA or RNA molecule in solution. For such a molecule to be stable in water at neutral pH, the bases will have to tuck themselves into the very center of some folded structure, so as to avoid the water; while the sugars and phosphates, both of which are soluble in water, will have to be on the outside. In fact, this is just what happens. If we take some measurements of the known structure of a DNA sugar–phosphate chain (determined by X-ray analysis, as described in Chapter 9), we see right away how DNA forms a spiral or helix on account of the low solubility in water of the bases. We can even do a first-order calculation to determine what kind of helix it makes.

We know from elementary chemistry that the distance between adjacent sugars or phosphates in the DNA chain is 6 Å (or 0.6 nm) in the usual case (Fig. 2.2). It cannot become much longer than 6.5 Å or shorter than 5.5 Å, or else the strong bonds between the atoms will strain too much. The thickness of the flat part of a DNA base is 3.3 Å, and this distance cannot change much either, because the bases are chemically rigid with strong, inflexible bonds between the atoms. This leaves us with a 'hole' of 2.7 Å between the bases, which some greasy object (and definitely not water) would have to fill, otherwise we will leave a vacuum. In brief, the bases are attached to a sugar–phosphate chain that is twice as long as the thickness of the bases themselves.

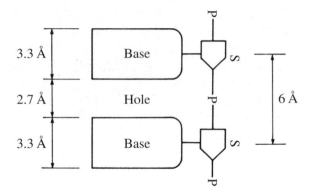

Figure 2.2 Two nucleotides in schematic form, showing key dimensions.

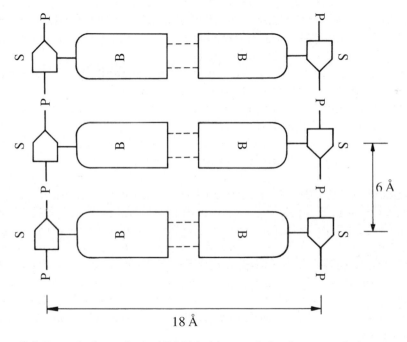

Figure 2.3 Part of a hypothetical DNA ladder, made by the cross-chain pairing of bases.

How can we tuck these insoluble bases into the center of a DNA molecule where they can avoid water, and at the same time be rid of the 'holes'? The most obvious form of DNA, as an assembly of two chains with the bases on the inside, would be a *ladder*. A segment of such a hypothetical ladder is drawn in Fig. 2.3. (A single chain of DNA could also fold back on itself to make a ladder, but this structure would be so similar to the one shown that it need not be considered separately.) In our ladder model, the two bases from opposite strands, joined in ways which we describe below, hold the phosphates 18 Å apart. Also, the two chains run in opposite directions, for reasons that

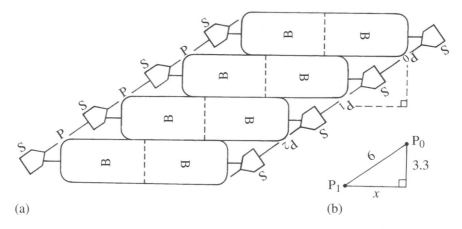

Figure 2.4 A skewed ladder, with no gaps between the paired bases. The plane geometry of this ladder is shown in part (b).

we discuss below. All seems satisfactory in that respect, but we are still left with many 'holes' between the bases within each strand. What can we do to remove them?

As shown in Fig. 2.4(a), one solution might be to skew the ladder strongly to one side. Once the sugar–phosphate chains tilt to an angle of about 30° from the horizontal, then the holes disappear, as shown. The key to the geometry is shown in Fig. 2.4(b), where we see a right-angled triangle that accurately describes the structure. The phosphates are connected along the hypotenuse of this triangle at a distance of 6 Å, while the bases proceed upward by 3.3 Å along the right-hand side. The sideways-shift of the ladder per base-step, left to right across the paper, is therefore given by

$$x = \sqrt{(6^2 - 3.3^2)} = 5.0 \,\text{Å}.$$

Our skew-ladder seems perfectly satisfactory as a way of closing up the bases so as to exclude water. But it is not quite the same as the conformation which DNA adopts in Nature. As we have said in Chapter 1, DNA takes the form of a spiral or helix. In fact, the DNA double helix is nothing more than a highly twisted ladder. It provides another, slightly different way of solving the same problem: how to separate the bases by 3.3 Å while leaving the phosphates 6 Å apart. Figure 2.5(a) shows a simplified view of 2 base-pairs from our skew ladder, while Fig. 2.5(b) shows that the bases can stack onto each other just as well, without gaps, if they twist about an imaginary vertical axis into the shape of a helix. The two chains climb from the horizontal at exactly the same angle as before – about 30° – but now they lie on the surface of a cylinder of diameter 18 Å; and the base-pairs within the helix are now arranged as in the treads of a spiral staircase.

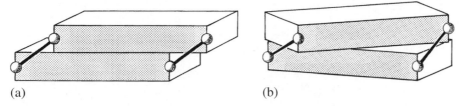

Figure 2.5 (a) Stacking of base-pairs as in the skewed ladder of Fig. 2.4; (b) stacking of base-pairs by means of helical twist.

Why does DNA prefer to form a helix rather than a skew-ladder? Our present model is too crude to answer this question convincingly. When we build a more accurate model which shows individual atoms, we find that the skew-ladder leads to many unacceptably close contacts between neighboring atoms, and so this model has to be abandoned. Nevertheless, the skew-ladder is a useful tool in thinking about the internal structure of DNA, because it has several important geometrical features of the real DNA helix, and yet lies in a plane and is thus easy to visualize.

Now when we go into three dimensions, and consider the shape of a DNA helix, the geometry is almost the same as in our skew-ladder above. We can take a series of the twisted, two-base-pair units shown in Fig. 2.5(b), and stack them on top of each other to get a proper, double-helical model of DNA. Figure 2.6(a) shows such a model schematically. Only the first two base-pairs are shown, but then we show all parts of the sugar–phosphate chains. These chains wrap as spirals around an imaginary cylindrical surface of radius 9 Å, and each sugar ring is represented by a dot. Figure 2.6(b) shows a side view of the cylinder for just one of the two sugar–phosphate chains. Here the phosphates, P_0, P_1, P_2, etc. – counting from the top – are drawn as open circles, and the same lengths of 6.0 Å, 3.3 Å, and 5 Å that were found for our skew-ladder characterize the path of these phosphates through space. Finally, a top view along the vertical axis of the DNA cylinder is shown in Fig. 2.6(c). Again, for the sake of simplicity, only one chain is shown, and the phosphates along it are labeled P_0, P_1, ..., P_{10}. Each successive phosphate in this view lies 3.3 Å further away from us than the one before. The chain is shown with a break between P_{10} and P_0, because P_{11} lies directly behind P_0 in this view: it is 11×3.3 Å $= 36$ Å further away from us, when we look down into the plane of the paper.

Simple geometry enables us to calculate the angle θ (*theta*) by which each phosphate turns relative to its neighboring phosphate along the helix. As shown in Fig. 2.6(c), the distance $x = 5.0$ Å is the base of an isosceles triangle, whose vertex lies at the center

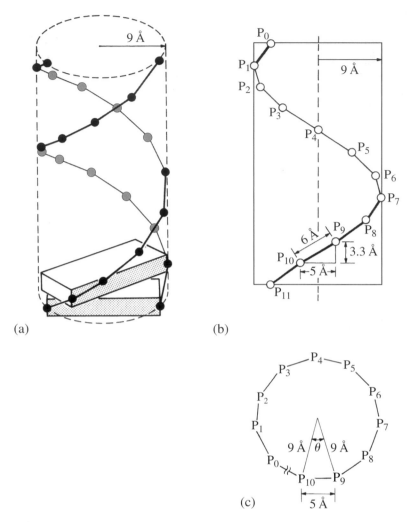

Figure 2.6 Sugar–phosphate chains wrapped helically around a cylinder: three views. In (a), sugar rings are drawn as shaded or filled circles, while phosphates are thin lines. In (b), phosphates are drawn as open circles, while sugars are thin lines. In (c), the view is down the long axis of the cylinder, looking along the dashed line in part (b).

of the cylinder at a distance of 9 Å from any phosphate. The value of θ can, therefore, be found by making a scale drawing, or else calculated as $2 \times \arcsin(2.5/9.0) = 32.3°$. Thus, each phosphate-to-phosphate rotation makes an angle of $32.3°/360° = 1/11$ part of a circle; and that is why we have put 11 phosphates in Fig. 2.6(c), to represent a complete turn of DNA.

This calculation, although relatively simple, tells us something which agrees closely with experiment: almost all DNA double helices have between 10 and 12 phosphates per turn of helix, within each

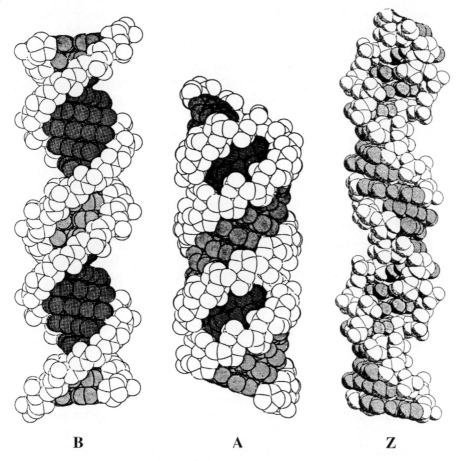

B   **A**   **Z**

Figure 2.7 Three well known (but highly idealized) forms of DNA. 'B' and 'A' are right-handed with 10 and 11 phosphates per helical turn, respectively, while 'Z' is left-handed with 12 phosphates per turn. Real right-handed DNA in solution averages about 10.5 phosphates per turn, or halfway between 'B' and 'A'. Pictures of 'A' and 'B' from C.J. Alden and S.-H. Kim (1979) *Journal of Molecular Biology* **132**, 411–34. Picture of 'Z' from H.R. Drew and R.E. Dickerson (1981) *Journal of Molecular Biology* **152**, 723–36 (with atoms shown somewhat smaller).

strand. For example, the well-known 'A' form of DNA (see Fig. 2.7) has 11 phosphates per turn, while the 'B' form has 10, and the 'Z' form has 12. These slight differences are significant in biology, and we shall discuss them later. But the crucial point here is that we have learned something important about the internal structure of the helix. Simply by studying the dimensions of the bases and of the sugar–phosphate chains, and knowing that the bases are insoluble in water (and so must stack directly onto each other), we have been able to determine that DNA will form a helix with about 11 phosphates per turn.

One point has been overlooked so far: how can we decide if our helix should be right-handed or left-handed? That is, as the phosphates spiral forward, should they go clockwise or counter-clockwise? There are, in fact, known examples of both right-handed and left-handed double helices: the 'Z' form (Fig. 2.7) should perhaps be assigned -12 phosphates per turn, rather than $+12$, because it is left handed rather than right-handed. It turns out that most DNA double helices are right-handed because of certain details of the chemical structure.

These details can be seen clearly when we build accurate, space-filling models of DNA: the atoms do not easily fit together if we try to build left-handed versions. Only very special combinations of DNA bases can become left-handed. Even then, the structures formed are so complicated and difficult to understand that they will not be considered much in this book.

Before proceeding further, let us look closely at some realistic pictures of DNA that include all of the atoms. Figure 2.8(a) shows a space-filling model of a very small part of right-handed DNA, corresponding roughly to the fragment shown more crudely in Fig. 2.5(b), or to one small part of the 'B' helix in Fig. 2.7. The atoms in this model are color coded as hydrogen (white), carbon (black), nitrogen (blue), oxygen (red), and phosphorus (yellow). The bases are joined at the center in 'pairs' (discussed below), while sugars and phosphates lie along the outside. This kind of model is widely used by scientists because it is so accurate. In fact, one could measure from the model a 3.3 Å separation of bases and a 6 Å separation of phosphates, and so check our previous calculation. Also, if you try to twist the base pairs backward in this model, from right handed to left handed, the bonds between atoms fall apart.

Other more detailed chemical representations of just one DNA chain are shown in Fig. 2.8(b) and (c). There the atoms (or groups of atoms) are drawn as letters or balls: for example, P = phosphorus, O = oxygen, C = cytosine, and A = adenine; while the bonds between the atoms are drawn as lines or sticks. The 5' to 3' direction of chains is also shown.

Let us now return to Fig. 2.7 to look more closely at the three forms of DNA shown there. Because we now know something about the internal structure of DNA, these models make sense to us. In all cases we can see that the bases compactly fill the centers of the double helices, where they successfully escape from contact with the surrounding water, and that the sugars and phosphates spiral around the outside of each helix at a rate of 11 ± 1 phosphates per turn. The 'Z' form also shows some jagged features in its sugar–phosphate chain that we shall not analyze here; they come from irregularities in the packing of bases on the inside of the helix. Using Fig. 2.8 as a guide, you can attempt to identify single bases, sugars, and phosphates in the pictures of Fig. 2.7, or even single atoms, if you have enough patience. It should be stressed that all of the pictures shown in Figs 2.7 and 2.8 are equally valid ways of representing the same thing; one kind of picture may be preferred over another, depending upon the level of detail at which you wish to visualize the DNA.

We have assumed throughout our exposition that the two chains of a DNA double helix run in opposite or 'antiparallel' directions

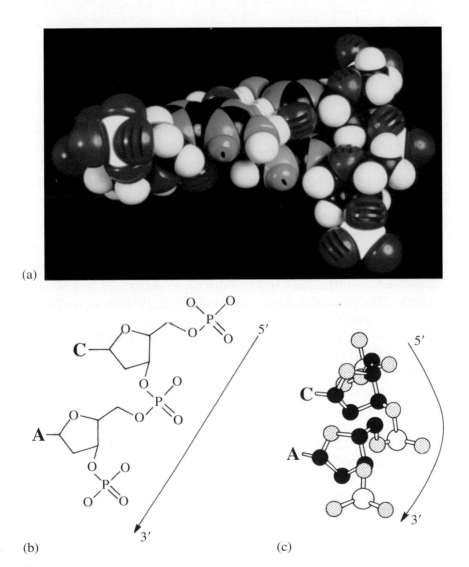

(a)

(b)   (c)

Figure 2.8 (a) Space-filling model of two stacked base-pairs, and their associated sugar–phosphate chains; (b) schematic, plane layout of the right-hand chain, with bases cytosine (C) and adenine (A) shown only as letters; (c) three-dimensional version of (b), in same configuration as (a).

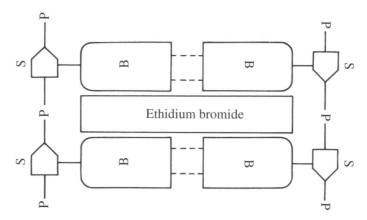

Figure 2.9 Ethidium bromide fills a gap between base pairs (cf. Fig. 2.3).

(Figs 1.7, 2.3, and 2.4), because 'parallel' double helices are not found in Nature. To a large extent, the interactions or pairings between bases at the core of the helix require that naturally occurring DNA be made from antiparallel chains. Some scientists have made short DNA molecules through chemical synthesis, in which the bases can only pair with one another when the chains run in a parallel sense: but these pairings are far less stable than when the chains run in the usual, antiparallel way, as found in Nature.

Another point before we go on: scientists sometimes add to DNA a substance called ethidium bromide, which fluoresces a bright red-pink when ultraviolet light shines on it, especially if it is bound to DNA. This provides a way of seeing exactly where the DNA is located in certain preparations, and so it constitutes an important tool for investigating DNA. How exactly does ethidium bromide stick to DNA? It is a greasy, mostly hydrophobic molecule, about the size of a base-pair. It can escape from contact with water by slipping between neighboring base-pairs along the chain, as shown in Fig. 2.9. Going back to Fig. 2.3, we must expect that the DNA will untwist locally to form an ordinary ladder before the ethidium bromide can fit in; and indeed it turns out that DNA plus ethidium bromide make a largely untwisted ladder. This feature, combined with the fluorescence mentioned above, makes ethidium bromide an important tool for scientists who investigate DNA.

It is the *flexibility* of sugar–phosphate chains which allows them to change from a spiral to a straight ladder when ethidium bromide is added. This flexibility is of an unusual and indirect sort which arises from features shown in Fig. 2.10. There we see in Fig. 2.10(a) that the phosphate group is essentially a rigid tetrahedron, having a phosphorus atom at its center and one oxygen at each vertex. Only when we go further along the chain from the phosphorus, as in Fig. 2.10(b),

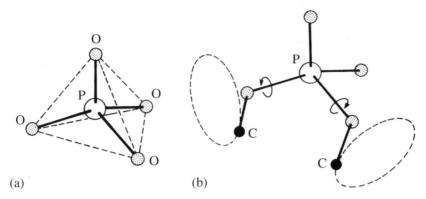

Figure 2.10 (a) Tetrahedral arrangement of phosphate group: atom types are phosphorus (P) and oxygen (O); (b) freedom of rotation for adjacent links of the chain with atom type carbon (C).

and attach two carbon atoms to two of the oxygens, can these carbon atoms swivel about the line of the phosphorus–oxygen bond. Imagine that your shoulder is the phosphorus, your elbow is the oxygen, and your hand is the carbon; obviously your hand and lower arm can swivel relative to your shoulder, but in DNA the angle at your elbow would be fixed. All parts of the DNA sugar– phosphate chain are rigid locally, but they have this kind of indirect rotational flexibility over several bonds.

We have almost finished our survey of the basic principles that determine the structure of DNA. All we have to do now is to learn how the bases adhere to one another in the central core of the double helix. James Watson and Francis Crick solved this problem in 1953, by putting forward a set of rules for base-pairing. They said that the most stable base pairs would be of the kind A–T or G–C, as shown in Fig. 2.11(a) and (b). One advantage of their scheme was that all four possible Watson–Crick base pairs, A–T, T–A, G–C, and C–G, were of the same size, and hence could fit easily into the framework of a regular double helix. Another advantage was that it explained how the genes in DNA could be duplicated (or stably inherited) on cell division. Whenever a cell divides, and needs to duplicate its DNA, it can do so simply by splitting the DNA into two separate strands; then certain enzymes will come along and use each of these old strands as a 'template' for the precise synthesis of a new strand, according to the Watson–Crick rules of base pairing: A with T and G with C. (More will be said about this in Chapter 4.)

Note that some of the interatomic connections within the A, T, G, and C rings are drawn as two lines, rather than as one: these are the 'double bonds', which give the base rings both their flatness and their rigidity. Also note that the CH_3 or methyl group on the

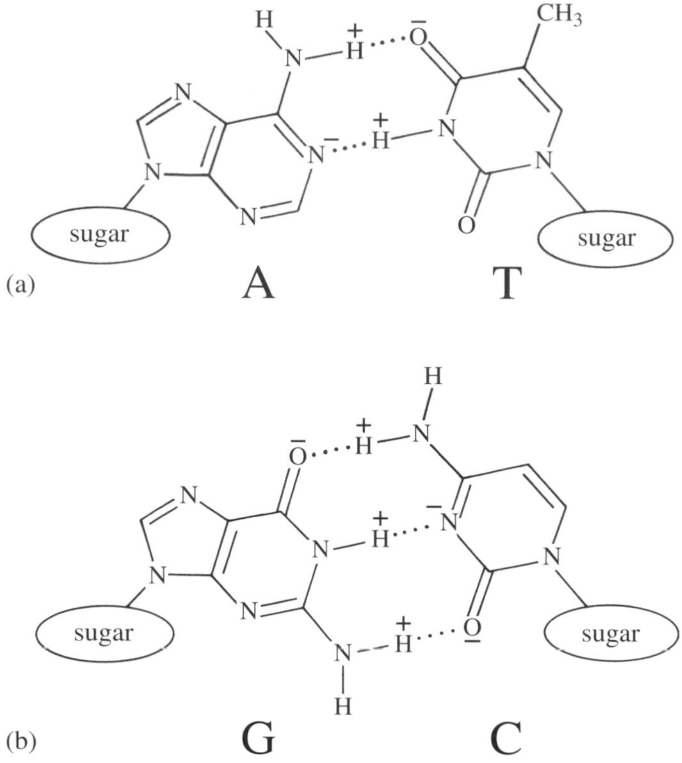

Figure 2.11 Watson–Crick base pairs showing hydrogen bonding: (a) A–T; (b) G–C. Atoms which are not labeled are carbon, while other atom types are hydrogen (H), nitrogen (N), and oxygen (O). Not all hydrogens are shown. Hydrogen bonds are represented by dotted lines. Symbols + and − here represent partial electric charges of about 1/3 electron or proton, which are typical for hydrogen bonds. A T–A pair is the same as an A–T pair, but turned over.

thymine ring would be absent in RNA, where the methyl-less base is called 'uracil' (see the right-hand part of Fig. 2.13 and its caption).

But what is the physical basis of these Watson–Crick rules of pairing? They are based on the simple fact that, within any DNA base, there is a small surplus of negative electric charge on nitrogen and oxygen atoms where they are *not* attached to hydrogen, while there is a small surplus of positive charge on these same atoms where they *are* attached to hydrogen. Thus, consider the base pair of adenine (A) and thymine (T), as shown in Fig. 2.11(a). Not counting the two nitrogens that are attached to sugars, there are three nitrogens on adenine and two oxygens on thymine that have a surplus of negative charge. On the other hand, one nitrogen on adenine and one nitrogen on thymine have a surplus positive charge. So all we have to do is to put the pluses and minuses together, thereby making the 'hydrogen bonds' which are shown here as dotted lines.

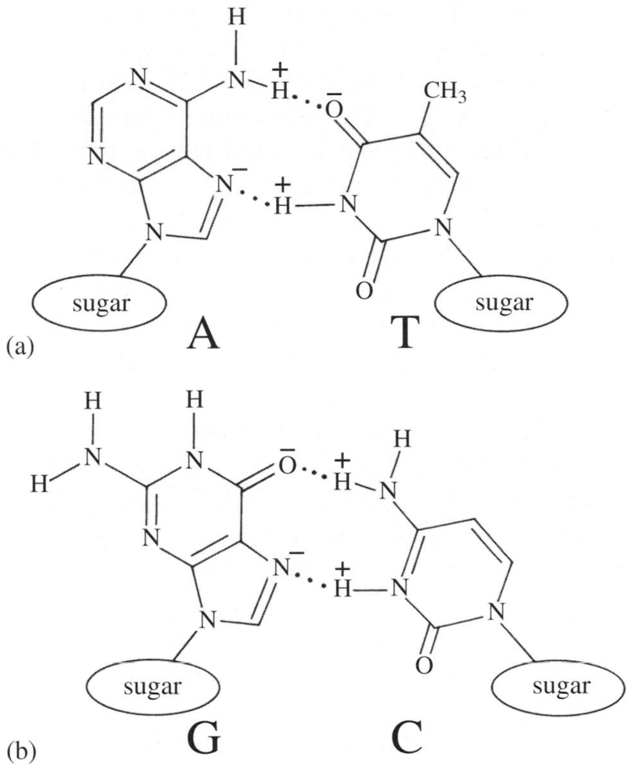

Figure 2.12 Hoogsteen base pairs, showing hydrogen bonding: (a) A–T; (b) G–C. Symbols + and − are partial charges, as in Fig. 2.11. The Hoogsteen G–C pair is stable only at mildly acidic pH (about 4–5), because it requires protonation of a cytosine nitrogen, that is, the adding of a hydrogen to it.

There are, in principle, two ways of doing this, consistent with making a double helix. These are called the Watson–Crick and Hoogsteen base pairs. The Watson–Crick A–T pair is shown in Fig. 2.11(a), while the Hoogsteen A–T pair is shown in Fig. 2.12(a). Both are roughly of equal stability; there are two hydrogen bonds which can be drawn between the bases in both cases. Note that the adenine base has to be rotated through 180° about the bond to the sugar in order to change between the two kinds of pairing – like rotating a tennis racquet from 'rough' to 'smooth'.

The history of these base-pairs is rather interesting. Watson and Crick found their A–T base-pair as part of a search for the double-helical structure of DNA by playing with paper cut-outs of bases. (Some details of this are given later, in Chapter 9.) Karst Hoogsteen tried 10 years later to confirm the Watson–Crick pair for adenine and thymine by heating up a solution of these two bases and letting it cool slowly in order to make a crystal; but he found instead a different kind of base-pair in his crystal.

Watson–Crick and Hoogsteen pairs for guanine (G) and cytosine (C) can also be drawn as shown in Figs 2.11(b) and 2.12(b). There are two things to note here. First, the Watson–Crick guanine–cytosine pair has three hydrogen bonds, rather than two as for adenine– thymine; so DNA double helices with guanine and cytosine should be more stable than those with adenine and thymine. Secondly, the Hoogsteen guanine–cytosine pair is only stable at low pH, since one of the nitrogens on cytosine must be protonated (i.e. be attached to a hydrogen) for this structure to form. The midpoint for protonation is pH 5, or more acidic than the normal pH 7 to 8 found in cells. That is the main reason why practically all DNA double helices contain Watson–Crick rather than Hoogsteen pairs; the Hoogsteen G–C pair is *not stable at neutral pH*. An additional reason is that a Watson– Crick G–C pair has more hydrogen bonds than a Hoogsteen G–C pair – three *versus* two. In principle, DNA containing only A and T bases should be relatively stable in either Watson–Crick or Hoogsteen forms, but few purely Hoogsteen-paired double helices have yet been detected with certainty. Also, it seems unlikely that both Watson– Crick and Hoogsteen pairs could be easily accommodated in close proximity within the same double helix, because the two kinds of base pair have different sugar-to-sugar distances.

Until recently it would have been enough to learn just about Watson–Crick base-pairs. But now it is important to learn about Hoogsteen pairs as well, because such pairs show up occasionally in complexes of DNA with anticancer drugs, and also in triple helices where a third strand of DNA joins the first two. The base 'triplet' then contains both Watson–Crick and Hoogsteen pairs: try to draw a triplet with one adenine and two thymines. Two scientists in California, Scott Strobel and Peter Dervan, have actually been able to cut a yeast chromosome cleanly in half by designing a third strand of DNA that binds only to one specific double-helical sequence within the whole chromosome, and then by attaching an iron atom to this third strand, so as to 'rust' (or oxidize) the DNA into pieces.

Why can't there be other stable base pairs, such as G with A, or C with T? Some of these are ruled out by the difficulty of making two or more hydrogen bonds. But others, such as guanine (G) with thymine (T) or uracil (U), as shown in Fig. 2.13, are not excluded for that reason. The hydrogen bonding produces a pair with similar overall shape to those in Fig. 2.11. In fact, a guanine–uracil pair is perfectly stable, and is used when transfer-RNA – the molecule that carries amino acids for protein synthesis – binds to messenger-RNA on the ribosome. You may recall from Chapter 1 that DNA makes a copy of itself in the form of RNA, and that this copy, called 'messenger-RNA', travels outside the nucleus to the protein-making machinery or

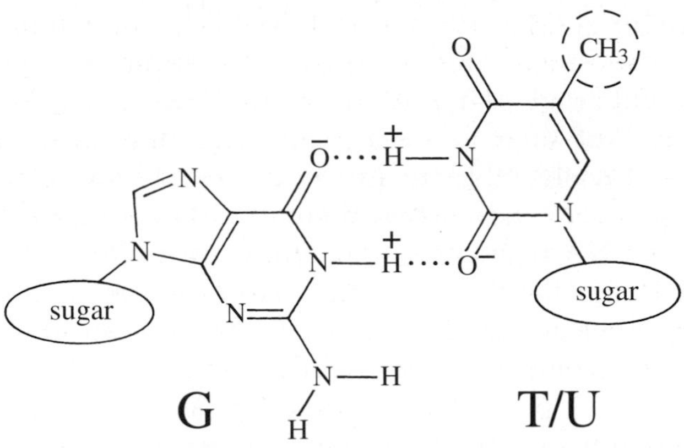

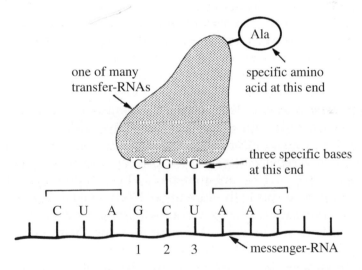

Figure 2.13 Pairing of G with T allows two good hydrogen bonds, plus an overall shape similar to that of the Watson–Crick pairs shown in Fig. 2.11(a) and (b). A closely related base-pair, that of G with U (uracil), is used routinely to specify amino acids for the synthesis of proteins, as shown in Fig. 2.14; as explained in Chapter 1, U is like T but without the CH_3 group here enclosed in a broken circle.

Figure 2.14 The transfer-RNA molecule that carries alanine can recognize its preferred triplet of bases on a messenger-RNA chain by using either a G–U or a G–C base-pair in position 3.

'ribosome'. There it becomes attached to a series of different transfer-RNA molecules, one of which is shown in Fig. 2.14. Each kind of transfer-RNA contains three specific bases at one end, and one specific amino acid at the other. If the three specific bases on transfer-RNA can form base-pairs with three neighboring bases in the messenger-RNA, then the amino acid being carried by the transfer-RNA is added to a growing protein chain.

Now, there are $4 \times 4 \times 4 = 64$ possible combinations of three bases in the messenger-RNA, and yet sometimes there are only 25 to 30 different kinds of transfer-RNA molecule in a cell. So several different sets of three bases in the messenger-RNA must have to 'share' a transfer-RNA in much the same way that two birds will share the same nest when there are not enough trees. In fact, the transfer-RNA is flexible, and will accept either of two possible base pairs in the third position. An example is shown in Fig. 2.14: here either an RNA sequence GCU with an 'unusual' G–U pair, or else a sequence GCC, with the 'normal' G–C pair, serves to determine one particular amino acid. A guanine–uracil pair is not the only unusual one to be used in this way; there are several others.

That brings us to a final question: if all of these different base pairings are stable enough to be used by transfer-RNA when it links with messenger-RNA, as in Fig. 2.14, why aren't they also found in DNA? If indeed they were used every time a cell divides, the sequence of bases in the DNA could change drastically. Yet this evidently does not happen, since it is now known that just a few G–C to G–T changes at critical places in the DNA program could cause cancer. In fact, Nature uses special proteins called 'proofreading enzymes' to prevent the occurrence of slight changes in sequence when DNA replicates. The enzymes that copy DNA to DNA, or DNA to RNA, are indeed very clever. They can sense at several stages during synthesis whether anything is going wrong; for example, if they have added or are about to add the wrong base, according to the Watson–Crick rules of pairing. Also, there are 'repair' enzymes that go around correcting occasional mistakes of copying or 'mismatches'. Thus, Nature goes to great lengths to avoid errors in the copying of DNA, even though the atoms in the DNA structure are actually quite tolerant of mismatch pairings. These enzymes are extremely efficient in doing their job, yet no one knows exactly how they work.

In summary, we have learned in this chapter how the insolubility of bases in water provides the driving force for DNA to form a double helix; and how the geometry at the core of the helix depends on subtle interactions between partial electrical charges on the bases. These subtle interactions alone are not sufficient for accurate copying of DNA from generation to generation, and so cells contain many enzymes that enhance the efficiency of copying. Yet unusual base pairs are a fact of life at other places in the cell, such as when proteins are made *via* the binding of transfer-RNA to messenger-RNA; so one has to be aware of all these subtle possibilities, in order to appreciate the internal structure of nucleic acids as they act in biology.

Further Reading

Abrescia, N.G., Thompson, A., Huynh-Dinh, T., and Subirana, J.A. (2002) Crystal structure of an antiparallel DNA fragment with Hoogsteen base pairing. *Proceedings of the National Academy of Sciences, USA* **99**, 2806–11. A short d(ATATAT) double helix which contains Hoogsteen A–T base pairs throughout.

Arnott, S. (1970) The geometry of nucleic acids. *Progress in Biophysics and Molecular Biology* **21**, 265–319. A summary of early X-ray studies on fibrous samples of DNA.

Chaires, J.B. and Waring, M.J., eds (2001) Drug-nucleic acid interactions. *Methods in Enzymology* **340**. An excellent volume which summarizes the latest techniques and results in drug-DNA work.

Dickerson, R.E., Drew, H.R., Conner, B.N., Wing, R.M., Fratini, A.V., and Kopka, M.L. (1982) The anatomy of A-, B- and Z-DNA. *Science* **216**, 475–85. A summary of the first work from X-ray studies of DNA in single crystals.

Matray, T.J. and Kool, E.T. (1999) A specific partner for abasic damage in DNA. *Nature* **399**, 704–7. The importance of correct size and shape for inclusion of bases within a double helix by polymerase enzymes.

Saenger, W. (1984) *Principles of Nucleic Acid Structure*. Springer-Verlag, New York. A comprehensive review of the published literature on DNA structure up to the early 1980s.

Schwartz, T., Rould, M.A., Lowenhaupt, K., Herbert, A., and Rich, A. (1999) Crystal structure of the Zα domain of human editing-enzyme ADAR1 bound to left-handed Z-DNA. *Science* **284**, 1841–5. Possible biological role for left-handed DNA as a substrate for the RNA-editing enzyme adenosine deaminase.

Wan, C., Fiebig, T., Schiemann, O., Barton, J.K., and Zewail, A.H. (2000) Femtosecond direct observation of charge transfer between bases in DNA. *Proceedings of the National Academy of Sciences, USA* **97**, 14052–5. A study of how electrons move between adjoining base pairs in DNA over very short time scales.

Wittung, P., Nielsen, P.E., Buchardt, O., Egholm, M., and Norden, B. (1994) DNA-like double helix formed by peptide nucleic acid. *Nature* **368**, 561–3. The DNA bases form double-helical structures even if attached to a chain of peptides, rather than of sugars and phosphates.

Bibliography

Hoogsteen, K. (1959) Structure of a crystal containing a hydrogen-bonded complex of 1-methylthymine and 9-methyladenine. *Acta Crystallographica* **12**, 822–3. Discovery of an alternative pairing-scheme for A with T.

Strobel, S.A. and Dervan, P.B. (1990) Site-specific cleavage of a yeast chromosome by oligonucleotide-directed triple-helix formation. *Science* **249**,

73–5. Use of a triple helix with iron (Fe) attached to 'rust' a yeast chromosomal DNA molecule into two pieces.

Watson, J.D. and Crick, F.H.C. (1953) A structure for deoxyribose nucleic acid. *Nature* **171**, 737–8. First proposal of the rules for base pairing of A with T and G with C, and of a two-stranded, double-helical model for DNA.

Wilkins, M.H.F., Stokes, A.R., and Wilson, H.R. (1953) Molecular structure of deoxypentose nucleic acids. *Nature* **171**, 738–40. First companion paper to Watson and Crick (above), giving theory of diffraction of helices.

Franklin, Rosalind, E., and Gosling, R.G. (1953) Molecular configuration of sodium thymonucleate. *Nature* **171**, 740–1. Second companion paper, showing a clear X-ray diffraction picture of 'structure B'.

Exercises

2.1a   On planet P all living things are found to contain a DNA-like double-helical molecule just like that found on Earth, except that the molecule on planet P consists only of the two nucleotides A and G. Thus, A–G is the only scheme for base pairing, and the large A–G base pairs impart a separation of 20 Å to the two sugar–phosphate chains. By adapting the calculations shown in Figs 2.4 and 2.6, estimate the angle of helical twist and the number of base-pairs per helical turn for the special DNA on planet P.

b   On planet Q the DNA molecule contains four nucleotides, namely A, G, C, and T, just as on Earth; but the sugar–phosphate chain is found to be 7 Å long between phosphates (compared with 6 Å on Earth) on account of an extra carbon atom in each nucleotide unit. Given that the bases pair in Watson–Crick style, how many base-pairs do you expect per double-helical turn in the DNA on planet Q?

c   On planet R the genetic molecule is exactly like terrestrial DNA, except that the oceans on planet R are slightly acidic (pH 4), such that the base pairings A–T and G–C are mostly in accordance with the Hoogsteen scheme (Fig. 2.12). In consequence, the sugar–phosphate chains spiral about an imaginary cylinder of diameter 16 Å. Estimate the number of base-pairs per helical turn of DNA on planet R.

2.2   Idealized 'B' form DNA has a helical twist of 36° and 'rise' of 3.3 Å, in the axial direction, per base-pair step (see Fig. 2.7). When a molecule of the intercalating drug ethidium bromide inserts itself into a step of DNA in the manner of Fig. 2.9, it increases the length of the DNA by 3.3 Å, and at the same time reduces the helical twist at the step by 26°, i.e. from 36° to 10°.

Find the overall length of a 100-bp segment of 'B' form DNA, and the total number of helical turns:

a   with no ethidium bromide;
b   with one ethidium bromide molecule for every 10 base-pairs;
c   with one ethidium bromide molecule for every 2 base-pairs.

(Case **c** corresponds to the largest possible uptake of ethidium bromide by DNA.)

2.3   In some circumstances, DNA can make a triple helix by forming planar hydrogen-bonded base triplets in place of the usual base-pairs. Construct such a triplet from one A and two Ts in two different ways, with each thymine T connected by two hydrogen bonds to the adenine A.

a   Begin with the A–T pair of Fig. 2.11(a), and add a second T to the 'unoccupied' upper edge of the A by moving the first T base around in the plane of the paper.
b   Begin with the usual A–T pair, as before; but now obtain the second T by flipping the first T over onto its other face before moving it around.

Which of the new pairings is equivalent to the Hoogsteen arrangement of Fig. 2.12(a)?

(Hint: Work with a copy of Fig. 2.11(a), and use tracing paper – either way up – for the second T. Note that either oxygen of the thymine can act as a hydrogen-bond acceptor.)

2.4   Investigate possible G–A base-pairings as follows: begin with the G base from Fig. 2.13, along with its hydrogen-bonding scheme; and use A bases with the hydrogen-bonding schemes of Figs 2.11(a) and 2.12(a), but 'flipped over' in each case. Work on tracing paper. Do the two different G–A pairings differ much in their sugar–sugar distances?

2.5   The three-base code in a DNA or messenger-RNA chain (see Table 1.1) is read with the assistance of various transfer-RNA molecules, as shown schematically in Fig. 2.14. There, one particular transfer-RNA molecule is shown recognizing the messenger-RNA sequence GCU (corresponding to GCT in DNA); but the same transfer-RNA can also recognize the sequence GCC, by using a standard G–C pair rather than an unusual G–U in position 3 of the message.

a   Suppose that a particular cell has 32 different kinds of transfer-RNA molecule, each with one of 20 possible amino acids attached at its distal end. On average, how many different sets of base triplets in the messenger-RNA chain would each transfer-RNA have to recognize, in order to make all the proteins necessary for cell growth?

b   What is the smallest number of different transfer-RNA molecules that the cell could use and still be viable? How many triplets would any particular transfer-RNA molecule then have to recognize, on average?

CHAPTER 3
Different Kinds of Double Helix

In the previous chapter we learned two things: why DNA forms a double helix, and how the bases interact with one another at the oily, water-insoluble core of the helix to form 'base-pairs'. The driving force for helix formation was shown to be the need for the bases to escape from water by joining with other bases at the core of the helix. Yet they cannot stack directly on top of one another while doing so: rather, they must twist around slightly, because they are attached to sugar–phosphate chains that are twice as long for each base as the thickness of the base itself.

All of these points are what might be called 'first-order' influences on the structure of DNA. Now you might be hoping that you will not need to learn about 'second-order' effects: you can perhaps manage by just knowing the first-order effects. That is a forlorn hope, however, because you will have to learn about the second-order effects before you can understand many of the roles of DNA in biology: for example, how promoters work, how DNA wraps into chromosomes, and even how DNA binds to the 'repressor' proteins, which influence how well promoters work. As you may recall from Chapters 1 and 2, the DNA within any gene makes an RNA copy of itself which then goes on to make protein; a 'promoter' is a short region of DNA near the gene that tells the cell how many RNA copies to make, and hence how much protein.

We shall limit ourselves to three themes in our study of second-order effects on the structure of DNA: first, how the bases undergo 'propeller twist' to make sure that as much as possible of their oily, flat surfaces escape from contact with water; second, how the base-pairs stack on one another in particular ways that depend upon the ordering or sequence of bases; and third, how certain small,

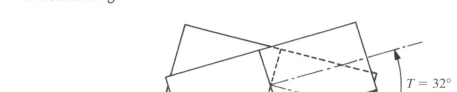

Figure 3.1 Two base-pairs with 32° of right-handed helical twist: the 'minor-groove' edges are drawn with heavy shading, as in Fig. 3.5. As in Fig. 2.11, the base-pairs are shown parallel to the paper.

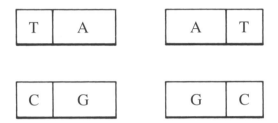

Figure 3.2 Four possible base-pairs of the Watson–Crick type, each of which joins a large purine (A, G) to a small pyrimidine (T, C), and in the same perspective as Fig. 3.1.

subtle motions of the base pairs can accumulate over a series of such base pairs, to make different kinds of double helix.

We said in the last chapter that the bases form ordered pairs at the core of the helix, leaving sugars and phosphates on the outside; and that each base-pair twists with respect to its neighbor by about 32°, as in the treads of a spiral staircase. The sense of this rotation is right handed, or clockwise going forward, as shown in Fig. 3.1: this is the same helical sense as in an ordinary corkscrew.

Almost always, the base-pairs are of a Watson–Crick kind, joining guanine (G) with cytosine (C), and adenine (A) with thymine (T): see Fig. 3.2. Bases A and G are called 'purines', and they are bigger than bases C and T, which are called 'pyrimidines'; yet the overall size of the base-pair is roughly the same in all four possible arrangements. The apparent simplicity of these arrangements once led scientists to conclude that the base sequence of DNA could not influence its three-dimensional structure, because all four kinds of base pair could slot into a perfectly uniform double-helical 'staircase'. This conclusion was not based on any firm evidence, however, and it has proved to be incorrect. For example, typical angles of base-pair twist in real DNA molecules, as determined from the many high-resolution maps of DNA structure collected since 1980, range from

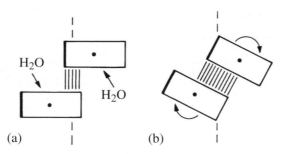

Figure 3.3 Propeller twist, as in (b), allows greater overlap of bases within the same strand and reduces the area of contact between the bases and water.

20° to 50° about a mean of 34°. The mean value of 34° is close to our prediction of 32° from the simple theory presented in Chapter 2; yet our first-order theory did not predict anything about a variation in twist away from 32° to other values. It certainly did not predict a broad range of twist from 20° to 50°, as is typically observed. Evidently, we must develop a theory that includes these second-order effects before we can claim to have any real understanding of the structure of DNA.

Our starting point for the second-order theory is rather subtle: it seems that because of the substantial twist between adjacent base-pairs, less than the entire surface of any base pair can escape contact with water. Thus, only the central overlapping portions of the base surfaces in Fig. 3.1 are protected from water, while the four overhanging triangular portions are not protected. When we view the two right-hand bases in Fig. 3.1 from the perspective of the right-hand margin of the page, edge-on in the plane of the paper, we see the arrangement shown in Fig. 3.3(a). Taken together, Figs 3.1 and 3.3(a) show that the overlap of consecutive pairs is good in the interior of the stack, but is only poor in the outer regions, which are exposed to the water.

What can we do to improve this situation? One solution would be to rotate each of the bases shown in Fig. 3.3(a) in a clockwise sense about its long axis, which points down into the plane of the paper in this view. Such a motion is shown in Fig. 3.3(b), where each base rotates slightly about its end-centerpoint, which is a black dot in the diagram. Stacking is improved, since the water is now excluded from a larger fraction of the surfaces of the two bases. If this motion is not clear to you, stretch out both of your arms in front of you, with your right hand above your left, both hands horizontal, and with the tips of the middle fingers vertically above each other and about 5 cm apart. Then move your right hand by about 5 cm in a left-to-right sense, so that it no longer lies directly over

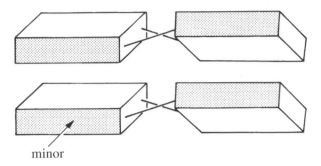

minor

Figure 3.4 Propeller-twisted base-pairs. Note how the hydrogen bonds between bases are distorted by this motion, yet remain intact. The minor-groove edges of the bases are shaded.

your left. Finally, rotate both hands clockwise about the wrists by around 20°; and you will find that your two hands 'cover' each other much better than they did before.

Now the two bases on the left-hand side of Fig. 3.1 can likewise rotate about their long axes to exclude water, but their sense of rotation must be counter-clockwise, when viewed from the perspective of Fig. 3.3, in order to achieve the same result. All four bases from Fig. 3.1 are shown together in Fig. 3.4. In this picture, each base pair looks somewhat like an old-fashioned airplane propeller, since the left-hand and right-hand bases twist in opposite directions. Hence, the overall motion is called 'propeller twist'. Its sense is that of a left-handed screw as one goes forward along the pair from one base to its partner.[1] Propeller twist obviously distorts the hydrogen bonds that hold the two bases together, which are shown schematically by two lines in Fig. 3.4; but these weak bonds can accept some distortion of that kind, provided the degree of propeller twist is not too great.

In general, such propeller twist tends to be higher than average in regions of double helix containing mostly AT base pairs, typically 15° to 25°; but lower in regions of helix containing mostly GC base pairs, typically 5° to 15°.

Also note that we have shaded one edge of each base in Fig. 3.4, and have labeled one of these with the term 'minor'. There is a convention to call one side of a base pair the 'minor-groove side', and the other the 'major-groove side'. Where do these names come from? As shown in Fig. 3.5(a), the two sugars to which a base-pair are attached lie closer to one side of the base-pair than the other. The edge which lies closer to an imaginary line drawn between the two sugars is called the 'minor-groove side', while the other edge is called the 'major-groove side'. By convention, we shall always

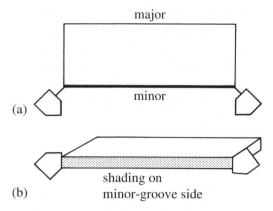

Figure 3.5 Two views of a base-pair, showing directions of the sugar–phosphate chains, just as in Figs 2.3 and 2.4. By our convention, the minor-groove edges are shaded. Here, the base-pair is shown as a single block.

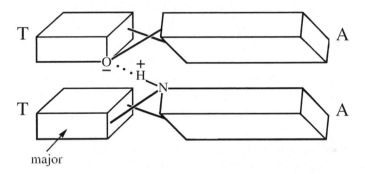

Figure 3.6 Propeller-twisted A–T pairs, showing an additional hydrogen bond between the base-pairs in the major groove, as proposed by Hillary Nelson (cf. Fig. 2.11(a)).

shade the minor-groove side of a base-pair in our drawings, as indicated in Fig. 3.5(b).

Why do we use the term 'groove' in this labeling convention? Early structural models of DNA showed a cylinder with two hollow, spiral grooves lying between the two sugar–phosphate chains. One of the grooves – 'minor' – was smaller than the other. You can see these two grooves in the 'B' model of Fig. 2.7. But for some of the helices which we shall study in this chapter, it turns out that the so-called minor groove is actually as large as or larger than the so-called major groove: for example, see the 'A' model of Fig. 2.7. We can get around this difficulty by talking instead about the minor-groove *side* of the base pairs themselves.

Of course, we can also view the base-pairs from the major-groove side. An example of this is shown in Fig. 3.6, where we see that the near edge, labeled 'major', remains unshaded. Furthermore, in this drawing we can see that the two bases A and T within either pair are of unequal size. In general, as we have said, purines A and G are larger than pyrimidines T and C. In several previous drawings, for example, Figs 3.1 and 3.4, we have omitted this feature for the sake of clarity; but when we start to consider the interactions of real DNA bases, such as those shown in Fig. 3.6, we have to make our drawings more accurate. Figure 3.6 also shows the detail of a possible hydrogen bond (N–H...O) between adjacent base pairs, from adenine in the lower pair to thymine in the upper. Such a hydrogen bond might be expected to increase the propeller twist, because the distance is right for it to form only when the base pairs are highly twisted along their long axes. Indeed, experiments show that regions of DNA with all adenine bases on one strand and all thymine bases on the other do have an unusually high propeller twist of about 20° to 30° as against 10° to 20° for other sequences.

You can imagine that if we were to study all kinds of two-base-pair arrangements in DNA, or indeed all kinds of three-base-pair arrangements, we should find a lot of unexpected but important contacts between the bases. For that reason it is very hard to understand, on an atomic scale, the behavior of any long DNA molecule such as those found in biological systems. We need some sort of simplified description of base-pair arrangements in DNA, at less than an atomic level of detail, if we wish to understand the many roles of DNA in biology.

The simplification which we make at this point is one which we have already used in some drawings: we construct an imaginary, flat plane that coincides as well as possible with the twisted surface of any base-pair. In other words, we shall pretend that the base-pairs shown in Figs 3.4 and 3.6 have no propeller twist, and treat them as rigid, rectangular blocks, such as those shown in Figs 3.5 and 3.7. Previously, of course, we have said how important it is to

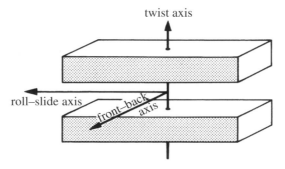

Figure 3.7 Local reference axes for an individual base-pair step, following the mathematics of Leonard Euler.

build propeller twist into a base-pair; but now it seems that we are going to ignore this twist entirely in a simplified model! You must simply remember that the propeller twist is always *there*, but that it is not always *shown* in the diagrams. This is a bit like drawing a pocket watch without showing the gears inside: they are there, but not visible. For some purposes we can understand things more clearly by thinking about a deliberately simple representation of a base-pair.

Our simplified model is based on the work of the famous Swiss mathematician, Leonard Euler (1707–83). He explained that if you have two rigid objects, such as the rectangular blocks shown in Fig. 3.7 – here representing two base-pairs – and you want to describe the position of one block relative to the other, then you will need to use six variables or 'degrees of freedom': three translations and three rotations. A translation[2] is a change of position without any rotation: imagine moving a cardboard box from one place in a room to another, so that every face of the box always moves parallel to itself. The position of the box in the room can be specified completely by the values of three 'coordinates', say x, y, and z, measured from a suitably chosen 'origin'. A rotation, on the other hand, involves a change of angle without a change in position: for example, you could pick up a cube from a table, turn it through 90°, and then put it back on the table in the same position as before.

It is easy to understand why you need six variables to describe completely the position of one solid block relative to another. Suppose that the upper base-pair or block shown in Fig. 3.7 were temporarily replaced by a *point*: then you would only need three translation coordinates, x, y, and z, to say where this point might be located relative to the lower block. Next let us build an upper block around the point, and ask how many kinds of rotation we need to orient the upper block relative to the lower block. The answer is exactly three, as indicated by the three axes of rotation drawn in Fig. 3.7. We call these the 'twist', 'front–back', and 'roll–slide' axes for reasons that will become clear soon.

So it seems that we need only six numbers to describe the local configuration of any two neighboring base-pairs from a mathematical point of view. That is not too bad. But when we look at real DNA structures, the situation becomes even more favorable. To a first approximation, only three of Euler's six possible degrees of freedom are actually mobilized in real DNA double helices. This fortunate simplification comes about because the base-pairs are attached, as we recall from the previous chapters, to sugars and phosphates which limit their range of maneuver in certain directions, notably along the front–back axis of Fig. 3.7. Also, the

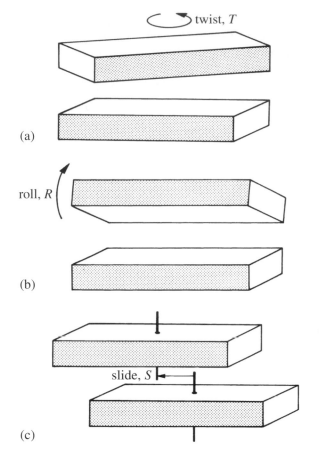

Figure 3.8 Twist, roll and slide motions at a base-pair step. Each drawing defines the positive sense of twist, roll, or slide, as used in this book.

base-pairs cannot be separated along their vertical twist axis without introducing water or a vacuum between them.

Thus, in practice, we need only three variables to describe the motions of any base-pair relative to its neighbor. Two of these are rotations, and one is a translation. All three are shown separately in Fig. 3.8. The first rotation, or 'twist', (Fig. 3.8(a)) is the same twist which we estimated as about 32° in Chapter 2, and which we showed in Fig. 3.1. We shall now give it a rigorous definition: it corresponds to a rotation about the local twist axis that runs vertically through, or near, the centers of any two neighboring base-pairs, as shown in Fig. 3.7. Note that if the DNA happens to be curved, or if the base-pairs stack on one another locally, such that they do not advance directly along the overall helix axis, then the *local* twist axis shown in Fig. 3.7 may not coincide with a *global* twist axis for the whole molecule, averaged over many steps; and the twist angles measured about these two different axes may also be slightly different. So it is important to define what kind of twist we mean, when we talk about the 'twist' of DNA.

The second kind of rotation, or 'roll' (Fig. 3.8(b)), describes the rolling open of base pairs along their long axes. Angles of roll vary from $+20°$ to $-10°$ in the usual DNA structures. By convention, we say that the roll is positive if base pairs open up towards the minor-groove side, as shown in the diagram. Actually, the surfaces of individual bases do not come apart from one another very much in the roll motion: they only appear to do so because we have drawn the two base-pairs as uniform, rigid blocks. This simple kind of drawing conceals details in much the same way that the cover of a pocket watch conceals the gears.

The last commonly observed kind of motion is a translation or 'slide' (Fig. 3.8(c)). It describes the relative sliding of neighboring base pairs along their long axes. Slide is defined as positive if the upper pair goes further to the left than the lower pair, when we look at the minor-groove edges. Typically, values of slide range from $+3\,\text{Å}$ to $-2\,\text{Å}$ in real DNA; the sugar–phosphate chains do not easily allow any further motion. Still, these sugar–phosphate chains do allow a great deal more flexibility about the roll–slide axis, as epitomized by 'roll' and 'slide', than they do about the front–back axis shown in Fig. 3.7.

In summary, there are three significant relative motions of the base pairs at any base-pair step. They are called roll, slide, and twist, and they may be abbreviated to R, S, and T, respectively. The positive sense of each is shown in Fig. 3.8(a), (b) and (c).[3] You cannot forget the series R, S, and T, although you have to remember which letter stands for which degree of freedom.

The propeller-twist motion, which we were talking about earlier, is a property of a single base-pair, and not of two base-pairs that lie over each other. Thus, we can imagine two short pieces of DNA that have different values of propeller twist but the same values of roll, slide, and twist. As shown in Fig. 3.9, if we stack two base-pairs directly on top of one another, at zero slide and zero twist for the sake of simplicity, then the roll remains zero no matter how much propeller twist we add equally to both pairs. At the level of the 'rigid block' drawings shown in Figs 3.7 and 3.8, both parts (a) and (b) of Fig. 3.9 would look identical, despite the change in propeller twist on going from (a) to (b). Bases on the right-hand strand tilt upward along their minor-groove edges by $10°$ and those on the left-hand strand tilt downward by $10°$; nevertheless, the two effects cancel when we calculate a mean plane for the entire base-pair.

Drawing base-pairs as rigid blocks, and then identifying roll, slide, and twist motions at the steps between base-pairs may seem to be a strange way of thinking about DNA; but it is the most successful way so far devised. Therefore let us continue, and see how the three parameters roll, slide, and twist might be useful for understanding different arrangements by which base-pairs stack onto each other in different types of two-base-pair steps. Even though we have pushed the idea of propeller twist out of the limelight for the moment, we must not forget about it, because propeller twist is partly responsible for certain relations between R, S, and T which are found in real DNA.

Previously, in Fig. 3.6, we noted one important feature of DNA structure, which was a hydrogen bond between neighboring base-pairs in sequences of the kind AA or TT. The high propeller twist of

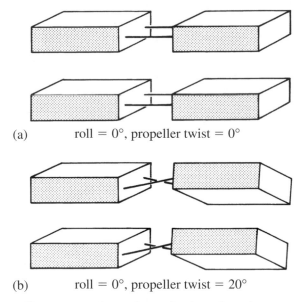

(a)   roll = 0°, propeller twist = 0°

(b)   roll = 0°, propeller twist = 20°

Figure 3.9 Propeller twist need not alter roll. The roll angle remains zero in part (b), because the mean planes of the base-pairs remain parallel.

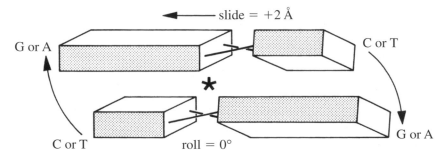

Figure 3.10 A pyrimidine–purine step with zero roll: positive slide is needed to avoid a steric clash at * if the base-pairs have propeller twist.

the two base-pairs allows a cross-chain link between an oxygen atom of T and a nitrogen N–H of A on the major-groove side. It also causes a close contact between an oxygen atom of T and a carbon atom of A on the minor-groove side. Those two cross-chain contacts help to hold the AA/TT step in a nearly fixed conformation, with $R = 0°$, $S = 0\,\text{Å}$, and $T = 36°$.

Another important feature of DNA structure can be found at steps which we shall describe as 'pyrimidine–purine'. Figure 3.10 shows an example of such a step. You should have learned from the many pictures shown so far that the two sugar–phosphate chains of DNA run in opposite directions. In Fig. 3.10 these sugar–phosphate chains are not shown, but their directions are marked by arrows. Suppose in a particular step that the upper base-pair is C–G and the lower one is A–T in going from right to left across the page, then the entire step can be described as 'CA/TG': coming down along the right-hand chain, we have the sequence CA, and going back up along the left-hand chain we have TG. The 'slash' between CA and TG indicates the 'jump' from one chain to the other. If you were to turn the book through 180° in its own plane, and use the same procedure for naming this particular step, you would get 'TG/CA' instead of 'CA/TG'. It does not matter which term you use: both describe the same collection of atoms in space.

The drawing in Fig. 3.10 is intended to represent all possible pyrimidine–purine steps, which we may now list as TG/CA (=CA/TG), CG/CG, and TA/TA. These steps are seen commonly with two different kinds of base-to-base overlap, as shown schematically in Figs 3.10 and 3.11.

In Fig. 3.10, the large purine bases G or A slide away from each other by +2 Å, in order to avoid too-close contact at the point shown by the 'star' in the picture. Some positive slide is needed here, because both of the bases concerned are the large purines,

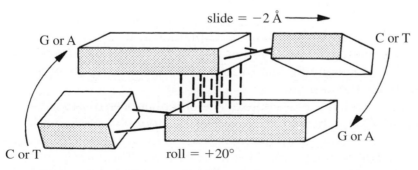

Figure 3.11 A pyrimidine–purine step in an alternative configuration having negative slide and positive roll, due to the cross-chain stacking of purines.

G or A. If the base-pairs were not propeller twisted, then contact of purines at the star would not be a serious problem, and the slide could be zero. Note that the roll angle remains near zero, since all bases remain parallel to their neighbors on the same strand.

In Fig. 3.11, the large purine bases slide on top of one another by -2 Å. In this conformation, the roll angle becomes large and positive, $+20°$, since the small pyrimidine bases must be inclined with respect to the large purine bases in either strand by $+20°$, in order to maintain the 20° of propeller twist. Apparently, the close stacking of pyrimidine on purine within either strand of real DNA is less significant than is the maintenance of high propeller twist.

There are two important points about Figs 3.10 and 3.11. First, values of slide lying between $+2$ Å and -2 Å might be expected to be less stable than the ones shown, because at intermediate values (say 0 to -1 Å) the two large purine bases will neither avoid one another fully, nor stack firmly on top of one another. One might expect, therefore, that pyrimidine–purine steps in DNA would be weakly 'bistable': that is, capable of adopting two extreme conformations but not always a continuous range; and this is indeed what is found. Second, by comparing the two structures shown in Figs 3.10 and 3.11, we can see that roll changes as we change the slide, on account of propeller twist. Thus, slide $= +2$ Å gives roll $= 0°$, while slide $= -2$ Å gives roll $= +20°$. This is very much like the motion of a bolt-action rifle: the bolt 'rolls' as it 'slides' forward to place a bullet in the chamber.

We have now seen two different ways by which the preferred close contacts of base pairs can influence the roll, slide, and twist values that are adopted at any step. First, as shown in Fig. 3.6, an AA/TT step can be held in a single, preferred position by a possible hydrogen bond on the major-groove side. Second, as shown in Figs 3.10 and 3.11, pyrimidine–purine steps can adopt at least two different types of stacking that cover a wide range of slide S and roll R. Both of these stacking effects are important in real DNA; but there is also a third stacking effect which is just as important as the other two, and which will be explained fully in Appendix 2 but will be mentioned briefly here. It has to do with how adjacent bases stack on one another, according to partial electric charges within the bases themselves.

The stacking of bases onto one another has been discussed so far in terms of only the van der Waals or hydrophobic effect, which provides for good overlap of any two bases in proportion to the area of contact of their flat, water-insoluble surfaces. The hydrophobic effect applies with equal strength to all bases or base pairs. It thereby contributes a tendency to propeller twist as mentioned above, and it can be discussed adequately by means of our current pictures in which the base-pairs are drawn as featureless 'blocks'.

However, the usual A–T and G–C base pairs also contain within themselves many precise distributions of partial electric charge, which are spread over their flat surfaces. These are analogous to the partial electric charges used to explain Watson–Crick base pairs in Fig. 2.11. Yet in addition to the partial charges shown in Fig. 2.11, which lie on nitrogen N or N–H or oxygen O atoms that are extended outward from the main parts of each ring, other distributions of partial electric charge are located near nitrogen N or oxygen O atoms within the rings themselves. Such additional electric charges were not shown in Fig. 2.11 for the sake of simplicity; yet they are just as real, and can have almost as much effect on the preferred stacking of bases vertically, as they do when they form the Watson–Crick hydrogen bonds which join one base to its partner.

To a first approximation, the G–C base pair contains a large plus-or-minus electrical 'dipole' along its long axis, specifically as plus on C but minus on G; whereas the A–T base pair contains only small patches of isolated plus-or-minus electrical charge along its long axis, which are relatively dispersed over the entire pair, and hence do not amount to a substantial dipole.

Let us now imagine how two base-pairs will move over one another at different values of roll R and slide S. What effect, if any, will the partial electrical charges have on preferred values of roll and slide for different sequences? In general, we know that 'unlike' electrical charges attract, whereas 'like' electrical charges repel; and so we expect the largest effects of an electrical kind to be observed for steps where two G–C base pairs stack onto one another, such as GG/CC, CG/CG or GC/GC. Much smaller effects due to electrical charge are expected for steps containing both a G–C and an A–T base-pair, such as CA/TG; and hardly any electrical effect is expected for steps containing two A–T base pairs, such as TA/TA.

Thus, for those steps containing two successive G–C base pairs, where each G–C base pair shows a partial plus charge on its C-ring, and a partial minus charge on its G-ring, one expects that the two base-pairs will repel each other near slide $S = 0\,\text{Å}$, in an arrangement where they are stacked directly on top of one another, due to 'like-to-like' repulsion. Hence, those two successive G–C pairs might prefer to lie slightly offset from one another vertically, at either positive or negative slide, in order to reduce the expected like-to-like charge repulsion. The effect of charge–charge repulsion for any step with two G–C pairs should, therefore, be similar to that shown in Figs 3.10 and 3.11 for pyrimidine–purine steps in general; yet it has a different chemical origin.

Our discussion of base-stacking arrangements in DNA seems to be getting rather complicated. We have three different situations to think about: (a) AA/TT steps, (b) pyrimidine–purine steps, and (c) steps with two G–C pairs. Yet the overall picture may be made clear by examining plots of roll R versus slide S for real DNA as shown in Fig. 3.12. All of the data shown in these plots come from structures of DNA oligomers, determined by X-ray crystallography, and are very precise.

First, Fig. 3.12(a) shows roll R and slide S values as observed for the AA/TT step, and collected from many different X-ray structures in a crystal. We can see there that the AA/TT step allows little variation of either roll or slide away from a mean of $R = 0°$ and $S = 0$ Å, as drawn previously in Fig. 3.6. Next, Fig. 3.12(b) shows the same plot for a CA/TG step, which is an example of the general pyrimidine–purine type. There we can see a wide range of roll and slide values, and also a strong connection between them as drawn schematically in Figs 3.10 and 3.11. Finally, Fig. 3.12(c) shows roll and slide data for the step GG/CC, which is broadly representative of the two G–C pair type. There we can see a clear 'forbidden zone' of slide S near 0 Å, where similar charges in adjacent G and G, or C and C bases, repel one another vertically. The experimental data, therefore, support all that we have said previously, and give us confidence that the structure of DNA can be understood in terms of ordinary chemistry, without any mysterious features!

Finally, to close our discussion of base stacking, could roll R and slide S be coupled with twist T in some general way? From first principles, we might expect to see a relationship between slide and twist if the sugar–phosphate chains were semi-rigid and of constant length: the relevant geometry is shown in Fig. 3.13. In this example, the twist angle decreases from $36°$ to $28°$ as the base pairs slide from 0 to -2 Å. There is a direct mechanical coupling involved here, because the sugar–phosphate chains are assumed to be rigid links of constant length. Actually, plots analogous to those shown in Fig. 3.12, but with twist plotted along the vertical axis instead of roll, do show a broad but definite tendency of the sort expected: low slide goes with low twist, while high slide goes with high twist. But the correlation of slide with twist is not so strong as the correlation of slide with roll, perhaps because sugar–phosphate chains are not actually as rigid as the picture of Fig. 3.13 suggests (see Appendix 2). In any case, it seems that roll, slide and twist in DNA are all related to one another.

Thus, having started with Euler's six degrees of freedom, and having eliminated three of these by introducing constraints on the base stacking due to various factors, we find finally that the three

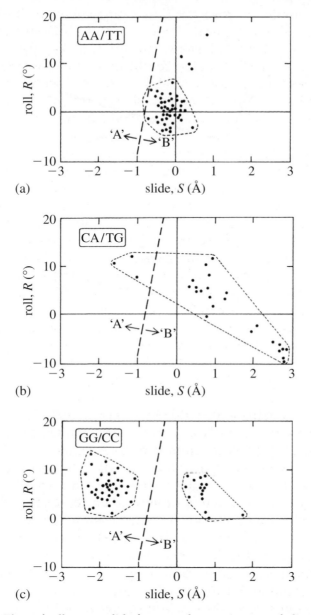

Figure 3.12 Plots of roll versus slide for many base-pair steps of oligomeric DNA as studied by X-ray diffraction. Separate plots are given for three of the ten distinct steps by sequence: AA/TT is a 'rigid' step, CA/TG is a 'flexible' step and GG/CC is a 'bistable' step. From El Hassan and Calladine (1997). *Philosophical Transactions of the Royal Society, A* **355**, 43–100.

remaining parameters R, S, and T are broadly related to each other in ways which depend both on the base composition of the step, and also on the general behavior of the sugar–phosphate chains, which connect the two base-pairs.

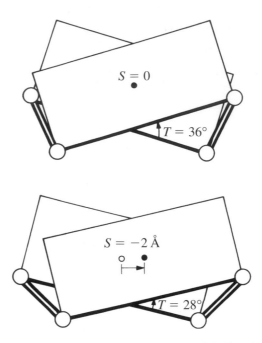

Figure 3.13 Schematic model of a base-pair step (cf. Fig. 2.5(b)) showing a possible mechanical linkage between slide and twist, which is confirmed weakly by experiment.

Now we have explained almost everything that is known today about the internal structure of DNA. In summary, the base-pairs adopt propeller twist to minimize their contact with water; this propeller twist prevents the otherwise flat base-pairs from sliding freely on one another's surfaces, and can sometimes 'lock' certain steps into a nearly rigid configuration. Some base-pairs also contain partial electric charges, which can prevent a step from adopting certain values of slide or can cause it to favor two separate values of slide. There are also some relations between roll, slide, and twist, that come from the connection of base pairs to sugar–phosphate chains of roughly constant length.

Having completed our study of the patterns of base stacking in DNA, our final task is to explain how different values of roll R, slide S, and twist T *generate different kinds of double helix* – as seen, for example, in Fig. 2.7. Thus, suppose that the same values of roll, slide, and twist are repeated over and over again along a significant length of DNA, what kind of double helix will be formed? Once we understand this relationship between the 'internal' variables R, S, and T, which describe the base-stacking relationships, and the 'external' form of the resulting helical structure, we shall be able to understand how different sequences of bases in DNA can generate

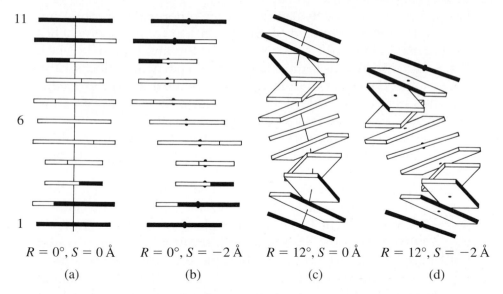

11

6

1

$R = 0°, S = 0\,\text{Å}$ $R = 0°, S = -2\,\text{Å}$ $R = 12°, S = 0\,\text{Å}$ $R = 12°, S = -2\,\text{Å}$

(a)   (b)   (c)   (d)

Figure 3.14 One complete helical turn of DNA having $T = 36°$, showing the effects of introducing uniform roll R or slide S at each step. Broadly, (a) corresponds to the 'B' form of DNA, while (d) corresponds to the 'A' form as shown in Fig. 2.7. Parts (b) and (c) correspond to structures intermediate between 'B' and 'A' which have, in fact, been seen recently in DNA crystals by X-ray diffraction.

different double-helical structures by favoring different values of R, S, and T at a local level.

Our analysis will consist solely of three-dimensional geometry. For the present, we shall assume that every step in a given helix has the same values of R, S, and T. An obvious way of proceeding might be to build some physical models of DNA, step by step, by means of a suitable home-made construction kit. That is exactly what we did ourselves in the first instance, when we were struggling to understand the geometry of double helices. From careful study of these models, we were able to derive the relevant equations that describe the geometrical form of the DNA in three dimensions, as functions of R, S, and T. Do not worry if you cannot follow the details of our presentation: many people have difficulty with the simpler two-dimensional geometry that can be drawn on a piece of paper! The crucial point to grasp is that the final results could be established firmly by means of a few hours of practical construction at a woodwork bench.

The various stages of our analysis correspond to the pictures shown in Fig. 3.14(a)–(d). Let us look first at Fig. 3.14(a), which shows the side view of a stack of 11 base-pairs, with $R = 0°$, $S = 0\,\text{Å}$, and $T = 36°$ at each of the 10 steps. When roll and slide are both zero, then the helical geometry is very simple: neighboring base-pairs remain exactly parallel to each other and horizontal, and exactly 10 steps are needed (with $T = 36°$) to complete one helical turn of 360°. Hence, the side views of the top and bottom blocks in Fig. 3.14(a) are just the same. For these two particular blocks, numbered 1 and 11, we get a full-frontal view of the minor-groove edge, which is here colored black. If we go along five steps from either end to base pair 6, we find a block which is seen with a full-frontal view of the white, major-groove edge; in between you should be able to make out the different edges of the blocks as they rotate about the vertical axis.

Next, let us start from the simple helix shown in Fig. 3.14(a), having $R = 0°$, $S = 0\,\text{Å}$, and $T = 36°$, and then introduce a slide of $S = -2\,\text{Å}$ at each step. The resulting helical geometry is shown in Fig. 3.14(b), where the base pairs spiral outwards from a central helix axis. To understand this motion more clearly, we can study a top view of the structure, as shown in Fig. 3.15. This diagram shows a half-turn of a helix in which the center of every block has moved outwards, radially from the axis by a distance r. By looking at the sideways displacement of the centers of the blocks, we can see that the outward motion is, in fact, associated with a negative slide at each step. Applying trigonometry to Fig. 3.15, we find that

$$r = (-S/2)/\sin(T/2).$$

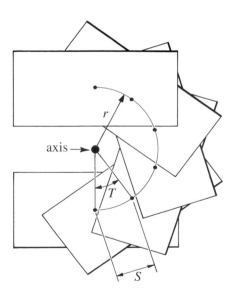

Figure 3.15 A top view of part of the helix shown in Fig. 3.14(b) illustrating the geometry by which radial displacement r is related to the magnitude of slide S and twist T. The small spots mark the centres of the blocks: to find which belongs to which, trace the full outline of a given block, and see which spot is at its centre.

Thus, given $S = -2\,\text{Å}$ while $T = 36°$, the base-pairs move outwards by

$\qquad r = (2/2)/\sin(36°/2) = 3.2\,\text{Å}.$

In summary, the top view of Fig. 3.15 shows that negative slide S makes a big hole in the middle of the helix, while the side-view of Fig. 3.14(b) shows that the stack of bases becomes wider than it was before.

Next, what happens if the base-pairs roll apart from each other by some angle R, while $S = 0\,\text{Å}$ and $T = 36°$? How will the picture of Fig. 3.14(a) change on account of the introduction of roll R? The resulting helical geometry is shown in Fig. 3.14(c), where each base pair tilts from the horizontal relative to its position in diagram (a). (In fact, each block rotates through a few degrees about its own 'front–back' axis, Fig. 3.7.) This kind of motion may well be a surprise to you, but if you look carefully at (c), you will see that, at every step, the gap between the minor-groove edges is larger than that between the major-groove edges, corresponding to the positive roll of $R = 12°$ which has been used for the drawing. The cumulative effect of many roll angles is to cause a *tilt* of the base-pairs with respect to the vertical axis. When the roll angles are small, it can be shown that

$\qquad \text{tilt} \approx (R/2)/\sin(T/2).$

Thus, when $R = 0°$, tilt $= 0°$; but when $R = 12°$ and $T = 36°$, then the tilt becomes

$\qquad (12°/2)/\sin(36°/2) = 19°,$

as shown in (c).

Also, roll and tilt cause the base-pairs to move out slightly from the axis. The best way to understand this is to think about the thin black 'rod' which connects the base-pairs in pictures Fig. 3.14(a) and (c), and passes through them at right-angles. In (a) it is straight but in (c), the roll at each step curves the rod; and since the direction of curvature is different for each step, the rod ends up as a gentle spiral. Calculations show that here the outward movement of base-pairs along the rod is approximately $3.3 \sin(\text{tilt})/2 \sin(T/2) = 1.7\,\text{Å}$ here, where $3.3\,\text{Å}$ is the distance between base-pairs along the stack locally.

So far we have considered combinations of just two parameters, S and T, to give distance from the axis as in Fig. 3.14(b), or R and T to give both distance from the axis and also tilt from the axis as in (c). But when all three parameters R, S, and T act simultaneously, something new happens: the helix gets shorter, because the negative sliding motion goes 'downhill' on account of the tilt. The relevant

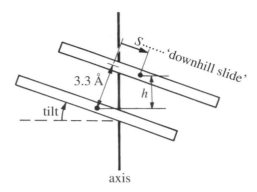

Figure 3.16 For base-pairs at a standard perpendicular separation of 3.3 Å, the 'rise' h in the direction of the helix axis depends both on tilt and slide S. This picture has been drawn with a negative value of slide, as in Figs 3.14 and 3.15.

geometry is drawn in Fig. 3.16, and its effect on the overall structure may be seen in Fig. 3.14(d). The vertical stacking distance at each step remains constant at 3.3 Å measured perpendicular to the blocks, but the rise h along the helix axis from the center of one base-pair to the next is changed to

$$h = 3.3 \cos(\text{tilt}) + S \sin(\text{tilt}).$$

For $R = 12°$, we know that tilt $= 19°$; and so for $S = -2$ Å, we have

$$h = 3.3 \cos(19°) - 2 \sin(19°)$$
$$= 3.1 - 0.6 = 2.5 \text{ Å}.$$

To summarize, one can calculate approximately the overall shape of a long helix made out of many uniform (R, S, and T) steps according to three formulas:

(i) tilt from axis $= (R/2)/\sin(T/2)$,

(ii) distance from axis $= (-S/2) \cos(\text{tilt})/\sin(T/2) +$
$\qquad\qquad\qquad\qquad\qquad 3.3 \sin(\text{tilt})/2 \sin(T/2)$,

(iii) length along axis $= 3.3 \cos(\text{tilt}) + S \sin(\text{tilt})$.

We are now ready to relate different values of R, S, and T, as determined by the base sequence of DNA, to the overall shape of a double helix. For example, in Fig. 3.6, we showed that AA/TT prefers to stack so that it can form an extra hydrogen bond in the major groove; values of (R, S and T) for an AA/TT step in its preferred arrangement are ($0°$, 0 Å, $36°$). Therefore, a double helix containing only AA/TT steps will resemble Fig. 3.14(a), except, of course, that the base pairs will be highly propeller twisted: remember that the drawings of Fig. 3.14 do not show the propeller twist.

As another example, consider some DNA consisting entirely of GG/CC steps, all having uniform slide $S = -2$ Å and roll $R = 6°$, as shown earlier on the left-hand side of Fig. 3.12(c). This DNA will resemble the models shown in Fig. 3.14(b) and (d); and will be intermediate in structure between them, since the roll angle of 6° is halfway between the roll angles of 0° and 12° used to construct those two models.

Finally, note that the models shown in Fig. 3.14(a) and (d) correspond broadly to the 'B' and 'A' forms of DNA as shown earlier in Fig. 2.7. Look carefully at the pictures in Fig. 2.7, and locate the base pairs. Then compare the locations of the base pairs with those shown in Fig. 3.14(a) and (d), especially with regard to tilt and distance from the axis.

Almost all of the external features of the 'B' and 'A' helices, such as the distance of base-pairs from an axis, the tilt of pairs with respect to an axis, and the rise along the axis, can be calculated from their (R, S, and T) values according to the formulas given above.

Why then do different sequences of DNA prefer either the 'B' or 'A' forms in crystals? As shown earlier in Fig. 3.12, AA/TT steps do not appear at all in the 'A'-form region, whereas CA/TG or GG/CC steps may be found in either the 'B'-form or 'A'-form region, with some space in the center between them near slide $S = 0$ Å. Hence, the preference of DNA for two general forms, whether 'B' or 'A', may derive in part from the tendency of steps CA/TG or GG/CC with two G–C base pairs, to favor either of two separate values for roll R and slide S, rather than a continuous range. In fact, the classification of a long double helix into either of the two categories, 'B' or 'A', becomes somewhat arbitrary and ambiguous, when many successive steps having different roll R and slide S are considered as a broad average.

These few examples of the use of (R, S, and T) values in understanding DNA have related mainly to physical measurements of DNA structure from X-ray studies, rather than to the role of DNA in biology. But we must start somewhere! In the following chapters, we shall explain how the roll–slide–twist model is indispensable for understanding many of the roles of DNA in biology, such as how promoters work, how DNA coils around proteins in a chromosome, and how DNA binds gene-regulatory proteins such as 'repressors' or 'activators'. We do not need to bother about the other three of Euler's six degrees of freedom, unless we are dealing with DNA that has been severely distorted by contact with a protein or drug: see Appendix 2. Do not worry too much about the details of this chapter, such as the various formulas and constructions, so long as you grasp the meanings of propeller twist, and of roll, slide, and twist.

You may be puzzled that we have not shown the sugar–phosphate chains in any of the pictures of this chapter, except in Fig. 3.13. The chains are there, of course, but they have not been shown in the drawings. This is analogous to the way in which we have not shown propeller twist in some of the diagrams, either. The really important point in the present chapter is that the outward features of DNA *all* depend strongly on base-stacking arrangements at the inner core of the molecule.

Notes

1. In this book, we regard the left-handed sense of propeller twist, as shown in Fig. 3.4, as 'positive'. This is opposite from the sign convention given in the 'Cambridge Accord' (Dickerson *et al.* (1989) *EMBO Journal* **8**, 1–4), but it should not lead to confusion, since almost all propeller twists are seen to be of the same sense as that shown in Fig. 3.4, which is positive according to our convention.
2. See Appendix 1.
3. We have used here the symbols R, S, and T for roll, slide, and twist, respectively, with the sign conventions shown in Fig. 3.8. In fact, many different symbols and sign conventions have been used for those quantities by different scientists, and we have chosen the present set for the sake of simplicity in a textbook. Our symbols can be translated into those of the 'Cambridge Accord' of the X-ray diffraction workers (see Note 1, above) as follows: $R = \rho$, $S = D_y$, and $T = \Omega$, without any change of sign.

Further Reading

Calladine, C.R. (1982) Mechanics of sequence-dependent stacking of bases in B-DNA. *Journal of Molecular Biology* **161**, 343–52. Steric consequences of propeller twist for the stacking of base-pairs in the 'B' form of DNA.

Calladine, C.R. and Drew, H.R. (1984) A base-centred explanation of the B-to-A transition in DNA. *Journal of Molecular Biology* **178**, 773–82. First full statement of the roll–slide–twist model for DNA, as generalized to all right-handed forms of the molecule.

Dickerson, R.E. *et al.* (1989) Definitions and nomenclature of nucleic acid structure parameters. *EMBO Journal* **8**, 1–4. A comprehensive listing of possible structural parameters for DNA. Note that propeller twist is reversed in sign there, as compared with many papers in the literature and in this book.

Dickerson, R.E. and Ng, H.-L. (2001) DNA structure from A to B. *Proceedings of the National Academy of Sciences, USA* **98**, 6986–8. A review of X-ray studies on intermediates in the B-to-A helical transition.

Hogan, M., Dattagupta, N., and Crothers, D.M. (1978) Transient electric dichroism of rod-like DNA molecules. *Proceedings of the National Academy of Sciences, USA* **75**, 195–9. Early solution data favoring the existence of propeller twist in DNA base-pairs.

Laughlan, G., Murchie, A., Norman, D.G., Moore, M.H., Moody, P., Lilley, D.M.J., and Luisi, B.F. (1994) The high resolution crystal structure of a parallel-stranded guanine tetraplex. *Science* **265**, 520–4. A detailed view of an unusual four-stranded DNA helix using only guanine-to-guanine base-pairs.

Levitt, M. (1978) How many base-pairs per turn does DNA have in solution and in chromatin? Some theoretical calculations. *Proceedings of the National Academy of Sciences, USA* **75**, 640–4. Early theoretical calculations favoring propeller twist in DNA.

Minasov, G., Tereshko, V., and Egli, M. (1999) Atomic-resolution crystal structures of B-DNA reveal specific influences of divalent metal ions on conformation and packing. *Journal of Molecular Biology* **291**, 83–9. Detailed X-ray studies of where magnesium or calcium ions reside around DNA.

Nh, H.-L. and Dickerson, R.E. (2002) Mediation of the A/B DNA helix transition by G-tracts in the crystal structure of duplex CATGGGCC-CATG. *Nucleic Acids Research* **30**, 4061–7. An A/B intermediate as induced by runs of guanine bases.

Vargason, J.M., Henderson, K., and Shing Ho, P. (2001) A crystallographic map of the transition from B-DNA to A-DNA. *Proceedings of the National Academy of Sciences, USA* **98**, 7265–70. Double helices which lie intermediate in structure between A and B may be induced in d(GGCGCC) by either methylated or brominated cytosine bases.

Wing, R.M., Drew, H.R., Takano, T., Broka, C., Tanaka, S., Itakura, K., and Dickerson, R.E. (1980) Crystal structure analysis of a complete turn of B-DNA. *Nature* **287**, 755–8. First direct structural evidence for propeller twist in DNA base-pairs, by X-ray diffraction.

Bibliography

Nelson, H.C.M., Finch, J.T., Luisi, B.F., and Klug, A. (1987) The structure of an oligo(dA).oligo(dT) tract and its biological implications. *Nature* **330**, 221–6. Direct observation of very high propeller twist in a series of A–T base pairs, and a postulate of an additional hydrogen bond in the major groove between adjacent pairs (see Fig. 3.6).

Exercises

3.1 Normally the DNA double helix is right handed, as shown schematically in Fig. 3.1. In this case, the provision of left-handed or counter-clockwise propeller twist (when looking along the long axis of any base pair), as shown in Fig. 3.4, can reduce the access of water to the bases, as shown in Fig. 3.3.

For a hypothetical *left-handed* double helix of DNA, with $T = -32°$, what sense of propeller twist would be required to reduce likewise the access of water to the bases?

(Note: the left-handed 'Z'- DNA shown in Fig. 2.7 has almost no propeller twist, because the bases there stack not only onto neighboring bases, but also onto neighboring sugars.)

3.2   In the models of the 'A' and 'B' forms of DNA shown in Fig. 2.7, the *major*-groove edges of the bases are shaded heavily – a convention which is opposite from that used elsewhere in this book. Identify the major and minor grooves which lie between the sugar and phosphate chains in these two models. In the 'B' form, which groove has the larger width? In the 'A' form, which groove is wider? Or are the widths about the same? In the 'A' form, which groove is *deeper*?

3.3   Using Fig. 3.5 as a guide, identify the major- and minor-groove edges of the base-pairs shown with atomic detail in Fig. 2.11(a) and (b).

3.4   Leonard Euler explained long ago that any one rigid block has six degrees of freedom of motion with respect to another rigid block. Each of these may be described in terms of a translation along, or a rotation about, any of the three axes which are labeled in Fig. 3.7. In practice, three of these six degrees of freedom are *not* mobilized significantly in the base-pair steps of DNA.

By use of a simple model involving blocks of wood, or cardboard boxes, confirm that translation or shortening along the 'twist axis', and rotation about the 'front–back' axis, are inhibited because of the close surface-to-surface stacking of the bases.

Also, examine the third unused degree of freedom, which involves translation along the front–back axis. It is not clear whether this motion is inhibited in real DNA by the chemical forces which influence base stacking, or by the action of sugar–phosphate chains, or both. The front–back motion could conceivably be favored for certain sequences in DNA, but we have at present few good examples, apart from a few steps which include two G–C base pairs, and so repel due to partial electric charge along their short axes, and a few CA/TG steps in DNA wrapped around protein spools.

Finally, confirm that the three allowed motions of roll, slide, and twist in DNA, as shown in Fig. 3.8, are also allowed in the wooden-block models.

3.5a   Make a simple physical model to illustrate the linkage between slide and twist shown in Fig. 3.13. (Expanded polystyrene foam can be cut easily into suitable blocks using a bread knife, and the sugar–phosphate chain links may be made from wires or paper clips, with their ends pushed into the blocks.)

b   Make a simple physical model of two adjacent propeller-twisted base-pairs, with the bases of unequal size as in Fig. 3.6. (Expanded polystyrene foam blocks may be held apart conveniently, in a propeller-twisted arrangement, by means of cocktail sticks.) Use these model base pairs, without any sugar–phosphate chains, to study the linked slide–roll motion which is shown in Figs 3.10 and 3.11.

c   Use the model of part (b) to investigate the 'locking' of both slide *S* and helical twist *T* in AA/TT steps by the additional hydrogen bond shown in Fig. 3.6 and the associated high propeller twist. Also demonstrate that the absence of these effects allows *S* and *T* to vary without hindrance.

3.6   The table below gives approximate, uniform values for roll, slide, and twist at the base-pair steps in three different well-known forms of DNA – 'A', 'B', and 'C' – which have been studied by X-ray diffraction of fibers:

	'A'	'B'	'C'
$R(°)$	+12	0	−6
$S(\text{Å})$	−1.5	0	+1
$T(°)$	32	36	40

Use the equations on p. 58 to calculate approximate values of the following parameters for each of the three forms 'A', 'B', and 'C':

a   the distance of the centers of base-pairs from the axis of the double helix;

b   the angle by which base-pairs are tilted from planes normal to this axis;

c   the length, or 'rise', of the molecule per base-pair along the axis. (Note: calculate the tilt angle first.)

CHAPTER 4

Twisting and Curving

In the last two chapters we have learned some rudimentary things about DNA. We have learned: (a) why DNA forms a helix, (b) how the bases make ordered pairs at the center of the helix, (c) how the bases twist like a propeller within any base-pair, and (d) how the overall shape of a helix depends on the local parameters roll, slide, and twist over a series of base-pair steps. This is the stuff of chemistry, not biology. When are we going to start talking about DNA in biology? That is exactly what we shall be doing in this chapter.

The two most fundamental actions of DNA in biology involve either the *twisting* or *curving* of a DNA double helix. First, we consider the twisting of DNA – or, to be more precise, its *un*twisting. There are two main instances where DNA has to untwist as it carries out its duties in a cell: first, when DNA is copied into the messenger-RNA that tells the cell how to make protein; and second, when DNA is copied into another DNA strand just before a single cell divides into two cells. The first of these processes is called 'transcription', and the second is called 'replication': see Fig. 4.1. In each case, the DNA unwinds into two separate sugar–phosphate strands. These pictures provide, of course, only a static representation of a brief instant in the life of a cell. You have to imagine that the unwound regions of DNA in pictures (a) and (b) are moving rapidly across the page, from right to left, in order to grasp the dynamic nature of DNA unwinding in living systems.

Often DNA unwinds only over a short region, say 15 to 20 base-pairs, when making RNA as in (a), because it takes a lot of energy to pull the base-pairs apart and expose them to water. The 'bubble' of unpaired bases can travel along the length of the DNA very rapidly, at about 100 base-pairs per second; but then, time doesn't mean

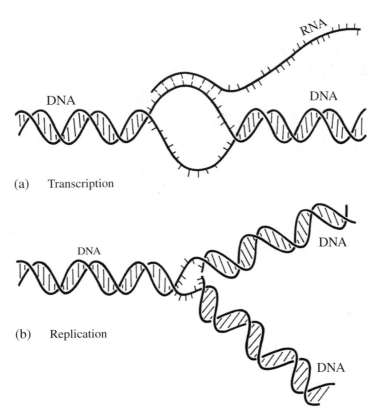

Figure 4.1 Schematic representations of transcription and replication of DNA. In each case, the DNA must unwind locally to let one strand serve as a template for the synthesis of a new strand, either of (a) RNA or (b) DNA.

much to these tiny molecules that we can hardly see by using a light microscope. When DNA gets copied into RNA, a copying protein or enzyme called 'RNA polymerase' attaches itself to one of the two DNA strands: see Fig. 4.2. Then the polymerase pulls nucleotides out of solution to match the bases it finds within the DNA chain. In Fig. 4.2 the enzyme is just about to add an RNA base C to a DNA base G. Another somewhat similar enzyme carries out the process of copying DNA into DNA (Fig. 4.1(b)); it is called 'DNA polymerase'. Any cell contains several varieties of RNA polymerase, and several varieties of DNA polymerase, to do different kinds of copying tasks.

These RNA polymerases in the cell always try to put C with G, G with C, A with T, and U with A to make Watson–Crick base-pairs, as shown inside the 'bubble' of Fig. 4.2. The DNA polymerases do the same, but add T instead of U to A. Some of the copying enzymes, in a test-tube, will incorrectly put T with G if you feed them a lot of T nucleotides and no C at all; but they add T to G much more slowly than C to G. These kinds of copying error do not happen very often

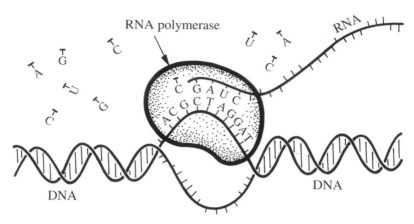

Figure 4.2 An RNA strand in the process of being made from single nucleotides by the enzyme RNA polymerase. One of the two DNA strands serves as a template for accurate synthesis of the new RNA strand, by the rules of Watson–Crick pairing. The actual length of the DNA–RNA hybrid, shown schematically here, is close to 8 base-pairs.

in living cells, where all four kinds of nucleotide are relatively abundant.

Now, you remember that DNA is helical, like a screw: so, as the RNA polymerase and its 'bubble' of unpaired bases move along the DNA, either the polymerase must screw *around* the DNA, or the DNA thread must screw itself *through* a stationary polymerase. Which of these alternatives will be favored in living cells? The two competing models are shown in Fig. 4.3, and it is not obvious on first inspection which might be right. Actually, James Wang and Leroy Liu have shown that the polymerase often remains somewhat stationary, as in Fig. 4.3(b), while the DNA screws through it. There are good reasons why this should be so. First, it will be hard for the big polymerase protein to move rapidly through the sticky, viscous fluid of a cell nucleus, in order to rotate about the DNA by 10 times per second; the thin, wiry DNA can rotate in the fluid much more easily. Second, the polymerase drags behind it a long RNA 'tail' as it goes about its duties, as shown in Fig. 4.2. It would be very hard for this long tail of perhaps 500 to 1000 nucleotides to follow the polymerase round and round the DNA, at a rate of 10 times per second. Indeed, some experiments suggest that in a test-tube the polymerase finds it easier to rotate about the DNA if the RNA tail is cut off by another enzyme (called RNAase) as it is being made.

If you have ever been out fishing, and the current twists your bait around the end of your line, you will know that there will be an unholy mess. Similarly, it would cause big trouble if Nature did not

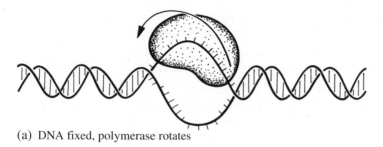

(a) DNA fixed, polymerase rotates

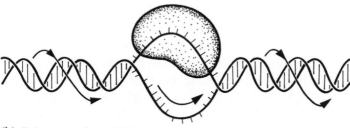

(b) Polymerase fixed, DNA rotates

Figure 4.3 Alternative schemes for the copying of DNA into RNA by the enzyme RNA polymerase. In (a) the polymerase screws around a stationary DNA, while in (b) the DNA screws through a stationary polymerase.

find some way to relieve the twisting stress of DNA during its passage through a polymerase. In fact, Nature has invented several different kinds of de-knotting enzyme, called 'topoisomerases',[1] to get rid of the excess DNA twist. These enzymes act in very subtle ways, and no one is certain how they work on a near-atomic scale. They do however cut either one or both of the two sugar–phosphate chains to allow some kind of motion in the DNA to relieve torsional stress; then they re-connect the broken parts and thus leave an intact but relaxed DNA. Repeated application of this process to a tangled piece of DNA will eventually unravel it. We shall discuss the tangling (or super-coiling) of DNA further in Chapter 6.

Many clinically-used anti-cancer drugs, for example 'doxorubicin', are directed against the de-knotting enzymes. The cancer cells, which are growing out of control, can't divide if you poison their de-knotting enzymes. Some of these anti-cancer drugs 'trap' the topoisomerases in the middle of their cutting–relaxing–rejoining cycle, so that they cannot re-connect the strands of the DNA. Yet apparently, those cancer cells which survive the first few rounds of treatment with these drugs somehow find a way to grow with only a very low level of de-knotting enzymes; or else turn on new genes which help to expel the drug from the cell; and so they are eventually able to resist further treatment.

So it certainly doesn't look very promising, if you are a biological scientist today, to look for more anti-cancer drugs that poison topoisomerases. In fact, the most successful new anti-cancer drug to be discovered recently is called 'taxol' or 'docetaxol'; and it works by preventing the long cellular structures known as 'microtubules' from disassembling after they are made during cell division, when they help to carry one copy of each chromosome to each new cell – the process of separation mentioned on p. 4.

Apart from doxorubicin and taxol, the third most promising anti-cancer drug is called 'cisplatin' or 'carboplatin'. It was found in 1969 by Barnett Rosenberg, who was studying how bacteria grow in an electric field. He found that certain electrodes cause bacteria to grow long and thin, like spaghetti. By lots of detective work, he eventually worked out that the platinum in his electrode was combining with ammonia in his buffer, to make a platinum–ammonia compound that prevented cell division. This compound had been sitting on people's shelves for over 100 years, but no one had ever thought it would cure cancer. Anyway, he tested it on people, and it worked in some cases. Apparently, the platinum atom uses two of its ligand-binding sites to cross-link purines in GG or GA steps; which causes the DNA to bend, and so prevents replication.

We mentioned above that DNA unwinds into a bubble of 15 to 20 bases, as RNA polymerase copies one strand of the DNA to RNA. How is this accomplished? It turns out that the hard step is to unwind the double helix to begin with, in just one location. After that has been done, the bubble of DNA can travel with relative ease to some other location along the length of the molecule. It is therefore of crucial importance to understand how the cell tells its DNA to unwind in certain, specific locations. These events are then responsible for the large-scale synthesis of RNA from many particular genes on the DNA; and so they are responsible, ultimately, for the kinds of protein which are found in any cell. A similar kind of unwinding takes place at the start-sites for replication, where both strands of the DNA are copied into DNA; but there the enzymes are different.

What would you do if you were faced with the problem of unwinding the DNA from a cell in many specific locations? Would you make a protein with a little shovel attached at one end, to dig or pry a hole in the DNA? Some DNA-unwinding proteins do have little 'shovels', in the form of flat, oily amino acids such as phenyl-alanine, tyrosine, and tryptophan, which can make holes and then insert themselves between the base-pairs, and thereby effect unwinding of the helix. They would act just like the ethidium bromide molecule shown in Fig. 2.9, which converts the DNA locally from a helix into a partially untwisted ladder. Other DNA-unwinding proteins perhaps have clefts on their surfaces into which only one of the two DNA strands can fit, so that they bind to a single strand in preference to a double helix, and so unwind the molecule that way. Finally, yet other proteins might curve the DNA around themselves in the form of a telephone cord or a coiled bed-spring, so that the DNA can unwind as it vibrates like a concertina in solution. There is some evidence for all of these kinds of unwinding in the cell; yet only recently has there been much knowledge at a detailed, atomic level of what happens in any case.

All of these unwinding events involve the binding of some protein to DNA. Much progress has been made recently at visualising the complexes of these proteins with DNA, using X-ray crystallography or other methods; yet still more is known about the role of DNA in such interactions than about the role of the protein. As a general rule, Nature makes the double-helical connections very weak in places where unwinding must begin, so that it won't take much energy to break open the DNA base-pairs, and thereby separate the two strands from one another. You will recall from Chapters 2 and 3 that the main determinants of the stability of a double helix are the number of hydrogen bonds between the bases in a base-pair, and the extent of base-to-base overlap. An adenine–thymine pair has only two hydrogen bonds, while a guanine–cytosine pair has three (see Fig. 2.11); so a series of adenine–thymine pairs will make the double helix less stable, as shown in Fig. 4.4(a). Furthermore, it has been shown by many careful experiments that pyrimidine–purine sequences have

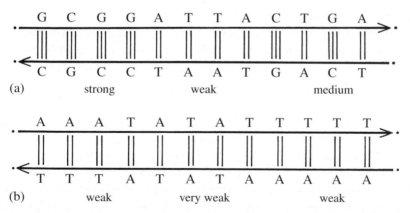

Figure 4.4 A two-part explanation for the ease of DNA unwinding at any TATA sequence. In (a), the double or triple cross-chain lines represent 2 or 3 Watson–Crick hydrogen bonds, respectively; and the double helix is less stable with 2 bonds than with 3. In (b), the TA steps (of kind pyrimidine–purine) cause a further weakening of the cross-chain connections, because they can easily unwind to yield low twist.

the least stability with regard to base-to-base overlap. These are the steps TA, TG, CA and CG. Why should this be so?

In the last chapter, we explained how to describe DNA in terms of the roll R, slide S, and twist T at any base-pair step. Pyrimidine–purine steps are special in this description because they can adopt either of two stackings: a high-slide form where the purines slide apart from one another (Fig. 3.10), or a low-slide form where the purines slide on top of one another (Fig. 3.11). Now the base-pairs are connected to sugar–phosphate chains in such a way that *low slide* leads to *low twist*. Thus, in the example shown in Fig. 3.13, a slide of $S = -2$ Å leads to a low twist of $T = 28°$ or thereabouts, as compared to $T = 34$ to $36°$ for other DNA. So it may be that the amount of energy required to unwind the DNA further, say to $T = 10°$ or $20°$, will be less for a pyrimidine–purine sequence than for other DNA, because the base-pairs there are already unwound to a significant degree; or perhaps they can unwind more easily under stress than at other sequences into a stable, low-twist form.

In any case, there seems to be little energetic barrier to low twist at a pyrimidine–purine step, for reasons that may not yet be fully understood.

Which, then, are the most easily unwound sequences in DNA? Simply those which combine the two characteristics we have been describing: few hydrogen bonds as in A–T pairs, and low twist as in pyrimidine–purine steps. In other words, DNA unwinds most easily at AT-rich regions that have many pyrimidine–purine steps, as shown in Fig. 4.4(b). The prototypic 'weak' sequence is something like 'TATATATA' or 'TAATAATAA', where TA is the pyrimidine–purine step of low twist.

This is not just a theoretical, hand-waving argument: there is actually strong experimental evidence for a low-twist intermediate in the unwinding of TATA-type sequences. It seems that such sequences unwind to 'cruciform' configurations more easily than other DNA. Now a cruciform is just a big 'bubble' of DNA which can be trapped and studied, because each strand of the bubble folds back and pairs with itself to form a double helix, as shown in Fig. 4.5. James McClellan and David Lilley set out to determine the reason for this odd behavior. They found that the unwinding of DNA into a cruciform at TATA-like sequences was being catalysed by partial unwinding of the DNA from 10.5 to about 12 base-pairs per turn, or from $T = 34°$ to about $T = 30°$, before the cruciform appeared. Their observations agree closely with what we expect from theory, as explained above.

Biologists identified the TATA-type sequences almost by accident, by determining the sequences of DNA in places where the double helix must unwind for transcription to begin. These broadly defined

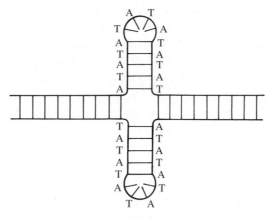

Figure 4.5 A 'cruciform' structure which has been 'extruded' from a suitable sequence of A and T bases. The DNA double helix untwists from 10.5 to 12 base-pairs per helical turn before the cruciform appears, owing to the easy unwinding of TATA sequences. As a consequence of this initial untwisting, the cruciform extrudes there more readily than at other sequences.

regions of DNA are called 'promoters', because they promote RNA synthesis from a nearby gene without being transcribed themselves. The biologists found in many cases a 'TATA' or similar sequence at or close to the required site of unwinding within each promoter. Then they drew a box around this sequence on their computer output, to show how important it might be. Hence, many textbooks call the weak region of DNA in some promoters a 'TATA box'. The majority of promoters contain a TATA or related sequence 'upstream' of the site where unwinding into separate strands begins; and this evidently plays an important part in starting the operation of the polymerase.

Thus, in most *bacteria*, an RNA polymerase protein binds to the TATA region: one part of that polymerase, the 'sigma initiation factor' actually contacts the TATA sequence, 10 base-pairs upstream from the start-point for making RNA. This sequence is where unwinding of the two strands starts.

But in the cells of our body the TATA-like 'box' performs a different function. Here, there is almost always an *additional* protein called 'TBP' (for 'TATA-binding protein') which binds to TATA-like sequences near the start of genes (and is sketched in Fig. 4.14, below). This protein is available in moderate amounts in pure form, and several groups of workers have found that it untwists the TATA sequence, as part of preparation for the polymerase, by about one-third turn. Here, the TATA-sequence preceeds the site of unwinding by about 25 base-pairs. We shall discuss the action of polymerases further in Chapter 6.

The TATA sequence is known to adopt at least two different helical forms: one with low twist and high roll, as seen when bound to TBP (see below), and some other proteins (also discussed below); and a second form with just 7.5 to 8.0 base-pairs per turn rather than 10 to 12 for normal DNA. That last helical form has been called the 'D' form of DNA in X-ray studies of DNA fibers, at low resolution. Until we establish the molecular structure of such a strange helix, we cannot even guess whether it might be used in biology.

As a last note on the subject of TATA-type double helices, we refer to the three triplets in the Genetic Code (Table 1.1) which code for 'stop', and are usually called 'stop-codons': TAA, TAG and TGA. The process of assembling amino acids into a protein chain comes to a halt at those particular triplets, because there are no transfer-RNA molecules which can recognize them. One of the stop codons is UAA in the messenger-RNA (like TAA in DNA), and if a transfer-RNA molecule for UAA were to exist, it would form a specially weak bond with this triplet (see Fig. 4.6) for precisely the reasons given above. We think that it is no accident that this specially weak-binding triplet does not correspond to a specific amino acid, for the hypothetical transfer-RNA would be unreliable at recognising it. The same considerations probably apply also to the two other stop-codons UAG and UGA, which are moderately weak in their capacity for base-pairing.

That gives an overview of twisting in DNA. Now let us look at the *curving* of DNA. In principle, DNA could curve either in a plane, like a banana, or else in three dimensions, like a coiled bed-spring. It is hard to learn about three-dimensional curvature at the first attempt, however; so we restrict ourselves in this chapter to the

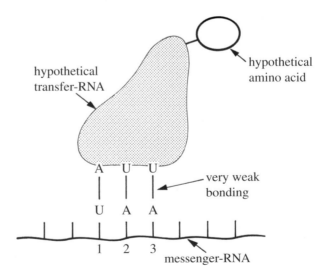

Figure 4.6 All of the 'STOP' signals in the Genetic Code (UAA, UAG, UGA, see Table 1.1) would correspond to very weak or moderately weak pairings between a hypothetical transfer-RNA and the messenger-RNA.

curving of DNA in a plane. The next chapter will deal with curving in three dimensions.

First let us forget about DNA for a moment, and think instead about our gardens. Many people have small, flowering trees in their suburban gardens, and they like to use slabs to build neat paths around them, as shown in Fig. 4.7. There are two ways to do this. One way is to buy special slabs that are made in a wedge-shape, with one end wider than the other. Such slabs will make a circle of a particular radius, just as stone 'voussoirs' will build a round arch of specific radius over a window or a door. The other way is to make the path out of rectangular slabs, and to fill the narrow triangular spaces between them with pebbles. An advantage of this scheme, of course, is that one can change the curvature of the path at will, making it circular, or straight, or of variable curvature as the plan requires, just as in the picture.

What do we mean by the *curvature* of a garden path? In Fig. 4.7 the path is obviously more curved where the angle between consecutive slabs is larger. Where this angle is zero, the path is straight; that is, it has zero curvature. It would therefore be sensible to define curvature in terms of the angle between successive slabs. Thus the curvature of the semicircular part of the path in Fig. 4.7 may be described as 15° per slab (or per step) because the path turns through 180° in 12 steps.

Now, if we lay a garden path with 15° between every pair of slabs, we shall eventually complete a circle. And the radius of the circle will be smaller if we make the angle of curvature larger, and vice versa. We can easily calculate the radius of the semicircle in Fig. 4.7, as

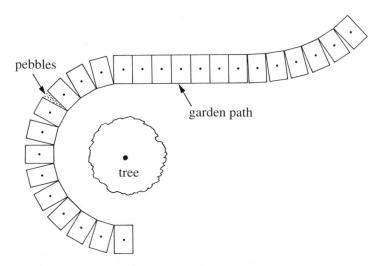

Figure 4.7 Curvature of a garden path in a plane, around a tree.

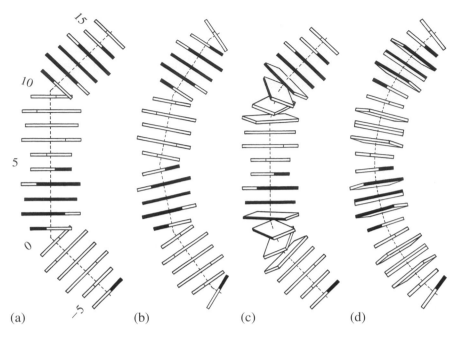

(a)   (b)   (c)   (d)

Figure 4.8 Two complete helical turns of DNA, with a curvature of 45° per turn, or 4.5° per step on average. Such tight curvature may be achieved, in principle, by any of the distributions of roll angle shown in parts (a) to (d).

follows. Suppose the width of each slab is 0.5 m. If the angle between slabs is 15°, there will be 360/15 = 24 slabs in a circle, with a total circumference of 24 × 0.5 = 12 m, and hence a radius of 12/2π = 1.9 m. This is the radius of the inner edge of the slabs; similarly we could find the radius of the center-line of the path, by using the center-to-center separation of the slabs, together with the angle between slabs.

The curvature of DNA is more complicated than that of a garden path, because DNA is three-dimensional; yet it follows broadly the same principles, because DNA is also made in discrete steps, being built up from base-pairs as we have seen. Thus, imagine that a piece of DNA of length 80 base-pairs, or 8 double-helical turns, has been bent into a 360° circle. That is just about the degree of curvature by which DNA wraps around proteins in the cell nucleus (Fig. 1.5). If you were to make the same circle from just eight tiny slabs, then the angle between slabs would have to be 360°/8 = 45°. This angle is precisely equivalent to the roll angle R in DNA, as described in Chapter 3. If we wish to make a circle from eight helical turns of DNA, we can do so by putting one roll angle of 45° in each double-helical turn: that will make a fine circle.

Such a scheme is shown in Fig. 4.8(a), which shows 20 base-pair steps that make up one-quarter of our 80-base-pair circle. In order to

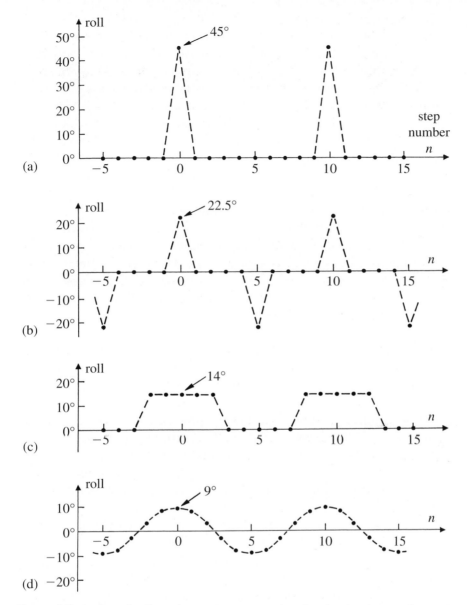

Figure 4.9 A plot of roll angle *versus* step number for the four cases shown in Fig. 4.8, each of which curves the DNA by 45° per helical turn.

convert straight DNA into curved DNA, we have introduced a roll angle of 45° at the two steps labeled 0 and 10, respectively, on the left-hand side of the drawing. These roll angles open up the minor-groove edges of the base-pairs, which are colored black, to yield a total curvature of $2 \times 45° = 90°$ over the 20 steps. The same type of curvature is shown schematically in Fig. 4.9(a), which gives a plot of roll angle *versus* step-number in the sequence: roll angles of 45° can be seen at steps 0 and 10 in this diagram, but at other steps there is zero roll.

It might be better, however, to spread these roll angles over a greater number of base-pair steps within the circle. While a single roll angle of 45° per helix turn is satisfactory from a geometrical point of view, it is not satisfactory from a physical point of view, because the opening of a single base-pair step by 45°, as in Fig. 4.8(a), would expose a lot of water-insoluble base surfaces to the solvent. A slight improvement might be made by dividing the 45° into two parts, and thus locating a roll angle of 22.5° at two places within each double-helical turn. This scheme is shown in Fig. 4.8(b), and it has also been plotted in Fig. 4.9(b): roll angles of +22.5° are located at steps 0 and 10, while roll angles of −22.5° are located at steps −5, 5 and 15. The sign of the roll angle changes from plus to minus every five base-pairs, because the helix rotates by 180° over that distance. So the two angles of 22.5° open up the same side of the helix, and make equal contributions to the curvature, despite the change in sign.

Another way of bending DNA into the required curve might be to have five steps with a roll of about 14° followed by five with zero roll, and then five more with $R = 14°$, etc. This new scheme is shown in Figs 4.8(c) and 4.9(c). In effect, we are converting successive half-turns of DNA into different uniform configurations; and it is not difficult to see that this will produce a curve. The model shown in Fig. 4.8(c) is sometimes called a 'junction' model for DNA curvature, because 'bends' seem to appear at the junctions between the successive portions of DNA, each having uniform but different roll: but it is important not to forget that curvature depends on the roll angles at all of the steps.

Perhaps the best scheme of all for making curved DNA would be to spread the roll over the entire helix, at every base-pair step, rather than placing big roll angles at only a few steps. Such an idealized model is shown in Fig. 4.8(d), and in the corresponding diagram of Fig. 4.9(d). If we let the roll angles vary as a cosine wave, starting at step 0 and ending at step 10, then we shall need to vary these roll angles by no more than plus or minus 9° to get a curvature of 45° per turn. You can see that there are much less abrupt changes in roll from step to step in Fig. 4.8(d) than in any of (a), (b), or (c). This may seem a bit like magic: how can we know what curvature will result from such an assortment of different roll angles as in this example?

The roll angles in Fig. 4.9(d) vary as a cosine wave of amplitude 9°, or as $R_n = 9° \cos(36°n)$, where $n = 0, 1, 2, \ldots, 9$ identifies each step in any double-helical turn. For example, the roll at step 0 is $R_0 = 9°$, while the roll at step 1 is $R_1 = 9° \cos(36°) = 7.3°$, and the roll at step 5 is $R_5 = 9° \cos(180°) = -9°$. This new arrangement is similar overall to the various schemes shown before in (a), (b), and (c), but it is more subtle in the way in which it assigns roll to different parts of the helix. Now we come to a crucial point: not all of the roll angles contribute fully to curvature, because some of them cause the DNA to bend in the wrong way, out of the plane of the paper in Fig. 4.8. Let us look again at the helix shown in Fig. 4.8(a). In this picture, a positive roll angle at step 2, between steps 0 and 5, would bend the DNA down into the paper, rather than to the right as desired. But a roll angle at step 0 bends the DNA to the right, while a roll at step 5, not shown in the drawing, would also bend the DNA to the left or to the right, as illustrated in Fig. 4.8(b). It can be shown that the contribution to rightward curvature by any step is its roll angle R_n, multiplied by the cosine of the total helix twist, or $R_n \cos(36°n)$. Thus, with arrangement (d) we get for step 0 a big contribution to curvature of $9° \cos(0°) = 9°$, and for step 5 we also get a big contribution of $-9° \cos(180°) = 9°$; but for step 2 we get only the small contribution of $9° \cos^2(72°) = 0.9°$, because the roll points mostly in the wrong direction.

It follows, then, that the overall curvature k due to roll at all steps $n = 0, 1, 2, \dots , 9$ can be found by adding together 10 different terms of the kind $R_n \cos(36°n)$, or in this case

$$\sum_{n=0,9} 9° \cos^2(36°n) = 9° + 5.9° + 0.9° + 0.9° + 5.9° + 9°$$
$$+ 5.9° + 0.9° + 0.9° + 5.9° = 45°$$

In summary, the roll angles in Fig. 4.9(d) vary as $9° \cos \theta$, where θ is the total twist (or sum of base-step twists T) relative to step 0; and then you have to multiply them by $\cos \theta$ again before adding up to get the total curvature, because not all steps point in the same direction. This simple scheme is not quite exact, geometrically, but it works well for small roll angles, of less than or equal to about 10°; and that covers most practical cases.

It is not difficult to show that if a constant roll angle is added to every step, so that $R_n = 9° \cos(36°n) + $ (constant), then the sum worked out above has exactly the same value as before. A constant change of roll at all steps will change the overall appearance of the DNA, just as in Fig. 3.14(c), but it imparts no curvature.

As we have seen, not every step contributes equally to curvature in our scheme. Steps 0, 1, 4, 5, 6, 9 (where the minor or major groove faces the center of curvature) contribute a lot, while steps 2, 3, 7, 8 don't contribute much at all. The reason for this behavior is that the

DNA is not flexible enough in a direction at right-angles to the roll axis (that is, about the 'front–back' axis of Fig. 3.7) to curve very much in that direction. Also, one might suspect that slide S and twist T vary along with roll R as the DNA curves, in accord with the relations among R, S, and T discussed above in Chapter 3; but this would make only small differences to the general pattern of behavior which we have described.

When studying the curvature of real DNA, which is made from a variety of base sequences, we cannot expect every kind of sequence to follow a smooth, wave-like pattern of roll angles exactly. Some steps, such as the pyrimidine–purine variety and others discussed in Chapter 3, are flexible enough to adopt either high roll or low roll with little difficulty. However, others such as AA/TT are known from studies of DNA in the crystal to prefer low roll, $R = 0°$; while steps such as GC/GC are thought to prefer high roll, $R = +5°$ to $+10°$. Essentially, what we must do for real DNA, when it is curved around a protein spool, is to find the best fit of a cosine wave to a given series of R values, at a period of 10 steps, even though some of these steps may not agree with the idealized cosine wave exactly. The spread of roll angles can range from $+9°$ to $-9°$, or from $+12°$ to $-6°$; it doesn't matter. Once we have obtained a 'best fit', then the amplitude of this best-fit cosine wave will be proportional to the absolute curvature of the sequence, while its left-to-right position or phase, as in Fig. 4.9(d), will tell which steps have the largest (or most positive) roll angle, and which steps have the smallest (or most negative). Where the roll angle is large and positive, the minor-groove edges of base-pairs will lie along the *outside* of the curved DNA, and where the roll angle is small (or most negative), the minor-groove edges will lie along the *inside* of the curve. This protocol is known to mathematicians as taking the 'Fourier transform' of the roll angles, to get the amplitude and phase of the curvature of the DNA helix.

You can see for yourself how this works by making a detailed plot along the lines of Fig. 4.9(c) for a series of steps such as those shown in Fig. 4.8(c). There, a batch of 5 steps $(-2, -1, 0, 1, 2)$ with $R = 14°$ is followed by a batch of 5 steps with $R = 0°$, and then by another batch with $R = 14°$, and so on. You will find that the cosine wave that can be drawn most closely over these points has an amplitude (or half-height) of about $9°$; which is just what is needed for a curvature of $45°$ per double-helical turn.

When we consider the DNA that is found in our chromosomes, we might expect to find AA/TT sequences where the DNA curvature requires low roll, and GC sequences where the curvature requires high roll. That is precisely what we do find, to a first approximation. Most of the DNA in our chromosomes curves strongly around protein 'spools' (Fig. 1.5) for almost two turns of 80 base-pairs each, or for about 160 base-pairs in all, into a flat, left-handed supercoil that resembles part of a telephone cord. The proteins that make up any spool are called 'histones', and they will be discussed in Chapter 7. When we examine the DNA sequences that reside in these tight coils that wrap around the protein spools, we find that AA/TT sequences have a higher probability of being in a low-roll position than other DNA, and that GC sequences have a higher probability of being in a high-roll position; but not every AA/TT goes to low roll and not every GC goes to high roll. The preferences of all 160 base-pair steps in this DNA have to balance against one another to attain an optimal positioning of roll angles for *most* of the steps in the sequence; and so some individual steps may not fit into the overall pattern.

Some actual data are shown in Fig. 4.10, concerning the preferred locations of different base sequences in curved, chromosomal DNA. To construct these plots, we isolated the DNA from over one hundred different spool–DNA complexes, and then counted how many times *N* each kind of sequence might be found at any possible location on the histone spool. During the isolation of this DNA, it was trimmed in length from 160 base-pairs to 145 base-pairs by a special enzyme, in order to mark more precisely where each spool might begin or end. The plots in Fig. 4.10 show data that have been averaged over positions 1 to 72 and 73 to 144 of the spool–DNA complex, about the center at 72.5, because these data were found to be much the same on both sides of the center.

Figure 4.10(a) shows the number *N* of AA/TT steps that were found at each possible location along the path of the long, curved DNA in our many examples. These AA/TT steps may be seen to be located preferentially at positions 5, 15, 25, 36, 46, and 56 from either end of the DNA; such locations are known to be positions of low roll, where the minor groove faces inwards, from an X-ray analysis of the spool–DNA complex. Around the center of the DNA at position 72.5, the periodic pattern is broken, and peaks for AA/TT appear at positions 62 and 72. It turns out that the preference of AA/TT steps for these central positions may not be due to DNA curvature, but perhaps to specific contacts between certain amino-acid side-chains on the protein and the exposed edges of the A/T base-pairs; the explanation is still not certain. But we need not worry about this detail: the bulk of the DNA shows roll-angle preferences in accordance with our general scheme.

Figure 4.10(b) shows, similarly, the number *N* of GC steps at each possible location along the long, curved DNA. These GC steps may be seen to be located preferentially at positions 11, 21, 31, 41, 52, or about 5 steps away from the preferred locations of AA/TT steps. In

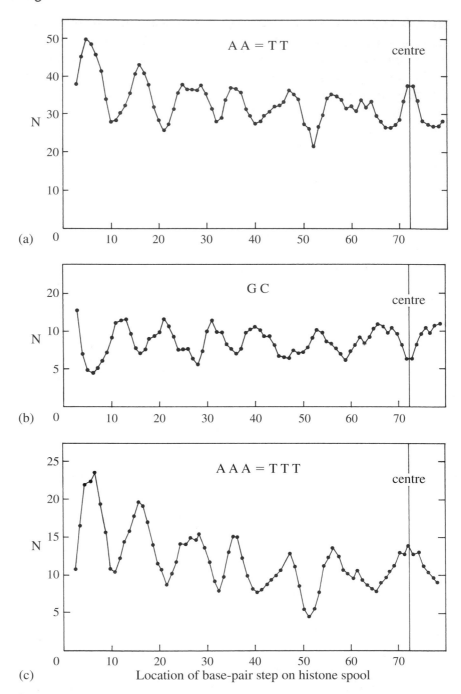

Figure 4.10 Number *N* of various short sequences in curved chromosomal DNA, plotted against their locations on the histone spool (see Fig. 1.5). The overall length of the DNA is 145 base-pairs, but the data could be averaged on both sides of the center at position 72.5, because they were found to be the same on both sides. Almost two hundred different DNA molecules were sequenced in order to construct these plots. From S.C. Satchwell *et al.* (1986) *Journal of Molecular Biology* **191**, 659–75.

other words, they prefer positions of high roll. The mean periodicity, or spacing of peaks, in both the AA/TT and GC patterns is close to 10.2 base-pairs, or one turn of double helix. The preferred helical periodicity of this DNA in other circumstances, when freed from the protein, is a slightly larger 10.6 base-pairs per turn.

Figure 4.10(c) shows an especially strong pattern of preferences for the trimer AAA/TTT: here two adjoining AA/TT sequences reinforce each other's individual preferences. Thus the AAA/TTT trimer has a strong preference to be located in a position of low roll rather than a position of high roll: the value of N differs by a factor of about 2 between these locations. This is quite a large factor, when you remember that the piece of DNA which is being curved around the histone spool consists of 144 steps or 145 base-pairs. Suppose this piece of DNA contains just one AAA/TTT trimer: then in 2 out of 3 cases, the AAA/TTT will be located in a position of low roll, due to its own influence in combination with the influences of other sequences. From this example, we can see that roll-angle preferences of particular short (i.e. two- and three-base-pair) sequences in DNA may be rather strong, provided they are not balanced by the opposing preferences of other sequences. Thus, certain AA/TT steps might predispose the DNA to adopt one position on the spool, whereas certain GC steps might prefer a different position. This is like the story of the husband and wife who always vote for different political parties, Republican and Democrat, or Conservative and Labour: in the end their votes cancel.

When studying pieces of DNA from a chromosome that are longer than the 145 base-pairs examined above, often we see that the spool–DNA complexes align themselves into an ordered array that may extend for 1000 to 2000 base-pairs. In such cases, the roll-angle preferences within any small region of the DNA, say 100 base-pairs, would have to be very strong to dominate the alignment of the other 900 to 1900 steps. Alternatively, these 100 steps could 'break' the array to yield a less regular structure; and many irregular models of chromosome structure have lately been proposed. In any case, the fit of roll-angle preferences in the DNA on one histone 'spool' to those of the DNA wrapped around neighboring spools is a matter of some interest, but it is not yet understood.

These alignments of DNA along the surfaces of histone spools turn out to be very important in biology, to decide for example where the AIDS or HIV virus will insert itself into human chromosomes. Such preferences for a particular rotational setting of the DNA are so strong, in fact, that some people are now using the HIV insertion enzyme as a way to probe the structures of chromosomes inside living cells! Thus in general, proteins such as the HIV insertion enzyme, that bind even to a very small region of DNA, seem to follow the same rules of DNA curvature.

As another example, a protein from a bacterial virus, known as the '434 repressor' (because it stops RNA from being made at a site of DNA unwinding in 434 virus), binds to DNA in the following way: it probes deeply into the grooves at either end of a 14-base-pair region, and curves the helix moderately in the middle, without making contact with it there, as shown in Fig. 4.11. The minor groove faces the protein at the center, and so the base-pair steps around bases 6, 7, 8, 9 must adopt a low or negative roll for the DNA to be bent by the protein in a correct fashion, so that the two parts of the protein can fit together properly. Indeed, if you put a low-roll sequence of four bases such as AATT in positions 6, 7, 8, 9 of the DNA, the protein binds 300 times more tightly than if you put a high-roll sequence such as GGCC. In chromosomes, such preferences are averaged out over a long region of DNA so there they appear to be weak; but in small regions of DNA, these preferences are not averaged out, and so they act strongly. In fact, you can calculate from the kind of bases which are found in positions 6 to 9 of the binding site how tightly this 434 repressor protein will bind to DNA, by using roll-angle preferences taken from the spool–DNA data discussed above. The same physical phenomenon of curvature-preference in DNA is operative in both cases.

But how does the 434 protein bind to DNA in positions 1 to 5 and 10 to 14, where it contacts the base-pairs most closely? Is the flexibility of DNA over positions 6 to 9 sufficient to provide enough specificity for the protein to carry out its function in the cell, of

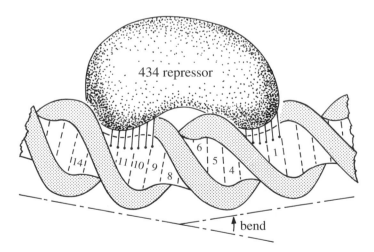

Figure 4.11 Schematic view of the binding of 434 repressor to DNA. This protein probes for the identities of base-pairs in positions 1 to 4 and 11 to 14 of its binding site, and also docks to the overall DNA conformation by testing the ease of curvature in positions 5 to 10. (More detail is shown in Fig. 8.2(a).)

repressing the synthesis of RNA from just one or a few genes on the DNA? For that purpose, the protein needs a specificity of binding to its particular target sequence over other sequences by a factor of 10 000, while the flexibility of DNA – as described above – provides a factor of 300 at most. The biological mechanism by which this protein blocks RNA synthesis is trivial: it simply binds to the same piece of DNA as that preferred by RNA polymerase, and so it physically prevents the polymerase from starting RNA chains. But the way in which the 434 repressor picks out a single DNA sequence (or just a few sequences) from many others in the bacterial chromosome is not easy to understand.

During the last few years, single crystals of many protein–DNA complexes have been analysed by X-ray methods, and this has made it possible to study in fine detail the close bonding of different parts of protein molecules to the bases of DNA. The general picture which has emerged from these studies is that the amino acids of a protein can make specific hydrogen bonds with exposed atoms on the sides of base-pairs, or along the 'floor' of the major or minor groove in the DNA. Such 'direct reading' of a DNA sequence by a small portion of protein will be a main subject in Chapter 8. But for the present, we may examine just one interaction of this sort as shown in Fig. 4.12, where a guanine base is making hydrogen bonds to an arginine amino acid. There are two contacts of hydrogen atoms on the arginine with oxygen or nitrogen atoms on the major-groove edge of the guanine ring. If the spatial patterns of electric charge on the two surfaces fit each other well, so that two or more hydrogen bonds can form, then there will be a highly specific local

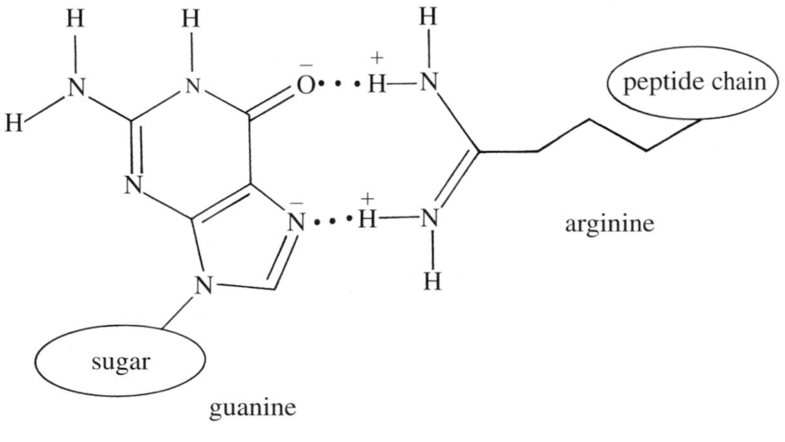

Figure 4.12 Hydrogen bonding between a guanine base in DNA and the arginine amino acid from a protein, as seen in many protein–DNA complexes (such as those shown in Figs 4.11 and 4.13). The paired cytosine base is not shown.

bonding arrangement between the amino acid and the base-pair. However, it is not completely true to say that the detailed patterns of amino acid-to-base hydrogen bonding, when summed over many different amino acids and bases, actually determine where a protein binds on DNA. For we have seen already that the flexibility of DNA plays a large part in its binding to the 434 protein, and to the histone–spool proteins, and to other proteins not discussed here; and this flexibility depends on the DNA sequence as well.

The picture we need in order to explain all aspects of protein–DNA recognition is one which involves consecutively two processes which we shall call 'docking' and 'probing'. By docking we mean the fitting together, on a large scale, of the protein and the DNA. If the fit at that scale is good, then the quality of hydrogen bonding between the contacting zones can be tested on a small scale; and only if this detailed probing is successful will the overall binding between the protein and the DNA be highly specific for some particular base sequence. In our example of the 434 repressor protein, the docking phase demands a capability of the DNA to bend in a certain way; if the DNA is too rigid, or prefers to bend in the wrong way, the docking cannot easily occur. But if the DNA has the right degree of flexibility, then the base-pairs in the contacting regions 1 to 5 and 10 to 14 can probe for the formation of enough hydrogen bonds with amino acids to let the second stage of binding take place.

Another example of the same two-stage process of recognition between DNA and protein is the cutting of DNA by the enzyme DNAase I. This enzyme is very useful for the study of DNA structure, as we shall see in Chapter 9. It can only dock with DNA if the width and depth of the minor groove lie within a certain range; and then only after it has docked can it begin to probe the configuration of the sugar–phosphate chain, and make a cut if it finds a specific geometry there. Thus the enzyme must recognize not only a *global* feature (the groove dimensions) but also a *local* feature (the phosphate configuration) if it is to succeed in cutting the DNA. These two features together provide typically a factor of 1000 in the specificity of cutting some bonds over others.

A third interesting example of protein–DNA recognition was provided by the X-ray analysis of a complex between a protein called Zif268 and the sequence to which it binds on DNA. Zif268 is part of a larger protein that is known as an activator of transcription, or a 'transcription factor', because it somehow stimulates the synthesis of RNA from certain genes close to where it binds on the DNA. No one today is certain how this might be accomplished.

Nevertheless, the structure of the complex between Zif268 and DNA is of interest in its own right. The protein contains a series of

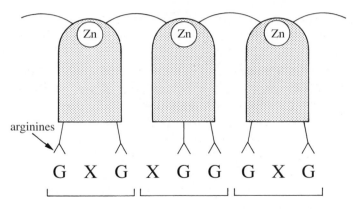

Figure 4.13 Schematic view of three 'zinc-fingers' recognising a particular DNA sequence, as in the X-ray structure of the Zif268 protein with DNA. Not shown are many contacts of the protein with DNA phosphates, or the moderately negative slide of the base-pairs. (More detail is shown in Fig. 8.6.)

small, modular units known as 'fingers', as shown in Fig. 4.13. Each unit contains a zinc atom which helps to fold the protein chain into a separate domain, and these domains are linked by short segments of the protein chain. Some proteins have as many as 10 to 20 fingers, but Zif268 has just three. When the Zif268 protein binds to DNA, it tries to attach itself to the DNA base-pairs through a series of arginine-to-guanine hydrogen bonds, indicated schematically in Fig. 4.13. Not shown in this diagram are many other hydrogen bonds which connect phosphate groups from one of the DNA sugar–phosphate chains to various other amino acids on the protein. In fact the three linked fingers form a spiral, which drapes itself along the sugar–phosphate chain on one side of the major groove, and thereby allows arginine amino acids to make firm contact with guanine bases in 6 out of 9 successive base-pairs. In other words, many contacts of the protein with the phosphates anchor it in place, so that it can contact the guanines.

How does such a zinc-finger protein 'dock' to the overall structure of DNA? In this example, the base-pairs of the DNA have moderately negative slide, and it may be that the negative slide imparts a particular spatial relationship to the phosphates and bases, which then constitutes a three-dimensional docking feature for any individual finger. Alternatively, the negative slide may act by increasing the distance of base-pairs from an overall helix axis, as shown in Figs 3.14(b) and 3.15, in which case the three fingers of the protein would have to follow a helical path of greater diameter than before. We shall return to the process of recognition of DNA sequences by zinc finger proteins in Chapter 8.

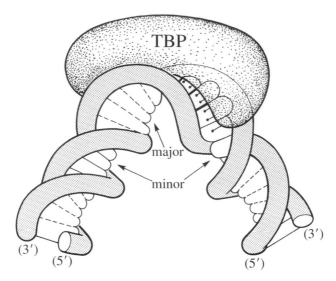

Figure 4.14 Schematic drawing of the TBP protein bound to a TATA-containing sequence in DNA, which it unwinds and bends sharply.

Studies of these zinc-finger proteins with DNA have provided some of our clearest insights into protein-DNA recognition. For example, N. Pavletich and C. Pabo have analysed by X-ray crystallography not only the three-finger complex of Zif268 with DNA as described above, but also a complex with five zinc-fingers from a human cancer-causing gene GLI, and two full turns of DNA to which it binds specifically. They find substantial variations in the slide and twist of this DNA which is recognized by the zinc-finger protein, that are also present in crystals of the same DNA without the protein. Thus, the GLI protein seems to recognize an inherent structural variation in slide and twist, as induced by the particular base sequence to which it binds.

In other studies, many workers have found systematic relations between the DNA bases recognized by any zinc-finger module, and the identities of amino acids in that finger module. These relations can be understood in terms of local patterns of hydrogen bonding, such as that shown in Fig. 4.12 and others of the same kind.

Much progress has also been made recently at determining the precise geometry by which the TATA sequence and other related sequences unwind. Three different crystal structures of the TBP protein (see above) with TATA-containing DNA show a highly untwisted double helix, as drawn schematically in Fig. 4.14. The base-step twist T decreases from about 34° outside the TATA region, to 20° within it; while roll R increases from about 0° outside the TATA region, to +20° within it. These changes untwist and open the minor groove by a large amount, without much change in slide S from near 0 Å.

Such DNA when bound to TBP is thus untwisted and bent somewhat, in the fashion of an accordion that has been bent into a right-angle shape. The TBP protein induces such a highly distorted DNA structure, by placing a large hydrophobic surface in the minor groove, to which the minor groove edges of A–T base-pairs of the sequence TATAAA can bind. There is little doubt that this untwisted structure facilitates transcription, probably by helping to orient the DNA into the active site of the polymerase, where the strands will be separated; and to serve as a 'scaffold' for the recruitment of accessory proteins (see Chapter 8). The specific untwisting of a TATA sequence by TBP has been measured in solution as one-half turn of DNA, by S. Hirose and colleagues or as one-third turn by J. Kahn, in good accord with the X-ray structure. The TATA sequence also unwinds in its complexes with various enzymes that carry out DNA strand-switching or recombination, often known as 'resolvase' proteins; but no one is yet certain how that slight untwisting leads to full separation of the two strands, in preparation for switching between different DNA molecules.

A similar untwisted helix has been suggested for the complex of a human sex-determining protein SRY with DNA, where the DNA has a related sequence CACAAA in contact with the protein (instead of TATAAA). The SRY protein also untwists the double helix, by protruding its surfaces into the minor groove; and in addition, it untwists by pushing or intercalating a hydrophobic isoleucine amino acid between two adenine bases of the sequence CAA, somewhat as for the intercalation of ethidium bromide described in Chapter 2. Therefore, such untwisted structural forms of DNA may be very common in living cells.

In summary, this chapter has been about the twisting and curving of DNA, and how DNA is recognized by proteins according to its twisting and curving and other features. In order for the bases of DNA to be exposed for copying into more DNA or into RNA, the double helix has to untwist. The places where DNA unwinds in Nature, to start the copying process, are often the weakest parts of the double helix. In order for DNA to wrap itself tightly into chromosomes, the double helix has to curve. The curvature spreads itself out over as many base-pairs as possible, in order to keep the roll angle at individual base-pair steps small, so that the bases will stack well onto each other. The sequence of the DNA influences how easily DNA can adopt any given curved shape, by the preferences of different sequences for different roll angles. These preferences apply not only to protein–DNA complexes as part of a chromosome, but also to protein–DNA complexes generally, such as those formed between the

DNA and 'repressor' proteins. In general, a protein recognizes certain sequences in DNA by first recognising some large-scale feature of the molecule such as the sugar–phosphate chains (in other words, by 'docking'), and then by probing the details of the bases.

In these first four chapters we have talked about curving DNA in a plane, but not in three-dimensional space; we have treated twisting and curving as separate subjects, whereas often they are related by the shape and thermal vibration of the DNA. We have not yet discussed how a naked DNA molecule is assembled into a mixture of protein and DNA to make a chromosome, nor how genes are activated within chromosomes by removal of some of the proteins. Nor have we shown how proteins recognize specific sequences in DNA. Finally, we have not yet explained any of the experimental techniques which scientists use to probe DNA structure and function, such as X-ray diffraction, nuclear magnetic resonance, and gel electrophoresis. The following chapters will deal with these and other topics.

Note

1. See Appendix 1.

Further Reading

Anderson, J.E., Ptashne, M., and Harrison, S.C. (1987) Structure of the repressor–operator complex of bacteriophage 434. *Nature* **326**, 846–52. X-ray crystal structure of a complex between the 434 protein and DNA, showing both 'docking' and 'probing' modes of binding.

Azizov, M., Ulyanov, A., Kuprash, D., Shakov, A., Gavin, I., and Nedospasov, S. (1992) Production and characterization of a library of mononucleosomal DNA from the chromatin of human cells. *Doklady Akademii Nauk* **322**, 415–20. Independent confirmation of DNA bending periodicities in the chromatin of human blood cells.

Bailey, K.A., Pereira, S.L., Widom, J., and Reeve, J.N. (2000) Archaeal histone selection of nucleosome positioning sequences and the prokaryotic origin of histone-dependent genome evolution. *Journal of Molecular Biology* **303**, 25–34. The AA/TT or GC sequence preferences seen for DNA curvature about eukaryotic histone proteins are seen also for DNA curvature about histone proteins from archaea.

Calladine, C.R. and Drew, H.R. (1986) Principles of sequence-dependent flexure of DNA. *Journal of Molecular Biology* **192**, 907–18. How roll angles in the DNA can change, to let DNA bend tightly about a protein.

Champoux, J.J. (2001) DNA topoisomerases: structure, function and mechanism. *Annual Review of Biochemistry* **70**, 369–413. A good review on various topoisomerase enzymes and their biological activities.

Chen, D., Bowater, R., Dorman, C., and Lilley, D.M.J. (1992) Activity of a plasmid borne *leu500* promoter depends on the transcription and translation of an adjacent gene. *Proceedings of the National Academy of Sciences, USA* **89**, 8784–8. Evidence for the local supercoiling of DNA in bacteria as induced by its motion through RNA polymerase.

Choo, Y., Sánchez-García, I., and Klug, A. (1994) *In vivo* repression by a site-specific DNA-binding protein designed against an oncongenic sequence. *Nature* **372**, 642–5. Specific contacts between single DNA bases and protein amino acids from zinc fingers are used to design new proteins, which bind to cancer-causing DNA sequences.

Drew, H.R., Weeks J.R., and Travers, A.A. (1985) Negative supercoiling induces spontaneous unwinding of a bacterial promoter. *EMBO Journal* **4**, 1025–32. The −10 region of a bacterial promoter, having a sequence of the kind TATA, unwinds more easily than other regions of the DNA at room temperature under torsional stress.

Fox, K.R. and Brown, P.M. (1996) Minor-groove binding ligands alter the rotational positioning of DNA fragments on nucleosome core particles. *Journal of Molecular Biology* **262**, 671–85. Small molecules which bind in the minor groove of AT-rich regions, can change the rotational setting with which DNA wraps around a histone octamer.

Kahn, J.D. (2000) Topological effects of the TATA-box binding-protein on minicircle DNA, and a possible thermodynamic linkage to chromatin remodeling. *Biochemistry* **39**, 3520–4. Unwinding of the TATA sequence by one-third of a double-helical turn upon addition of TBP protein, similar to that seen in the crystal.

Kim, J.I., Nikolov, D.B., and Burley, S.K. (1993) Co-crystal structure of TBP recognizing the minor groove at a TATA element. *Nature* **365**, 520–7. Unwound DNA at TATA sequence as induced by binding of the TBP protein in the minor groove.

Kim, Y., Geiger, J.H., Hahn, S., and Sigler, P.B. (1993) Crystal structure of a yeast TBP/TATA-box complex. *Nature* **365**, 512–20. More unwound DNA as induced by the binding of the TBP protein.

King, C.-Y. and Weiss, M.A. (1993) The SRY high-mobility-group box recognizes DNA by partial intercalation in the minor groove: a topological mechanism of sequence specificity. *Proceedings of the National Academy of Sciences, USA* **90**, 11990–4. Description of how the sex-determining SRY protein inserts a hydrophobic isoleucine side-chain between bases in the minor groove.

Koudelka, G.B., Harrison, S.C., and Ptashne, M. (1987) Effect of non-contacted bases on the affinity of 434 operator for 434 repressor and Cro. *Nature* **326**, 886–8. Quantitative measurements on how tightly the 434 and Cro repressors dock to different sequences in DNA.

Luisi, B.F., Xu, W.X., Otwinowski, Z., Freedman, L.P., Yamamoto, K.R., and Sigler, P.B. (1991) Crystallographic analysis of the interaction of the glucocorticoid receptor with DNA. *Nature* **352**, 497–505. The first crystal structure analysis of a hormone-receptor protein bound to DNA.

Malhotra, A., Severinova, E., and Darst, S.A. (1996) Crystal structure of a sigma-70 submit fragment from *E. coli* RNA polymerase. *Cell* **87**,

127–36. RNA polymerase from bacteria contains a protein structure that helps unwind DNA at TATAAT, which is different from that used by TBP.

Miller, J., McLachlan, A.D., and Klug, A. (1985) Repetitive zinc-binding domains in the protein transcription factor IIIA from *Xenopus* oocytes. *EMBO Journal* **4**, 1609–14. The earliest correct proposal of a repeating structure for the zinc-finger proteins.

Murakami, K.S. and Darst, S.A. (2003) Bacterial RNA Polymerases: the wholo story. *Current Opinion in Structural Biology* **13**, 31–9. Summarizes current understanding of initiation of transcription by bacterial RNA polymerase.

Murakami, K.S., Matsuda, S., Campbell, E.A., Muzzin, O., and Darst, S.A. (2002) Structural basis of transcription initiation: an RNA polymerase holoenzyme-DNA complex. *Science* **296**, 1285–90. How the sigma sub-unit of bacterial polymerase contacts the promoter region, and the transcription bubble is stabilized.

Nikolov, D.B., Chen, H., Halay, E.D., Usheva, A.A., Hisatake, K., Lee, D.K., Roeder, R.G., and Burley, S.K. (1995) Crystal structure of a TFIIB-TBP-TATA element ternary complex. *Nature* **377**, 119–28. Both TBP and another protein, TFIIB, visualized as they bind to the DNA in order to start transcription.

Ohndorf, U.-M., Rould, M.A., He, Q., Pabo, C., and Lippard, S.J. (1999) Basis for recognition of cisplatin-modified DNA by high-mobility-group proteins. *Nature* **399**, 708–12. The specific binding of cis-dichloro-diamino-platinum to one of its biological targets.

Pavletich, N.P. and Pabo, C.O. (1991) Zinc finger-DNA recognition: crystal structure of a *Zif268*–DNA complex at 2.1 Å. *Science* **252**, 809–17. Three-dimensional structure of a transcription factor somewhat like TFIIIA bound to DNA: the structure shows the probing of guanine bases by arginine amino acids.

Pruss, D., Reeves, R., Bushman, F.D., and Wolffe, A.P. (1994) The influence of DNA and nucleosome structure on integration events directed by HIV integrase. *Journal of Biological Chemistry* **269**, 25031–41. The insertion enzyme for the HIV virus recognizes the rotational setting of the DNA on histone spools, as it inserts into chromosomes.

Pryciak, P.M., Sil, A., and Varmus, H.E. (1992) Retroviral integration into mini-chromosomes in vitro. *EMBO Journal* **11**, 291–303. A recombination enzyme also recognizes GC *versus* AT bending preferences of DNA in chromatin.

Satchwell, S.C., Drew, H.R., and Travers, A.A. (1986) Sequence periodicities in chicken nucleosome core DNA. *Journal of Molecular Biology* **191**, 659–75. Different base sequences occupy different locations within DNA as it curves, owing to their preferences for different kinds of roll angle (see Fig. 4.10).

Werner, M.M., Gronenborn, A.M., and Clore, G.M. (1996) Intercalation, DNA kinking, and the control of transcription. *Science* **271**, 778–84. An excellent review of how proteins such as TBP and SRY bind to DNA.

Yang, W. and Steitz, T.A. (1995) Crystal structure of a site-specific recombinase or resolvase complexed with a 34 bp cleavage site. *Cell* **82**, 193–207. First vizualisation of a specific recombinase enzyme as bound to DNA.

Bibliography

Liu, L.F. and Wang, J.C. (1987) Supercoiling of the DNA template during transcription. *Proceedings of the National Academy of Sciences, USA* **84**, 7024–7. A classic paper describing what might happen if RNA polymerase could not rotate freely about the DNA during transcription.

McClellan, J.A., Palacek, E., and Lilley, D.M.J. (1986) $(A–T)_n$ tracts embedded in random-sequence DNA: formation of a structure which is chemically reactive and torsionally deformable. *Nucleic Acids Research* **14**, 9291–309. The unwinding of TATA-like sequences to a twist of 12 basepairs per turn before a cruciform appears.

Pavletich, N.P. and Pabo, C.O. (1993) Crystal structure of a five-finger GLI-DNA complex: new perspectives on zinc fingers. *Science* **261**, 1701–7. Analysis of five zinc-fingers bound to two turns of DNA, with many structural variations in the DNA being recognized by the protein.

Rosenberg, B., VanCamp, L., Trosko, J.E., and Mansour, V.H. (1969) Platinum compounds: a new class of potent antitumour agents. *Nature* **222**, 385–6. Testing of platinum compounds against cancer in mice, with positive results.

Tabuchi, H., Handa, H., and Hirose, S. (1993) Underwinding of DNA on binding of yeast TFIID to the TATA element. *Biochemical and Biophysical Research Communications* **192**, 1432–7. First detection of DNA unwinding by the TBP (or TFIID) protein in solution.

Exercises

4.1 Make a physical model to show some of the problems that occur when a polymerase molecule runs along double-helical DNA, as in Fig. 4.3.

To do this, take two equal lengths of rubber tubing, and tie them together firmly at one end. Then, working from that end, arrange the two tubes in the form of a fairly loose right-handed double helix; and when you get to the other end, tie the tubes together there also. The two tubes represent the two sugar–phosphate chains of a piece of DNA.

Now insert a stick between the two tubes, near the mid-point of the double helix. Holding an end of the double helix in one hand and the stick in the other, push the stick along the double helix. Observe how the helices tighten up ahead of the stick, and relax behind it – unless, that is, the stick is allowed to rotate as it moves forward along the model.

(This demonstration was suggested by Maxim Frank-Kamenetskii.)

4.2   The base sequence of a single strand of double-helical DNA is given below:

(5′) ACTTAAGGCCCTATATACCTAGACTCGGCGGTAAATTT (3′)

a   Underneath it, write out the base sequence of the complementary strand.
b   Identify AT-rich and GC-rich regions of the molecule.
c   Identify pyrimidine–purine steps.
d   Hence identify strong, medium, weak, and very weak regions of cross-chain bonding, in the manner of Fig. 4.4.

4.3   The base sequence of a single strand of double-helical DNA is given below:

(5′) GCGCCTAGAAATAATACTAGTATTATTTCTAGCCGG (3′)

a   Underneath it, write out the base sequence of the complementary strand.
b   Find a region which can make a 'cruciform' formation, as in Fig. 4.5; and draw a picture to show the new pairing.

4.4   A semicircle of garden path around a tree (cf. Fig. 4.7) is made from 10 rectangular slabs, with an individual angle of 20° at each of the 9 'steps' between consecutive slabs.
a   How many slabs, arranged in the same pattern, are needed to make a complete circle? Given that the center-to-center spacing of the slabs is 0.6 m, what is the circumference of the circle? And what is the radius of the circle?
b   Now do a different calculation on the same curved path, in order to find its radius. First, convert the curvature of 20° per slab into units of radian per meter by the use of the conversions: 1 radian $= 180°/\pi = 57.3°$; 1 slab $= 0.6$ m. Then calculate the radius of the circle from the formula (which we have not mentioned before, but which is easy to prove): radius of curvature $= 1/$(curvature).

4.5   Imagine that the four DNA molecules shown in Fig. 4.8 have been made with a curvature of 30° per helical turn, instead of 45° per turn.
Sketch out the corresponding version of each of the four plots (a) to (d) in Fig. 4.9, and list the maximum and minimum values of roll angle R in each case.

4.6   A particular DNA molecule has $T = 36°$ at every step.
a   The pattern of roll angles along the length of the molecule is somewhat similar to that shown in Fig. 4.8(c) and Fig.

4.9(c), except that the batches of non-zero roll are only three steps long, and the individual roll angles are smaller. Specifically

$$R_n = (10°, 10°, 0°, 0°, 0°, 0°, 0°, 0°, 0°, 10°)$$

where $n = 0$ to 9, and the pattern repeats many times.
Compute the curvature of the molecule (in units of degrees per helical turn) by evaluating the sum of 10 consecutive terms $R_n \cos(36°n)$, $n = 0$ to 9.
To do this, make a table of 10 rows and 4 columns, with the following entries:
Column 1, values of n;
Column 2, values of $\cos(36°n)$;
Column 3, values of R_n;
Column 4, values of $R_n \cos(36°n)$.
Obtain the required answer by taking the sum of all entries in Column 4.

b Repeat the calculation for

$$R_n = (0°, 0°, 0°, 0°, 10°, 10°, 10°, 0°, 0°, 0°)$$

c Repeat for

$$R_n = (0°, 0°, 0°, 0°, -10°, -10°, -10°, 0°, 0°, 0°)$$

d Repeat for the case (which is somewhat similar to that of Figs 4.8(d) and 4.9(d), and to Exercise 4.5):

$$R_n = 6° \cos(36°n), n = 0 \text{ to } 9$$

e Repeat for

$$R_n = 6° \cos(36°n) + 4°$$

4.7 Fig. 4.12 shows the hydrogen bonding between a guanine base of DNA and an amino acid arginine on the peptide chain of a nearby protein. The diagram below shows an amino acid asparagine in a similar way. Would asparagine form hydrogen bonds better to guanine or to adenine, in the major groove of DNA, as in the style of Fig. 4.12? See Fig. 2.11 for chemical formulas.

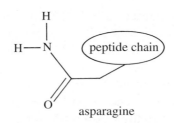

asparagine

CHAPTER 5

Curving in Three Dimensions

So far in this book we have said a lot about the structure of DNA as a double helix, and the way in which different kinds of double helix can result from the different geometries with which base-pairs stack on one another. But in most chapters our DNA, seen as a long rod or thread, has been straight. Only in Chapter 4 did we consider a molecule of DNA which follows a curved path; and the curve there was in a plane, like a garden path. We have now reached the stage where we must explain the curvature of DNA in three dimensions.

A DNA helix often proceeds through space as a three-dimensional spiral, rather than as a straight line or as a plane curve. The spiral path of DNA is usually described as a 'superhelix' or 'supercoil'. This is because we know that the DNA thread itself has a local helical, twisted structure even when it is straight. The qualitative aspects of DNA curvature in three-dimensional space are not difficult to understand, although we shall need to be careful to distinguish between the twisting of the path of the DNA as a whole, and the twisting of sugar–phosphate chains on a local scale. But you may find that some of the quantitative and mathematical aspects of the subject are hard to grasp. Do not worry if you find the mathematics rather heavy going: be content to appreciate the qualitative aspects of three-dimensional curvature.

In Chapter 4, we explained how DNA can curve in a plane to make a circle. There, curvature k was defined as the angle turned for a given length of DNA. For example, k might be $1°$ per base-pair or, equivalently, $10°$ per double-helical turn of 10 base-pairs. When DNA curves in three dimensions, our simple definition of curvature k still holds for any small segment of its path, such as one helical turn or 10 base-pairs; but in practice the DNA often departs from the local plane of curvature after two or more turns. In such cases, the plane of curvature twists by some angle t as the DNA advances through space, and so the DNA coils into three dimensions.

There is an easy way to understand this point. Consider a little man in an airplane (Fig. 5.1). In Fig. 5.1(a) he flies loop-the-loops, which are circular paths in a vertical plane, like the path of a car in a fairground or carnival ferris wheel; and he does so by pointing the nose of his 'plane either up or down, using his joystick to tilt the elevators on the tail of the 'plane. This is pure curvature k, since his path through the air curves by some angle k per unit length of loop. In Fig. 5.1(b), on the other hand, he flies straight ahead but in a narrow spiral, by pushing the joystick either to the left or to the right, and so activating the ailerons on the wings. The path of the airplane is now one of pure twist t, since the 'plane rotates about its long axis by some angle t per unit length of advance through the air. The twist t can be either plus or minus, depending on whether he turns the wheel to the right or to the left, respectively. After a while, the pilot gets dizzy from flying in a straight, twisted path, and he decides to combine the two motions of k and t. So he turns the nose of his 'plane both up and to the left at the same time; and he now flies in a broad spiral through the air, like the path of the wire in a coiled bed-spring. It is the *combination* of k and t that provides for a broad spiral path;

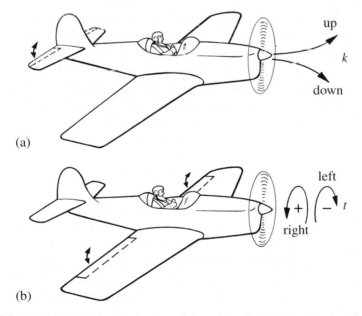

(a)

(b)

Figure 5.1 Our friend in the airplane explains to us about curvature k and twist t. Curvature k alone as in (a) makes the 'plane fly in vertical loop-the-loops. Twist t alone as in (b) makes the 'plane fly in a narrow spiral. Only the combination of curvature k and twist t produces a broad spiral.

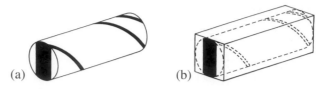

Figure 5.2 (a) One complete turn of double-helical DNA, showing two sugar–phosphate chains along the outside, and a black rectangle as the first base-pair: the whole structure has been drawn as a cylinder. (b) The same as in (a), but now the DNA has been encased in a semitransparent solid block.

k alone gives a plane circle, while t alone gives a narrow spiral like a twisted ribbon.

This story gives us a good initial understanding of how curvature k and twist t influence the path of DNA through space. Let us now ask, first, how k and t influence the shape of any small segment of DNA; and second, how these small segments can be joined together to make different paths through space, like those flown by our friend in the airplane.

We can represent any small segment of DNA as in Fig. 5.2(a) by a short cylinder with the sugar–phosphate chains along its outside. This particular segment contains exactly one double-helical turn of 360°, or 10 base-pair steps. The first and last base-pairs, numbers 0 and 10, can be drawn as black rectangles on the two ends of the cylinder, to demonstrate that they are parallel to one another, and are vertical in the drawing. Only one of these two black rectangles can be seen in the perspective of Fig. 5.2(a), so you have to imagine that there is another one on the far end of the cylinder.

An even simpler version of the same thing is shown in Fig. 5.2(b), where the single helical turn of DNA is now represented by a solid block. The two ends of the block are perfectly aligned, because the DNA has twisted by 360° along its axis in going from one end to the other; and the block is straight as in (a). You can imagine that the cylinder of (a) is converted into the block of (b) by first glueing squares to the ends of the cylinder, and then filling in the space between the squares with some sort of semi-transparent jelly which enables us to see, darkly, the sugar–phosphate chains buried inside. From now on we shall mainly disregard the twisting of these strands within the block, and will usually draw the block as in Fig. 5.3(a), without any hint of what is actually hidden inside it. In this figure both the curvature k and the twist t of the block are precisely zero, and all six faces of the block are plane squares or rectangles.

Figure 5.3(b) shows a block that has curved by 10° to the left, or equivalently by $k = 1°$ per step over 10 steps. It looks something like a banana. The upper surface is obviously curved, while the shaded

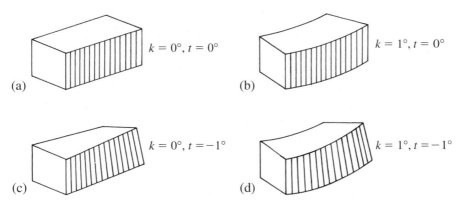

Figure 5.3 Three possible ways by which the solid block shown in Fig. 5.2(b) can change its shape, by the application of uniform curvature k and twist t. In this figure, curvature k and twist t are given in units of degrees per base-pair. Since there are 10 base-pair steps, the total curvature is 10° in (b) and (d), while the total twist is $-10°$ in (c) and (d).

surface might form part of a cylinder. Presumably, the roll angles R of the DNA within the block have changed on going from (a) to (b), so that the best-fit cosine wave to these roll angles, as described in Chapter 4, now sweeps out an angle of 10°. The twist t is still zero, because the two ends of the block remain aligned: note that the closely spaced lines on the shaded surface are all strictly parallel to one another.

Figure 5.3(c) shows a block that has twisted in a negative or left-handed sense by 10° from end to end, or equivalently by $t = -1°$ per step over 10 steps. Presumably, the local twist angles T between successive base-pairs in the DNA have decreased from $T = 36°$ to $T = 35°$ on average, in going from (a) to (c). Thus, the 10 steps of hidden DNA now twist only by a total of 350° rather than 360°, leaving a deficit of $350° - 360° = -10°$; and this is what shows in the picture. In (c) there is no curvature: the top surface, though now twisted, is still rectangular.

Lastly, Fig. 5.3(d) shows a block that is both twisted and curved; it looks like a slightly twisted banana. Both roll R and twist T have changed at the base-pair level to produce this result, in the same ways as described for parts (b) and (c) previously; but now R and T have both changed simultaneously.

We have used the word 'twist' in two senses here, with two symbols t and T. Lower-case twist t is the difference in total twist from 360° after one unit-length of helix. It amounts to $-10°$ after 10 steps in (c), as explained above. Upper-case twist T is the twist of any base-pair step locally. For example, $T = 36°$ in both (a) and (b), but

$T = 35°$ in (c) and (d). It follows, therefore, for any unit of n base-pair steps with uniform twist T:

$$t = nT - 360°.$$

Here $t = (10 \times 35°) - 360° = -10°$ in both (c) and (d) of Fig. 5.3. What do we mean by a 'unit of n base-pair steps'? Basically, we mean that the internal structure of the DNA is identical in each successive set of n steps – as it is likely to be, for example, if the *sequence* of the DNA repeats every n steps.

In summary, curvature k and twist t can describe the shape of any small segment of DNA on a local scale. These values of k and t come from variations in base-step roll R and twist T, respectively. We explain below how to calculate k and t from given values of roll R and twist T, for the base-pair steps of practically any segment of DNA.

Now you must use your imagination. What happens if we take many identical blocks of the kind shown in any one part of Fig. 5.3, and join them together, end-to-end, over a long distance? Using blocks all of type (a), we would get a straight, untwisted rod; using blocks of type (b), we would get a plane circle; while the blocks of type (c) would build a straight but twisted rod. Finally, the blocks of type (d) when joined together would make a broad spiral that we might call a 'superhelix', if we remembered the DNA hidden within it. Such a spiral will rotate counterclockwise as it goes forward, and so be left-handed, because the sign of twist t is negative. If t were positive, we would get a right-handed superhelix. This would happen, for example, if the local twist T became 37°, so that $t = nT - 360° = (10 \times 37°) - 360° = +10°$.

In fact, we do not have to rely entirely on our imagination for these constructions. We can *calculate* the path of the DNA through space for different values of k and t, by using certain geometrical formulas. Therefore, let us now derive these simple formulas.

A typical left-handed superhelix is shown in Fig. 5.4(a). This superhelix might be made by joining together many blocks like the one shown in Fig. 5.3(d). It winds uniformly like a ribbon about a vertical cylinder of radius r, at an angle α with respect to the horizontal at any point. It makes a left-handed spiral as it winds round the cylinder. It turns by 360° around the vertical axis of the cylinder after a vertical distance p, which is known as the superhelical 'pitch'. Finally, it has a contour length, not indicated in the picture, of N^* base-pair steps in one 360° turn. If an ant were to crawl along the path of the superhelix for any full 360° turn around the cylinder, and then measure or count over how many steps it had crawled, that would be N^*.

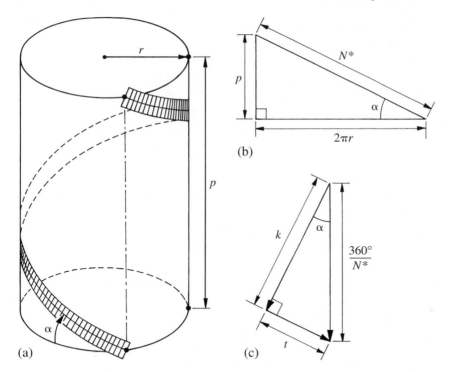

Figure 5.4 Curvature and twist in a spiral ribbon. In (a), a left-handed spiral ribbon of pitch angle α and radius r goes through one turn every p units of distance along the cylinder. In (b), the surface of the cylinder has been unwrapped on a smaller scale, to show geometrical relations among α, r, p and contour length N^*. In (c), the vector sum of curvature k and twist t, on a very small scale, gives the amount by which each step of the ribbon rotates around the axis of the cylinder. Note that triangle (c) is similar to triangle (b).

The usual way to analyse these parameters r, α, p, and N^* is to imagine that we can 'unroll' the cylinder onto a piece of paper, in the same way that you can unroll the cardboard cylinder on which paper towels are wrapped. This is called a 'cylindrical projection', and it is shown, to a smaller scale, in Fig. 5.4(b). The path of the superhelix becomes the diagonal of a right-angled triangle, while the pitch p and the circumference $2\pi r$ make up the other two sides; they are related by the angle α. It is a simple matter to find all of N^*, p, r, and α from any two of these four parameters; but we shall not do so here.

All of this is very well-known geometry; but the next part is not so well-known. Consider any small part of the superhelical path, say just 2° of the total 360° for each turn around the cylinder. For example, when $N^* = 180$ base-pairs, any single base-pair makes up $360°/N^* = 2°$ of rotation about the circumference, as seen in a view along the axis of the cylinder. In this case, the values of curvature k and twist t for the spiral ribbon are related to $360°/N^*$ as shown in

Fig. 5.4(c). Here, the total rotation of 2° about the axis is represented by the vertical arrow: so curvature k is the cosine component of it, while twist t is the sine component. This result follows from a branch of mathematics known as 'differential geometry', for the special case of a ribbon wrapped about a cylinder. In simple terms, the curvature k tells how far the superhelix should *curve* around the cylinder in the manner of Fig. 5.3(b), and t tells us how much the ribbon should *twist* up the cylinder in the manner of Fig. 5.3(c). The sum of the two effects (it is a vector sum, so is represented properly by the triangle) gives the total rotation about the axis of the cylinder.

Several interesting equations follow from the triangle of Fig. 5.4(c):

$$\tan \alpha = t/k$$

$$k = (360°/N^*)\cos \alpha$$

$$t = (360°/N^*)\sin \alpha$$

$$(360°/N^*)^2 = k^2 + t^2$$

We can now relate the diagrams in Fig. 5.4(b) and (c) to each other, and thereby get the overall shape of the superhelix from any given values of k and t. By similar triangles we have

$$p/N^* = t/(360°/N^*)$$

Substituting for N^* from the last of the four equations above, we obtain

$$p = 360°t/(k^2 + t^2).$$

Similarly,

$$2\pi r = 360°k/(k^2 + t^2).$$

Let us consider as a specific example a case where $k = 1°$ and $t = -1°$ per base-pair. Then $\alpha = -45°$, meaning that the superhelix is left-handed as in Fig. 5.4(a), since t is negative, and climbing at 45°. Also, contour length $N^* = 360°/\sqrt{2} = 255$ base-pairs, while pitch $p = 180$ base-pairs. Finally, the circumference of the cylinder $2\pi r = 180$ base-pairs, so radius $r = 29$ base-pairs. The final form of the superhelix is similar to that shown in Fig. 5.4(a), except that the angle α is larger.

In this particular example k and t both had the same magnitude. In a case where the magnitude of t is small compared with k, the formulas give us a low value of α; this means simply that the superhelical coil is almost flat, like a circle going round and round the same path. However, when k is small compared with t, the formulas give $\alpha \approx 90°$. This means that the superhelix takes the form of a highly extended spiral, almost like a straight line but with a small superhelical 'wobble'.

All of this seems very straightforward, once we have learned how to use the formulas. But how can we get k and t from the sequence of the DNA? Unless the base sequence is very regular, k and t will vary from one double-helical turn to the next. In that case, the path of the DNA will not describe a regular superhelix, and so the formulas which we have just derived will be useless. But if the sequence of the DNA does repeat exactly – or almost exactly – once every double-helical turn, then the DNA will form a regular superhelix and our formulas will hold good.

Fortunately, scientists have done many experiments on the structure of 'repeating-sequence' DNA, because they can make it rather easily. They synthesize chemically one small part of the DNA, say 10 base-pairs of a defined sequence, and then they join these units together to make a long polymer, just as we imagined when we were studying Fig. 5.3. Having made such repeating-sequence DNA, they can study its structure by the methods of electron microscopy and gel electrophoresis. Electron microscopy shows the shape of the DNA directly, although at low resolution, while gel electrophoresis measures indirectly the 'apparent volume' of the DNA cylinder. We shall discuss this method of analysis in Chapter 9, but for the present we may say that the curved DNA shown in Fig. 5.4(a) could be enclosed in a cylinder of larger volume than for straight DNA of the same length; and so in a gel, curved DNA would come into contact with more gel fibers, and hence go more slowly. That is precisely what is observed in experiments with gels.

Still, it is nice to see the DNA directly, without having to worry about indirect measurements of its volume by gel electrophoresis. For that reason, it is sensible to study repeating-sequence polymers by electron microscopy. Some pictures of curved, repeating-sequence DNA as obtained by this method are shown in Fig. 5.5. The first five frames in Fig. 5.5(a) show a collection of similarly curved DNA molecules of size about 1000 base-pairs. They lie in various shapes on the support 'grid', having been flattened onto two dimensions in preparation for microscopy. The last frame in Fig. 5.5(a) shows two DNA rings or 'plasmids' of length 3000 base-pairs, of ordinary (i.e. not repeating-sequence) DNA, as controls for the appearance of DNA with little or no curvature; they were prepared for microscopy under identical conditions. Figure 5.5(b) shows the outlines of curved DNA molecules from (a). Each of the curved DNA molecules has a wiggly, snake-like appearance compared with the two DNA plasmids. The 'wiggle' of the curves on the photographs can be fitted by a sine-wave of contour length $N^* = 533$ Å and pitch $p = 385$ Å, as an

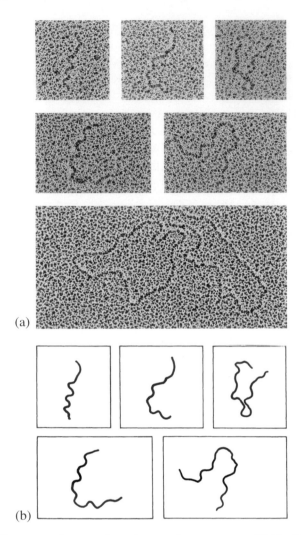

Figure 5.5 Electron micrographs of repeating-sequence DNA. The first five frames in (a) show an assortment of pictures of repeating-sequence, curved DNA of size about 1000 base-pairs. The last frame in (a) shows two DNA plasmid or ring molecules of size 3000 base-pairs, that have been included as controls. In (b), approximate tracings of the curved DNA molecules from (a) are presented. Courtesy of Margaret Mott.

average over many molecules. The amplitude of the 'wiggle' is obviously unreliable as a measure of the original three-dimensional structure, because the DNA superhelix was flattened significantly when placed onto the grid for microscopy; but the values of N^* and p should be representative of the original conditions in solution. From the values for N^* and p, and the triangle shown in Fig. 5.4(b), we can calculate that $\alpha = \pm 46°$, while radius $r = 59$ Å, before flattening onto the grid. The length per base-pair in these pictures has been measured as 3.0 Å, so we can express the dimensions of the DNA superhelix in units of base-pairs as $N^* = 178$ base-pairs, $p = 128$ base-pairs, and $r = 20$ base-pairs.

Figure 5.5 is a good piece of experimental evidence about the shape of curved, superhelical DNA. But can we calculate the values of N^*, p, and r independently, directly from the base sequence of the DNA, by using a suitable theory? It is easy to determine from the superhelical parameters listed above that $k = 1.40°$ per base-pair and $t = \pm 1.45°$ per base-pair; but how can we derive k and t values from the base sequence? Note that the electron microscope pictures do not indicate whether the original superhelix was left-handed or right-handed, and so we do not know from the experiment whether t is positive or negative.

The sequence which was used to make these polymers was of the kind …AAAAAANNNNAAAAAANNNN…, where N = C or G, mainly. This sequence repeats once every 10 base-pairs, as AAAAAANNNN. Now, it is known that the average local twist T for an AA step is close to $T = 35°$, while $T = 34°$ for other steps such as NN, NA, or AN. The slight difference in twist possibly comes from an extra hydrogen bond across the major-groove side of the AA step, as shown in Fig. 3.6. Therefore, we can calculate the overall twist t as $(5 \times 35° + 5 \times 34°) - 360° = 345° - 360° = -15°$ per 10 steps, or $t = -1.5°$ per base-pair. Not bad! We have already come close to the value for t of (plus or minus) 1.45° which was obtained by electron microscopy. Both of the values for T at particular steps, which were used in the calculation above, had been determined by gel electrophoresis or other techniques several years before the electron microscope pictures were taken.

But how can we calculate the curvature k from the base sequence? The roll angles R are thought to be close to 0° for an AA step, as against $R = +3.3°$ for the others as a broad average: again, these values come from gel electrophoresis, X-ray crystallography, or other techniques. Such approximate values of base-step roll are plotted in Fig. 5.6 against the step number, for this particular repeating sequence. As explained in Chapter 4, we must fit a cosine wave to the values of R plotted in Fig. 5.6, in order to determine the curvature k. We could draw a 'best-fit' cosine wave over the points in Fig. 5.6 by eye, in order to get a satisfactory approximate solution (see the broken line in the diagram), but really we would prefer to have a more systematic way of doing the calculation.

The most accurate method is to take what is known as the 'Fourier transform' of the roll angles over any period of steps 0 to 9 in the base sequence. This is just like what we did in Chapter 4, except that here we shall show the full mathematics. It doesn't matter

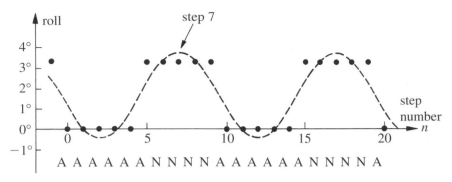

Figure 5.6 A plot of expected roll angles for the repeating-sequence, curved DNA studied by electron microscopy in Fig. 5.5. The curvature k may be calculated as $10.7°$ over any A_6N_4 repeat by taking a Fourier transform of these roll angles, as indicated by the broken line. Roll angles are most positive in the Fourier transform at step 7, where the minor groove of the DNA lies along the outside of the curve.

which step you assign to be step 0, or which step you assign as step 9, so long as all 10 steps of the repeated sequence (0, 1, 2, ..., 9) are counted only once.

First, you must evaluate two sums, which are the correlations of the roll angles R with either a sine or a cosine wave; for this purpose it is convenient to write R_n for the roll R at any step n:

$$\text{first sum} \quad = \quad \sum_{n=0,9} R_n \sin(36°n)$$

$$\text{second sum} = \quad \sum_{n=0,9} R_n \cos(36°n).$$

Thus, using the numbering of steps shown in Fig. 5.6, we find that:

$$\begin{aligned}
\text{first sum} \quad &= 0° + 0° + 0° + 0° + 0° + 3.3°(0.0) + 3.3°(-0.59) \\
&\quad + 3.3°(-0.95) + 3.3°(-0.95) + 3.3°(-0.59) \\
&= -10.2°
\end{aligned}$$

$$\begin{aligned}
\text{second sum} &= 0° + 0° + 0° + 0° + 0° + 3.3°(-1.0) + 3.3°(-0.81) \\
&\quad + 3.3°(-0.31) + 3.3°(0.31) + 3.3°(0.81) \\
&= -3.3°
\end{aligned}$$

Then the curvature k is simply given by the square root of

$$\begin{aligned}
k^2 &= (\text{first sum})^2 + (\text{second sum})^2 \\
&= (-10.2)^2 + (-3.3)^2 = 115; \quad \text{so } k = 10.7°.
\end{aligned}$$

Our calculated value of $k = 10.7°$ for 10 steps, or $1.07°$ per base-pair, compares fairly well with the value of $k = 1.4°$ per base-pair as determined from the electron microscope pictures of the overall shape. We could also calculate from this theory that our repeating-sequence

DNA should make a left-handed superhelix of contour length $N^* = 195$ base-pairs and radius $r = 61\,\text{Å}$, using the theoretical values of k and t. These are again fairly close to the experimental values of $N^* = 178$ base-pairs and $r = 59\,\text{Å}$, given above.

Before we go on, let us consider two more features of the Fourier transform: its amplitude and its phase. First, the amplitude of the best-fit cosine wave in most of our examples is equal to the curvature k divided here by 5.0, which is the sum of $\cos^2\theta$ over steps 0 to 9. Thus, in Fig. 5.6, the amplitude or half-height of the dotted line is simply $10.7/5.0 = 2.1°$. We can say then that the total variation in roll R, for the best-fit wave, is $2 \times 2.1° = 4.2°$. This wave goes from a peak at $R = +3.75°$ to a trough at $-0.45°$ about a mean of $+1.65°$. Second, the phase of a Fourier transform tells us where the best-fit cosine wave is located in a left-to-right sense, relative to the numbering of steps in the sequence. It can be calculated as the

arctangent of the ratio (first sum/second sum).

That gives, in the case above, $\arctan(-10.2/-3.3) = 252°$. This shows us that the origin of the best-fit cosine wave, where its value is most positive, lies $252°$ to the right of step 0 in the sequence, or in Fig. 5.6 at step $(252°/36°) = 7$ (see the arrow). In other words, the base-step roll R is the most positive at step 7, and least positive (or most negative) at steps 2 and 12, which are $180°$ out of phase in either direction. In molecular terms, this means that the minor groove of the DNA lies along the outside of the curve at step 7, and along the inside of the curve at steps 2 and 12.

Now suppose we had made another, slightly different DNA molecule with repeating sequence, but this time with a repeat once every 11 base-pairs rather than every 10. For example, AAAAAANNNNN would have an 11-base-pair repeat, as compared with the 10-base-pair repeat for AAAAAANNNN studied above. Then we could calculate the overall twist t by adding up the local twist angles T over 11 steps, to yield $t = (5 \times 35° + 6 \times 34°) - 360° = 379° - 360° = +19°$. Our new superhelix would thus be right-handed, since t is now positive. We could calculate the curvature k by taking the Fourier transform of the roll angles over 11 steps, at intervals of $360°/11 = 32.7°$, using cosines and sines of angles $(32.7°n)$, to yield $k = 11.6°$. Here, we have calculated both t and k for a complete 11-step repeat. In smaller units of base-pairs, t and k would be $t = +19°/11 = +1.73°$ and $k = 11.6/11 = 1.06°$. Finally, from these values of t and k in degrees per base-pair, we could calculate the dimensions of the right-handed superhelix, by use of the formulas of the preceding section.

Table 5.1 lists values of k and t as calculated for a series of DNA molecules of repeating sequence, which includes the two sequences

Table 5.1 Calculated superhelical parameters α, N^*, and r (see Fig. 5.4) for DNA of various repeating sequences

Sequence	Repeat	k (°/bp)	t (°/bp)	α (°)	N^* (bp)	r (bp)
A_6N_2	8	1.00	−10.38	−84.5	34.5	0.5
A_6N_3	9	1.06	−5.44	−79.0	65.0	2.0
A_6N_4	10	1.07	−1.50	−54.5	195.4	18.1
A_6N_5	11	1.06	+1.73	+58.5	177.4	14.8
A_6N_6	12	1.03	+4.42	+76.9	79.3	2.9
A_6N_7	13	0.99	+6.69	+81.6	53.2	1.2

The curvature and twist (k, t) were calculated from the base sequence according to angles of roll R and local twist T given in the text.

already mentioned, namely A_6N_4 and A_6N_5. These calculated values of k and t have then been used to compute the superhelical parameters α, N^* and r for each sequence. Looking first at the two central columns of the table for k and t, we can see clearly that if we had made a 9-base-pair repeat such as A_6N_3, or a 12-base-pair repeat such as A_6N_6, then twist t would have become much larger in magnitude than curvature k. In the case of A_6N_3, you would get $t = (5 \times 35° + 4 \times 34°) − 360° = 311° − 360° = −49°$ per repeat, which gives $t = −49°/9 = −5.44°$ per base-pair. But the curvature k is still only 1.06° per base-pair, and so the pitch angle α = arctan (t/k) becomes arctan($−5.44/1.06$) = $−79.0°$. This means that A_6N_3 would form a highly elongated, left-handed superhelix of contour length $N^* = 65$ base-pairs and radius $r = 2$ base-pairs.

Those calculations lead us to the conclusion that we can only expect to find DNA molecules in the form of a *broad* spiral when the DNA sequence-repeat is close to 10 or 11 bases, so that the twist t is low. Of course, if twist $t = 0$ exactly, then the DNA forms a plane circle of contour length $N^* = 360°/k$; and indeed many naturally occurring DNA molecules form plane circles.

How could we calculate k and t if the base sequence was more complicated than A_6N_4 or the related sequences discussed above? Previously we assigned roll R and twist T values only to AA/TT and 'other' steps. Thus, we set $T = 35°$ for AA/TT steps but $T = 34°$ for 'others'. This model is clearly too simple to account for all of the possible arrangements of base-pairs in DNA, but it serves as a first approximation for the case of free DNA in solution, or on an electron-microscope grid. When more data have been obtained about the precise shapes of DNA in solution or on a microscope grid, it should be possible to assign more accurate values of R and T to every kind of step. For example, experiments by many workers have shown that the sequences GGC/GCC and AGC/GCT have much higher roll angles R than the average, perhaps $+10°$ to $+15°$ over the two steps. Those high roll angles for certain GC steps seem especially pronounced in solutions that contain magnesium ions, at a concentration similar to that found in living cells. Recall that the GC step was also implicated earlier, in Chapter 4, to assist in the curvature of DNA about various proteins.

So far, we have been investigating the geometry of supercoiled DNA as if the molecule were a long, rigid object. That is satisfactory for short pieces of DNA, but not if the DNA is long. In general, a short piece of DNA will behave like a rigid body, whereas a long piece of DNA will behave rather flexibly, as it is buffeted in 'Brownian movement' by fast-moving water molecules in solution. Now the terms 'short' and 'long' are meaningless unless we have a standard length for purposes of comparison. Scientists have supplied exactly such a comparison length, and they call it the 'persistence length' of DNA. They define it as the length at which the time-averaged angle made between the two ends of an intrinsically straight DNA molecule is equal to one radian, or $57°$. The persistence length, as so defined, may then be deduced by several different experimental protocols such as electron microscopy, viscosity or light-scattering.

For many years, the accepted value for the persistence length of mixed-sequence DNA was near 140 base-pairs, or 450Å. Thus, a piece of DNA 100 base-pairs long was expected to behave more-or-less as if it were rigid, whereas a piece 200 base-pairs long was expected to behave more-or-less as if it were a flexible string. Recent studies, however, have shown that previous estimates for the persistence length of DNA reach only to about one half of its true value, which is near 240 base-pairs or 800Å. The mistake in previous assays was to assume that mixed-sequence DNA would be perfectly straight, whereas in fact it has a small but significant local curvature on account of its base sequence, near $1°$ to $3°$ per helical turn. So now the persistence length of DNA is taken as about 240 base-pairs or 800Å, for a thermally-induced deflection of $57°$. Thus, the contour length of a single turn of DNA supercoil, of size near 200 base-pairs, may be regarded as rigid to a first approximation – as seems plain, in fact, from electron micrographs such as those shown in Fig. 5.5.

It should be emphasized that free DNA in solution has, in general, little intrinsic curvature. Such curvature probably amounts at most to $15°$ or $20°$ per helix turn; and this level of curvature seems to be reached only for certain base sequences such as those listed in Table 5.1. In close association with positively charged proteins, however, DNA normally curves much more strongly than this, often by as much as 40 to $50°$ per helix turn, as described in Chapter 4. The

DNA has to deform to fit into the shape required by the protein, since this high level of curvature requires a systematic variation in roll angles of about plus or minus 9°, as we found earlier. In such cases of enforced curvature, a typical dinucleotide step which adopts $R = +3°$ under stress-free conditions in solution might be required to change to $R = +9°$ or $-9°$, when the DNA wraps tightly around a protein.

Yet in principle, if we are given a DNA sequence, and we have a table of allowable ranges of roll R for each type of dinucleotide step, we can still test for the ability of DNA to wrap tightly around a protein, by trying to establish a best-fit cosine wave of amplitude 9° through the allowable range. Figure 5.7(a) shows a hypothetical

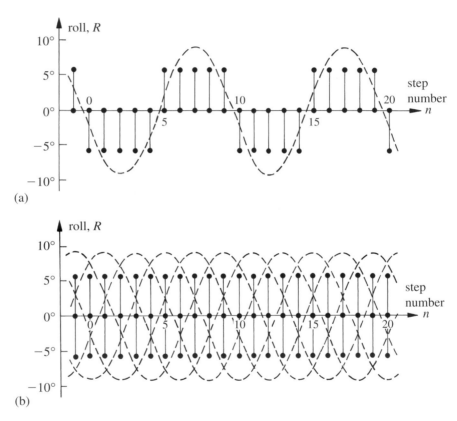

(a)

(b)

Figure 5.7 The ability of DNA to curve tightly about a protein depends on the allowable ranges of roll R at its base-pair steps. In (a), allowable ranges are $R = 0°$ to $-6°$ for steps 0, 1, 2, 3, 4, but $R = 0°$ to $+6°$ for steps 5, 6, 7, 8, 9. These roll angles may all remain equal to 0° for DNA free in solution; but when the DNA curves about a protein, many of them must switch to $-6°$ or $+6°$, in order to match the cosine wave of amplitude 9°, shown as a broken line, which coincides with the Fourier term of period 10. This DNA can bend only in one direction, since the best-fit cosine wave has a unique phase. In (b), allowable ranges of R are $-6°$ to $+6°$ for all steps. Now the DNA can bend in any direction, since the same cosine wave can fit the new roll angles with any left-to-right phase.

example of this sort, where a DNA molecule is perfectly straight in solution, but can curve around a protein when it is required to do so. The DNA in this case contains two kinds of dinucleotide step; one with an allowed range of roll $R = 0°$ to $-6°$, and the other with an allowed range of $R = 0°$ to $+6°$. In free DNA, the roll angles are all equal to zero for both kinds of step; and so a horizontal line can be drawn through these points, which indicates that the curvature k must be zero. But when such DNA is forced to bend around a protein, its sequence of roll values will approximate a cosine wave of amplitude 9°. In Fig. 5.7(a), a best-fit cosine wave of amplitude 9° has been drawn as a broken curve; it is clear that most of the steps must adopt their extreme roll values of $\pm6°$ in order to fit the curve. The DNA steps whose minor-groove edges lie along the inside of the curve are those in the middle of each batch (at steps 2 and 12) where $R = -6°$. Any other phasing of this cosine wave would not fit the allowable roll values so well.

In summary, this particular kind of DNA, which is typical of many DNA molecules found in biology, can either be straight, or else it can curve in one specific direction when wrapping around a protein. In this respect, a typical DNA molecule is just like your finger or your knee: it can only bend in one direction and in one plane. On the other hand, if we want to construct a piece of DNA that can bend with equal ease in all directions, we need to include an allowable range of R values from $-6°$ to $+6°$ at every step (or, for example, from $-2°$ to $+10°$). Then the required cosine wave can be fitted at any phase, as shown in Fig. 5.7(b), and the DNA can bend in any direction quite easily.

How can we find out which kinds of dinucleotide step have high or low allowable ranges of roll R, under the bending stress applied by a protein? Some experimental data on this subject have already been presented in Fig. 4.10. Steps such as AA or TT, which prefer to occupy the low-roll locations of DNA when it is wrapped around a histone spool, presumably find it difficult to adopt a high-roll configuration. The preference for low roll is seen to be even stronger for sequences such as AAA; and this fits in with the idea that such sequences are relatively rigid, with R near 0°. Indeed, it is well-known that DNA molecules containing more than about 40 AA or TT steps in a row cannot easily wrap around a histone spool, because they cannot curve to the degree required. Similar considerations apply also to the step GC, which prefers to occupy the high-roll locations of DNA wrapped around a histone spool; it might be assigned an allowable range of R from $+5°$ to $+10°$. Finally, it is thought that steps TA and CA, among others, are relatively flexible to changes in roll, and that they can adopt the full range of R from $-9°$ to $+9°$.

Thus, sequences such as TATATATA can wrap with any phase (or angular setting) around a histone spool.

Although we have used the same geometrical ideas in relation both to the intrinsic curvature of repeating-sequence DNA in solution, and also to the enforced curvature of DNA around a protein spool, it cannot be stated too strongly that the curvature of DNA about a protein is very different in magnitude from its curvature when it is free in solution. Most DNA sequences have only small intrinsic curvature, when they are free in solution; but they often adopt preferred, highly curved shapes when they bend around a protein, owing to the different kinds of roll-angle flexibility at different sequences.

In Fig. 5.7 we were discussing the simple two-dimensional curvature of DNA in a plane, which is well-understood. But how does the three-dimensional curvature of DNA into superhelical shapes relate to its role in biology? In principle, there are two ways by which this could happen. First, the three-dimensional shape of the DNA might affect how well it binds to different proteins in the cell. For example, one protein may prefer that the DNA should wrap around it as a left-handed superhelix, while another might prefer a right-handed superhelix. Again, one protein may prefer to bind the DNA in the form of an extended superhelix, while another might prefer a flat superhelix. Furthermore, the three-dimensional shape of DNA might affect how it vibrates in solution, in response to the thermal motion of water molecules. It is well-known that things vibrate differently according to their shapes. For example, small molecules such as carbon dioxide (CO_2) and nitrogen dioxide (NO_2) vibrate differently because one is linear whereas the other is bent; and so they show different infra-red spectra. Also, the nature of these vibrations or fluctuations may be influenced by any local patches of high flexibility within the DNA, which are present on account of particular base sequences, as shown in Fig. 5.7(b).

Unfortunately, although the theoretical principles are straightforward, chemists and biologists have not yet collected many clear experimental data on how the three-dimensional shape of DNA, over a large scale, affects its binding to proteins or its thermal fluctuations. The two-dimensional curvature of DNA into a plane has been implicated in many processes, but not yet the three-dimensional curvature into superhelices. All that we know today about the three-dimensional curvature of DNA is what we have learned from physical studies such as electron microscopy and gel electrophoresis. Here is a field where there is plenty of room for improved understanding of biological function.

Further Reading

Bednar, J., Furrer, P., Katritch, V., Stasiak, A.Z., Dubochet, J., and Stasiak, A. (1995) Determination of DNA persistence length by cryo-electron microscopy: separation of the static and dynamic contributions to the apparent persistence length of DNA. *Journal of Molecular Biology* **254**, 579–94. Half of the bending of DNA as seen by electron microscopy is due to its base sequence and half is due to its thermal motion.

Brukner, I., Susic, S., Dlakic, M., Savic, A., and Pongor, S. (1994) Physiological concentration of magnesium ions induces a strong macroscopic curvature in GGGCCC-containing DNA. *Journal of Molecular Biology* **236**, 26–32. Metal ions strongly influence DNA curvature in solution at GGGCCC-type sequences.

Brukner, I., Balmaaza, A., and Chartrand, P. (1997) Differential behaviour of curved DNA upon untwisting. *Proceedings of the National Academy of Sciences, USA* **94**, 403–6. Influence of the intercalator ethidium bromide on DNA supercoiled structures in solution.

Calladine, C.R., Drew, H.R., and McCall, M.J. (1988) The intrinsic curvature of DNA in solution. *Journal of Molecular Biology* **201**, 127–37. All relevant equations for the shape of curved DNA in three dimensions, in terms of its curvature and twist.

Calladine, C.R., Collis, C.M., Drew, H.R., and Mott, M.R. (1991) A study of electrophoretic mobility of DNA in agarose and polyacrylamide gels. *Journal of Molecular Biology* **221**, 981–1005. Pictures of long, superhelically curved DNA such as those shown in Fig. 5.5, by electron microscopy.

Dlakic, M. and Harrington, R.E. (1995) Bending and torsional flexibility of GC-rich sequences as determined by cyclization assays. *Journal of Biological Chemistry* **270**, 29945–52. The sequence GGGCCC helps DNA to curve into small circles, as much as does AAAAA but in an opposite helical phase, with high roll rather than low or zero roll.

Goodsell, D.S., Kopka, M.L., Cascio, D., and Dickerson, R.E. (1993) Crystal structure of CATGGCCATG and its implications for A-tract bending models. *Proceedings of the National Academy of Sciences, USA* **90**, 2930–4. The curved structure of a GGCC sequence as seen in a crystal with magnesium ions.

Lavigne, M., Kolb, A., Yeramian, E., and Buc, H. (1994) CRP fixes the rotational orientation of covalently closed DNA molecules. *EMBO Journal* **13**, 4983–90. Even when a DNA molecule is not much bent by its base sequence, the binding of a protein such as CRP that curves DNA locally, can extend the phase of curvature for hundreds of base-pairs in either direction.

Murphy, C.J. (2001) Photophysical probes of DNA sequence-directed structure and dynamics. *Advances in Photochemistry* **26**, 145–217. A good summary of DNA structure and its measurement by light-based molecular probes.

Nishikawa, J., Amano, M., Fukue, Y., Takana, S., Kishi, H. *et al*. (2003) Left-handedly curved DNA regulates accessibility to *cis*-DNA elements in chromatin. *Nucleic Acids Research* **31**, 6651–62. The artificial placement of left-handed supercoiled DNA, upstream from a gene in human cells, attracts a nucleosome to that curved location; and thereby places the downstream binding-site for TBP protein in an exposed location between nucleosomes, where it can act readily to initiate transcription.

Revet, B., Brahms, S., and Brahms, G. (1995) Binding of the transcription activator NRI to a supercoiled DNA segment imitates association with the natural enhancer: an electron microscopic investigation. *Proceedings of the National Academy of Sciences, USA* **92**, 7535–9. A left-handed super-coil with a sequence repeat of 10 base-pairs binds specifically to a protein that activates transcription, and may be seen clearly as a well-defined superhelix on the grid.

Rivetti, C., Walker, C., and Bustamente, C. (1998) Polymer chain statistics and conformational analysis of DNA molecules with bends or sections of different flexibility. *Journal of Molecular Biology* **280**, 41–59. The local bend-angle induced by d(AAAAAA) was estimated as 13.5° by atomic force microscopy (see Chapter 9).

Roychoudhury, M., Sitlani, A., Lapham, J., and Crothers, D.M. (2000) Global structure and mechanical properties of a 10-bp nucleosome positioning motif. *Proceedings of the National Academy of Sciences, USA* **97**, 13608–13, The local curvature of d(TATAAACGCC) was estimated as 13° by rates of ligase-mediated cyclization.

Sampaolese, B., Bergia, A., Scipioni, A., Zuccheri, G., Savino, M., Samori, B., and De Santis, P. (2002) Recognition of the DNA sequence by an inorganic crystal surface. *Proceedings of the National Academy of Sciences, USA* **99**, 13566–70. C-shaped or S-shaped DNA molecules will adhere to a mica surface in precise ways.

Tchernaenko, V., Radlinska, M., Drabik, C., Bujnicki, J., Halvorson, H.R., and Lutter, L.C. (2003) Topological measurement of an A-tract bend angle: comparison of bent and straight states. *Journal of Molecular Biology* **326**, 737–60. The local bend-angle induced by d(AAAAAA) was estimated by topological methods to be 26° at 4°C, or 17° at 37°C, by comparison with a high-temperature straight state. (Differences between these values and those of Rivetti *et al*. are probably attributable to the different concentrations of magnesium ions.)

Exercises

5.1 Any ribbon with uniform curvature k and twist t will generate a spiral or helix. The characteristic shape of the helix may be calculated from k and t by use of the formulas given on p. 100 for pitch angle α, radius r, pitch p, and contour length N^* of one complete turn (see also Fig. 5.4). By convention, k, r and N^* are always positive; while t, p, and α are all positive for a right-handed helix, but negative for a left-handed helix.

helix	k (°/bp)	t (°/bp)
a	2	−2
b	2	2
c	2	0.2
d	0.2	2

Compute values of α (in degrees) and r, p, N^* (in base-pairs) for the four sets of k and t specified above. Which helix has the smallest diameter, and which has the largest? Which helices are left-handed, and which are right-handed?

5.2 The helical geometry of a uniformly curved and twisted ribbon (Fig. 5.4) may be used in a different context to provide an approximate calculation for the situation shown in Fig. 3.14(c), where the addition of uniform roll at every base-pair step tilts the base-pairs with respect to an overall helix axis, which is vertical in the picture.

The key to the calculation is to regard the central 'wire', which is shown connecting base-pairs in models (a) and (c) of Fig. 3.14, as a 'twisted ribbon'. Since the blocks are attached locally perpendicular to the wire, tilt $= 90° - \alpha$.

In this application, t becomes the local twist T (°/bp) between successive base-pairs, while k becomes the roll angle R (°/bp). N^* is the number of steps per complete helical turn; and so $360°/N^*$ corresponds to the global twist T_G (°/bp) measured with respect to an overall helix axis, i.e. an imaginary vertical line in Fig. 3.14(c). Formulas from p. 100 may thus be re-written in the present context:

$$\tan \alpha = T/R, \text{ so } \tan(\text{tilt}) = R/T; \quad T_G^2 = T^2 + R^2.$$

a Given the T_G and tilt values shown below for idealized 'A,' 'B', and 'C' forms of DNA from fibers (see Chapter 9), use the formulas above to obtain the local twist T and the roll R in each case.

Fiber model	T_G (°/step)	Tilt (°)
'A'	32.7	+20
'B'	36.0	0
'C'	40.0	−10

b Inspection of Fig. 3.14(c) shows that it has global twist $T_G = 36°$. Use the formulas above to obtain the local twist T, for roll $R = 12°$. (Note that in Chapter 3 it was stated that $T \approx 36°$ for this model: thus the slight difference between values of global twist T_G and local twist T was overlooked there.)

5.3a   In Exercise 4.6 you were asked to compute the quantity which is now described on p. 104 as the 'second sum', for five different repeating sequences R_n of roll angles. Confirm by direct calculation (suggestion: extend the table of Exercise 4.6 to five columns) that the 'first sum' is precisely zero in all of these five cases. For ease of calculation, we fixed the roll angles in Exercise 4.6 always to give a first sum of zero; but this will not generally be the case.

b   Perform a calculation as set out on p. 104 for the case

$$R_n = (0°, 10°, 10°, 10°, 0°, 0°, 0°, 0°, 0°, 0°)$$

in order to find the curvature k and its phase. Is the first sum equal to zero in this example? Which step has the maximum roll R in the Fourier wave?

Note on the computation of $\arctan(f/s)$, where f = first sum and s = second sum. Suppose, for example, $f = 1.2, s = -1.6$. Then $f/s = -0.75$. If you ask your calculator for $\arctan(-0.75)$ it will probably give $-36.9°$, although another equally valid answer is $+143.1°$; for there are always two angles in the 360° circle, separated by 180°, whose tangents are identical. In the present case we must choose only one of these two angles. The rule which we need here is that if θ is the correct answer, $\sin\theta$ and $\cos\theta$ have the same signs as f and s, respectively. Here, $\sin -36.9° = -0.6$ and $\cos -36.9° = +0.8$,   while $\sin 143.1° = +0.6$ and $\cos 143.1° = -0.8$; and so it is the second answer, 143.1°, which is correct.

5.4   Consider a long DNA molecule having a repeating sequence of the kind $A_4N_6A_4N_6A_4N_6$ ..., or $(A_4N_6)_m$, i.e. a 10-base-pair repeat of the kind A_4N_6, where N stands for any base other than A.

a   Taking $R = 0°$ for steps AA, and $R = 3.3°$ for all other steps, convert this sequence repeat into the corresponding roll angle repeat R_n for $n = 0$ to 9, where $n = 0$ corresponds to the first AA step in the repeat. Compute the first and second sums as on p. 104, and hence evaluate k; and finally express this in units of $°/bp$. Which step has the largest value of roll R in the Fourier wave?

b   Taking twist values $T = 35°$ for step AA, and $T = 34°$ for all other steps, calculate the overall twist t for a 10-step repeat (i.e. sum the T values and then subtract 360°); and express this also in units of $°/bp$.

c   Using the values of k and t as calculated above, find the radius r of the superhelical curve which is made by this repeated-sequence DNA, and the number N^* of base-pair steps in a complete superhelical turn. Use the formulas on p. 100; and note that r is given here in base-pair units: $1\,\text{bp} \approx 3.3\,\text{Å}$.

5.5 Repeat Exercise 5.4, but using instead sequence $(A_6N_4)_m$.

5.6 Repeat Exercise 5.4, but using the sequence $(A_6N_5)_m$. Here, in working out k it will be necessary to use multiples of $360°/11 = 32.7°$ for $n = 0$ to $n = 10$;

i.e. first sum $= \displaystyle\sum_{n=0,10} R_n \sin(32.7°n)$, etc.

The value of k will be in degrees per 11-step repeat; but this should be expressed in $°/\text{bp}$ for use in the formulas on p. 100.

Which of the sequences $(A_4N_6)_m$, $(A_6N_4)_m$, and $(A_6N_5)_m$ form left-handed superhelices, and which form right-handed superhelices?

CHAPTER 6

DNA Supercoiling

In Chapter 4 we explained how a DNA molecule must twist – or untwist – and curve, in order to carry out its various functions in biology. For example, DNA has to untwist near the start-sites of all genes, often at or near TATA sequences, so that RNA polymerase can unwind or separate the strands and construct new RNA strands according to the rules of Watson–Crick base-pairing. Similarly, DNA must untwist at all origins of replication, so that DNA polymerase can construct a new DNA strand in readiness for cell division. Lastly, DNA must curve around a variety of proteins in the cell. Some of these proteins help to package the DNA into a compact form, while others help to control the activity of some particular gene.

In Chapter 5 we went on to explain how DNA can twist and curve at the same time, so as to form long, regular supercoils or spirals. Some DNA molecules are intrinsically twisted and curved, on account of their base sequences; but others become twisted and curved only when they bind to certain proteins.

In the present chapter we shall take the subject of supercoiling one step further, by describing DNA molecules which are not curved either on account of their base sequences, or as a consequence of binding to proteins, but which coil through space nevertheless on account of *torsional stress*. And in order to provide a simple introduction to what is admittedly a rather difficult subject, let us first see how torsional stress can affect the shapes of some familiar household objects.

Figure 6.1(a) shows a piece of telephone cord. This sort of cord has a regular coiled shape because it has been manufactured with inbuilt curvature and twist. It coils through space naturally, like the intrinsically curved DNA molecules described in Chapter 5. Figure 6.1(b) and (c), on the other hand, show a piece of ordinary electric power

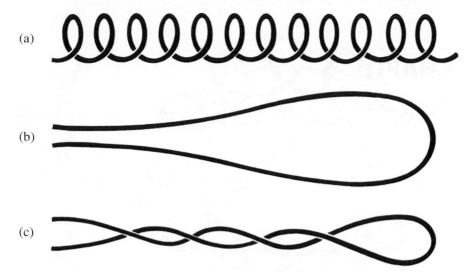

(a)

(b)

(c)

Figure 6.1 Everyday models for the supercoiling of DNA. In (a), a telephone cord coils through space naturally because it has inbuilt curvature and twist. In (b), an electric power cord has no inbuilt curvature or twist; but if you wind a few turns of twist into it, as in (c), it will cross over itself approximately once for every 360° turn of twist that you introduce.

cord. This kind of cord is manufactured without any inbuilt curvature or twist, and so it generally prefers to lie straight across a floor or table, or else in the kind of gently curving, broad loop sketched in Fig. 6.1(b). But if you take one end of this electric power cord in each hand, and wind a few turns of twist into it, then it will adopt the kind of shape shown in Fig. 6.1(c): the cord will cross over itself roughly once for every 360° turn of twist that you introduce. If you then release either end of the cord from the constraints of your hand, and shake it a bit, the free end will rotate in a reverse sense to eliminate the added twist; and the cord will return again to the uncrossed shape of Fig. 6.1(b).

Most DNA molecules, even though they are not intrinsically curved, can coil through space in the manner of the electric power cord shown in Fig. 6.1(b) and (c). It is very easy for a long DNA molecule to lose or gain a few turns of twist. This can happen, for example, by its binding to certain proteins, or by its unwinding locally during the synthesis of RNA; and if the two ends of the DNA are not free to rotate, then even a small change in twist can cause the path of the DNA to coil through space. Furthermore, there are several ways by which the two ends of a DNA molecule can be prevented from rotating. For example, the two ends can be joined with one another to form a closed circle of DNA, as shown in Fig. 6.2(a) and (b). Here, the ends of our telephone cord and electric cord have been joined into

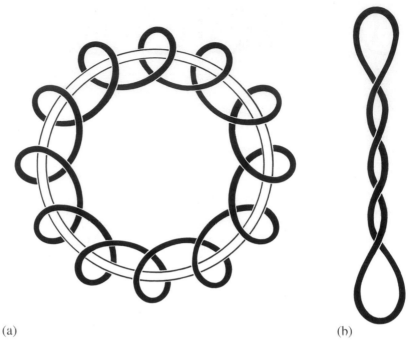

(a)   (b)

Figure 6.2 Two general varieties of DNA supercoil. In (a), the DNA coils into a series of spirals about an imaginary toroid or ring (shown here by open lines); and so this kind of wrapping is known as 'toroidal'. In (b), the DNA crosses over and under itself repeatedly; and so this kind of wrapping is known as 'interwound'.

shapes which are complex and yet 'circular', in the sense that they are endless. There are many examples of circular DNA in living cells.

Another example of end-restraint is shown in Fig. 6.3. Here, a long piece of linear DNA has been divided into a series of loops, and the two ends of each loop are constrained from rotation where they attach to some sort of supporting structure at the base of every loop. This kind of looped-linear arrangement is thought to be typical of the chromosomal DNA molecules which are found in higher organisms. We shall say more about this in Chapter 7. Both kinds of DNA, the circular and the looped-linear, will form supercoils upon any internal change of twist, because their ends are not free to rotate. In fact, there exists in our cells a variety of enzymes called 'topoisomerases', as mentioned in Chapter 4, that can cut the DNA temporarily so as to make a free end, and thereby relieve some of the effects of supercoiling. Without such enzymes, the DNA would get tangled about itself during normal cell function, and so it could hardly act as the genetic material. A full discussion of these highly complex enzymes is beyond the scope of our book, however; so the

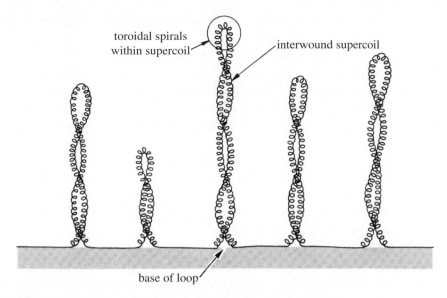

Figure 6.3 The division of a long, linear DNA molecule into loops generates end-restraint at the base of every loop, if the two ends are attached to some support or 'scaffold'. This kind of looped-linear arrangement is thought to be typical of the chromosomal DNA found in higher organisms.

reader should consult some of the references listed at the end of this chapter to learn more about them.

In any case, once we know that the ends of a DNA molecule are fixed, then we can identify two different kinds of supercoiling, which are epitomized by the two shapes shown in Fig 6.2(a) and (b). The circular DNA in (a) consists of a series of open spirals that wind around an imaginary ring, or toroid;[1] this kind of supercoiling is known as 'toroidal'. The circular DNA in (b), in contrast, winds above and below itself several times, and this kind of supercoiling is called 'interwound'. In practice, real DNA supercoils may contain portions of both the toroidal and interwound geometries. Thus, where certain parts of the DNA are highly curved, on account of either the base sequence or due to wrapping around a protein, one may find toroidal structures, since the DNA in a toroidal supercoil is highly curved throughout. Alternatively, if such curved portions of the DNA are not very long, they may locate themselves at the two strongly curved end-loops of an interwound supercoil, as shown at the top and bottom in Fig. 6.2(b). Sometimes the interwound and toroidal geometries may occur together, as in the looped-linear DNA which is shown schematically in Fig. 6.3. On a small scale, within any loop, the coiling is toroidal on account of the wrapping of DNA around protein spools; but on a large scale, over the full length of any loop, the structure is interwound. You often see this kind of arrangement in telephone cords, if people habitually rotate the handset.

In general, supercoiled DNA has the shapes seen in Fig. 6.2 because it either has more turns of twist, or fewer turns of twist, than the underlying, relaxed, right-handed double helix from which it is made. DNA with more than the natural number of turns is known as *overwound*, while DNA with fewer than the natural number of turns is known as *underwound*.

Is there any way in which we can tell, by looking at pictures such as those shown in Fig. 6.2, whether the supercoiled DNA is overwound or underwound? And can we say by how many turns the DNA is overwound or underwound? Oddly enough, the second question is easier to answer than the first. Recall that we said above, in relation to the electric power cord of Fig. 6.1(b) and (c), that the cord crosses over itself approximately once for every 360° turn of twist which is wound into it. This idea is rather general, and it may be applied also to both of the supercoils of Fig. 6.2. Thus the supercoil (b), which contains four crossovers, is either underwound or overwound by about four turns, while supercoil (a) contains 12 crossovers, and is underwound or overwound by about 12 turns. Yet it is not so easy to decide whether these supercoils are overwound or underwound; before we can say anything on that point, we must learn more about the theory of supercoiling, as described below.

When circular DNA molecules are isolated from the cells of our bodies, or from bacteria, they are generally found to lack one turn of twist for every 17 turns of stable, right-handed double helix. Thus, they are said to be underwound, or negatively supercoiled, by about 6%. DNA molecules which contain *extra* turns are not found in Nature, except under special circumstances; and we shall explain below why that is the case.

It might seem unlikely that one could ever arrive at a single scheme of explanation that would account for the shapes of all possible varieties of DNA supercoil, whether toroidal or interwound. Nevertheless, such a scheme has been invented by the mathematicians James White and Brock Fuller. It is based on the branch of mathematics known as 'topology', which concerns itself with how things change shape when they go from one form to another. That is just what we want here, because DNA supercoils have lots of different shapes even though they are all made from the same kind of right-handed double helix.

We shall explain the application of topology to DNA supercoiling in two steps. First we shall deal with interwound structures, and then with toroidal ones. You will see when we have finished that we have actually used the same scheme in both cases.

Figure 6.4 shows a series of five, closely related, circular[1] DNA molecules. Two are in the form of open circles or simple rings, while three are in the form of interwound supercoils. In all cases, the DNA has been drawn as if its relaxed, stress-free form were a long rubber rod of square cross-section – recall Fig. 5.2(b) – with one face painted black. For example, the open circle in Fig. 6.4(a) is entirely black on one side. This means that it has exactly the same twist as relaxed, linear DNA of the same length; it is neither underwound nor overwound. Above each molecule in Fig. 6.4 we have added two symbols: Tw for 'twist' and Wr for 'writhe'. To a first approximation, Tw tells by how many turns the rubber rod twists as it goes once around the circle, and Wr tells how many times the rod crosses over itself within any molecule. Thus, Tw = 0 for the open circle in Fig. 6.4(a), because there we can see the black face of the rod everywhere; and Wr = 0 because the rod does not cross over itself.

Figure 6.4(b) shows an open circle with three extra turns of twist. This twisted configuration can be obtained from the open circle of Fig. 6.4(a) by cutting the rod in one place, inserting there three full turns of twist and then closing it up again. It is easy to confirm that the open circle (b) contains three extra turns of twist, because the black face of the rod now shows at three locations as you go around the circle. You can also see that this new twist is right-handed, because the rod rotates in a clockwise sense, like a corkscrew, as you go around the circle. Accordingly, open circle (b) has been labeled as Tw = +3: the positive sign indicates that the extra twist is right-handed, i.e. in the same sense as that of the underlying DNA double helix. Wr = 0 again here, of course, since the rod in Fig. 6.4(b) does not cross over itself.

Now, if you make a model of the twisted circle (b) from a rubber rod, or from a long leather belt, you will find right away that this shape is unstable. Thus, if you hold the twisted ring firmly down onto a table it will look like (b), but if you let go it will collapse quickly into one of the other forms, usually into (d) or (e).

In each of the forms (c), (d), and (e), the DNA crosses over itself at least once, and the value of Wr shown above each drawing records the number of crossovers: 1 for (c), 2 for (d), 3 for (e). Note that for each new crossover, the circle loses one turn of twist. Thus, Tw decreases from +3 in (b), to +2 in (c), to +1 in (d), and finally to 0 in (e). You can confirm the value of Tw recorded in each drawing by counting the number of times the black face appears and disappears as you go around the circle. It is this decrease in Tw which provides the driving force for the rod to collapse spontaneously from shape (b) to (e). For most rubber rods, or leather belts, or DNA molecules,

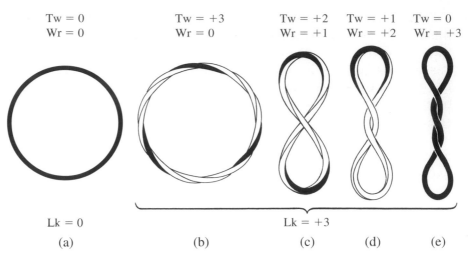

Tw = 0	Tw = +3	Tw = +2	Tw = +1	Tw = 0
Wr = 0	Wr = 0	Wr = +1	Wr = +2	Wr = +3

Lk = 0   Lk = +3

(a)   (b)   (c)   (d)   (e)

Figure 6.4 Five closely related circular DNA molecules: (a) and (b) show open circles, while (c), (d) and (e) show interwound supercoils. The DNA in its stress-free, relaxed form is drawn as a rubber rod of square cross-section, with one face painted black.

it is easier for the rod to cross over itself repeatedly than to alter its twist: the rod responds to the overwinding by *writhing* rather than by *twisting*. One could imagine, perhaps, that a very special rod could be constructed that would behave in the opposite way, preferring instead to remain in shape (b) rather than going to shape (e). Such a rod would be easy to twist, but hard to bend. However, most real rods and DNA molecules prefer configuration (e) to (b).

You may have noticed that the sum of Tw and Wr remains constant at +3 for all of shapes (b), (c), (d), and (e) in Fig. 6.4. This is no accident; and indeed it exemplifies a result of general significance. It can be shown to hold true for supercoiled DNA of any shape or size, provided that the meanings of Tw and Wr are defined rigorously (see below), and that the DNA double helix remains uncut and intact on both strands. For example, the sum (Tw + Wr) changed from 0 to +3 when we cut open the circle (a) and added three turns of twist; but then it did not change after that, in any of (b), (c), (d), or (e), because we did not cut the circle again.

Because the sum of Tw and Wr does not depend on the exact shape of the circle, but only on the intactness of its two DNA strands, mathematicians have given this quantity a special symbol, Lk, meaning 'linking number':

$$Tw + Wr = Lk.$$

Lk is known as a linking number because it is closely related to the number of times that the two sugar–phosphate chains of DNA wrap around, or are 'linked with', one another. In this chapter, we have taken DNA in its relaxed state as the reference point for counting Lk. Thus Lk = +3 tells us that the DNA has three *more* double-helical turns than it would have in a relaxed, open-circular form. In general, Lk measures the total excess or deficit of double-helical turns in the molecule. So, when we say that a DNA molecule is 'overwound by three turns' we mean, precisely, 'Lk = +3'. Note, in particular, that Lk can only be an integer, because the DNA can only join to itself by some integral number of turns. (However, if a collection of DNA molecules in solution happens to adopt a series of integral values of Lk such as −19, −20, −21 and −22, which are closely related in energy to one another, then one could say also that the mean Lk = −20.5 for the group as a whole.)

Do Tw and Wr have to be integers also, or can they be real numbers such as +0.5 or +2.5? All of the pictures in Fig. 6.4 show values of Tw and Wr that are integral, but this is not necessarily so in all cases. For example, one could make an interwound supercoil with Tw = +0.5, Wr = +2.5 by rotating the lower lobe of supercoil (e) through an angle of 90° relative to the upper lobe. Then it would lie midway in shape between (d) and (e), which is satisfactory physically but hard to draw. A further untwisting of the lower lobe by an additional 90° would convert Fig. 6.4(e) all the way to (d), and yield Tw = +1.0, Wr = +2.0. So we see that neither Tw nor Wr needs to be an integer.

We have reached a point where we need more precise definitions of twist and writhe than those which we have been using so far. Let us consider twist first. In Chapter 5 we introduced a special twist t, measured in units of degrees per base-pair, to account for the long-range coiling of DNA through space. Thus $t = 0$ for a plane curve, but t is negative for a left-handed spiral, or positive for a right-handed spiral. The twist Tw as described here is very closely related to the twist t of Chapter 5. To find the value of Tw, we can simply take the sum of t values for all of the base-pair steps in the circular DNA, and then divide the total by 360°, in order to express the result in units of helical turns. Sometimes we do not have to use this complex definition if we wish to find the value of Tw. For example, we can evaluate Tw by inspection for each of the drawings in Fig. 6.4. But for more complicated shapes, in which the value of Tw is not an integer, it would be necessary to do the calculation in full.

Next let us consider *writhe*, which we have thought of in terms of the number of times the rod crosses over itself. The crucial point about Wr is that it is a measure of the shape of the DNA as a three-dimensional curve through space. Previously we counted the number of crossovers of the DNA in a single view in order to estimate Wr. All we need to do to get Wr accurately is to count the number of crossovers that can be seen in many different randomly chosen views of the structure, and then take the average of all of these to get the actual value of Wr. This is not a hard concept to grasp, if we think of taking a large number of snapshots of the DNA as it tumbles randomly through space, due to thermal motion. In practice, however, this may not be such a straightforward procedure, for in some views there can be many crossovers, some of which will cancel each other out: see Exercise 6.3. We shall avoid such difficult ideas in the main part of this chapter, so that only the specialist need worry about them.

Now the diagrams in Fig. 6.4 that we have been studying so far were drawn for positive Lk, i.e. for overwound DNA. You may recall that DNA in living cells is normally not overwound, but rather is underwound, and so its value of Lk is negative. Therefore, in Fig. 6.5, we have provided a corresponding set of pictures for negatively supercoiled, underwound DNA. You can see by comparing the two figures that the twist is now left-handed; for example, Tw = −3 and counterclockwise in Fig. 6.5(b). Also, the way in which the DNA crosses over itself (or Wr) is subtly different in Figs 6.4 and 6.5. Thus, the DNA crosses over itself in a left-handed fashion in Fig. 6.4(e) to give Wr = +3, but in a right-handed fashion in Fig. 6.5(e) to give Wr = −3. In fact, the handedness of the crossovers in any interwound supercoil enables you to say definitely whether the DNA is underwound or overwound, simply by looking at a picture. For open circular DNA, where there are no crossovers, you have to look at the twist of the rod to see whether the DNA is overwound (clockwise twist) or underwound (counterclockwise twist). The drawings in Figs 6.4 and 6.5 should serve as reliable references for further study.

At the beginning of this chapter we described two general classes of supercoil, known as interwound and toroidal. So far we have investigated only the interwound form. But now that we are familiar with the meanings of Lk, Tw, and Wr, it should be a relatively easy matter to analyze the shape of a toroidal supercoil, such as that shown in Fig. 6.2(a).

The best way of proceeding might be to examine in detail one small portion of a toroidal supercoil, say a segment of two superhelical turns. Figure 6.6(a) and (b) show how two such toroidal, superhelical turns can be generated by the collapse of a highly twisted piece of DNA. In diagram (a) we have drawn a twisted ribbon, with one of its sides shaded. It represents a piece of DNA which has been underwound by two turns. The ribbon is attached at either end to a block, so that its two ends cannot rotate: this device enables us to consider a small piece of circular DNA in isolation. For this piece of DNA you can see that Tw = −2 and Lk = −2, since Wr = 0: there are no crossovers in any perspective.

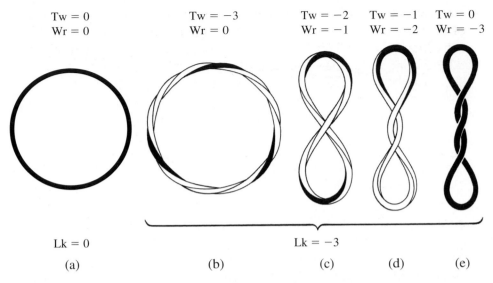

Tw = 0   Tw = −3   Tw = −2   Tw = −1   Tw = 0
Wr = 0   Wr = 0   Wr = −1   Wr = −2   Wr = −3

Lk = 0   Lk = −3

(a)   (b)   (c)   (d)   (e)

Figure 6.5 Five circular DNA molecules as in Fig. 6.4, but now with a linking number Lk of the opposite sense. These DNA molecules are underwound, with Lk negative, while the ones in Fig. 6.4 were overwound, with Lk positive.

In Fig. 6.6(b) we have moved the two blocks closer together, so that the twisted ribbon can collapse into part of a toroidal coil, like the coil made by a snake as it lies in the grass. The ribbon now makes two flat, left-handed turns: you can verify that they are left-handed by doing a simple experiment with a short piece of a leather belt, or a strip of paper. In this configuration both coils are almost planar, so Tw ≈ 0, and the linking number has not changed from its previous value of −2. Thus we find that Wr = Lk − Tw = −2. The main point which emerges from this exercise is that two flat, left-handed toroidal coils have a writhing number of −2. Thus, in going from Fig. 6.6(a) to (b), the twist of the DNA has been transformed into writhe.

Let us now return to the toroidal circle of Fig. 6.2(a). Its coils are certainly left-handed, but they are not so flat as those of Fig. 6.6(b). If we imagine a simple transformation by which the entire toroid is pushed inwards, so that its radius becomes smaller, then the coils will become flatter. In such a case the writhing number can be assigned as −2 for each pair of coils, just as in the example above; and so Wr = −12 for the entire molecule. On the other hand, if the supercoil were to be pulled out to a much larger radius, so that it became a simple, open circle, then it should have Wr = 0, and Tw = −12. In the configuration shown, Wr will lie somewhere between these two extremes, and closer to −12 than to 0; so perhaps Wr = −10.

You may have noticed in our pictures that a *left-handed* coil in the toroidal form of Fig. 6.2(a) gives negative writhe, while a *right-handed*

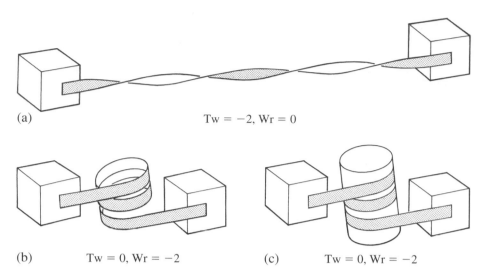

(a) $Tw = -2, Wr = 0$

(b) $Tw = 0, Wr = -2$ (c) $Tw = 0, Wr = -2$

Figure 6.6 A highly twisted ribbon will collapse spontaneously into part of a toroidal supercoil. In (a), the two ends of the ribbon are held apart by their attachment to blocks, so that $Tw = -2$. In (b), the blocks move together so that the ribbon can collapse to $Wr = -2$. In (c), a cork or protein spool stabilizes the shape of the ribbon shown in (b).

coil in the interwound form of Fig. 6.2(b) also gives negative writhe. Is there some sort of mistake here? Surely the rule of crossovers should be consistent among different types of supercoil, such as interwound and toroidal? In fact, there is no mistake at all; the two sets of pictures have been derived unambiguously from physical experiments with underwound rubber rods or leather belts. The key point here is that we need a way of allocating a *sign* to any given crossover which we can see, in order to say unambiguously whether it contributes positively or negatively to writhe. This is not a trivial exercise, and so we have consigned it to Exercise 6.3.

We have now described two different kinds of supercoiling for DNA – toroidal and interwound. But what are the relative stabilities of these two forms? In other words, when will a DNA molecule be interwound, and when will it be toroidal? The interwound shape is usually very stable, and most underwound or overwound DNA molecules will naturally adopt an interwound shape, in the absence of other forces. But the proteins that associate with DNA in living cells can sometimes change the situation dramatically, and favor the toroidal over the interwound form by wrapping the DNA around themselves.

For example, consider the cork which has been inserted between the two turns of ribbon shown in Fig. 6.6(c). This cork represents a typical protein 'spool' around which the DNA can wrap, and around which it does wrap in a left-handed sense in the chromosomes of most higher organisms on Earth. If the DNA or ribbon in Fig. 6.6(c) were to be cut free from the two blocks at either end, it would stay wrapped around the 'sticky' protein spool; whereas if it were cut free in the absence of a spool, as in Fig. 6.6(b), it would immediately spring back into a straight configuration. When we isolate DNA in the laboratory in pure form from any kind of cell or cells, at some point in the procedure we must strip off the proteins around which the DNA was originally wrapped, without breaking either of its two double-helical strands. In other words, we must remove the cork from the arrangement shown in Fig. 6.6(c), without cutting the DNA free from either of its two end-blocks. Naturally the 'naked' DNA will first spring out to the highly twisted form shown in Fig. 6.6(a), and then it can collapse into an interwound supercoil as shown in Fig. 6.5(e), because it has lost the curvature which stabilized the toroidal form. Therefore, we can expect to see highly interwound supercoils in the preparations of pure DNA which we make from living cells, after removal of various proteins. Incidentally, this is why DNA supercoils in Nature are usually underwound rather than overwound: the DNA always coils around proteins in the cell nucleus in the form of a left-handed toroidal spiral, giving negative Lk.

Some typical preparations of purified, protein-free DNA from bacteria are shown in Fig. 6.7(a), exactly as they appear in the electron microscope. These DNA molecules are circular forms of length 7000 base-pairs, and they are underwound on the average by 40 turns, i.e. Lk $-$ -40. A linear, relaxed double helix of the same length would contain $7000/10.6 = 660$ turns of right-handed DNA, given a typical helical twist of 10.6 base-pairs per turn; so these DNA molecules contain just 620 turns and are underwound by $40/660 = 6\%$.

Note that the supercoils shown in Fig. 6.7(a) have a branched formation of the kind indicated schematically in Fig. 6.7(b). This does not in itself have much effect on either the topology or the energetics of the structure. Indeed, the branched structure could 'migrate' smoothly into an unbranched form that would be much longer overall, if conditions were right.

Assuming that these DNA molecules are all of the interwound form, can we determine their Lk, Wr, and Tw? A value for the linking number of Lk $= -40$ can be determined independently of the electron-microscope pictures, by studying the mobility of such DNA during gel electrophoresis: this method will be described in Chapter 9. An approximate value for the writhing number, Wr $= -36$, can be determined by counting the mean number of crossovers per DNA molecule, as seen in many pictures from electron microscopy such as those of Fig. 6.7(a). Finally, by subtraction, Tw $=$ Lk $-$ Wr $= -4$. Thus, the preferred interwound structure is

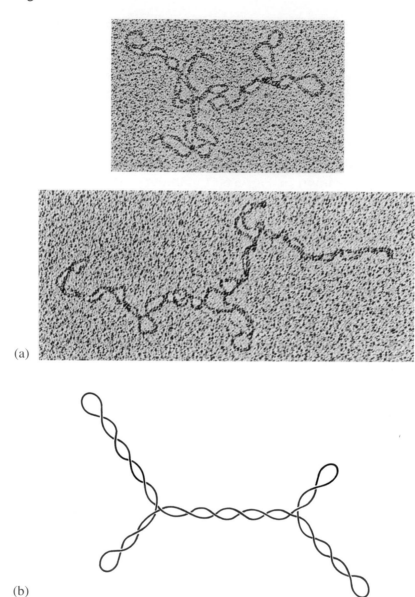

(a)

(b)

Figure 6.7 (a) Electron micrographs of negatively supercoiled, interwound DNA as prepared in pure form from *E. coli* bacteria. Each DNA plasmid or ring is 7000 base-pairs long, and has a mean Lk = −40. Courtesy of Christian Boles, Nick Cozzarelli, and James White; and from *Journal of Molecular Biology* (1990) **213**, 931–51. (b) Branched path of the interwound DNA shown in (a), in schematic form. Such branching has little effect on the parameters Lk, Tw, and Wr.

somewhat similar to the idealized shape shown in Fig. 6.5(e), since Wr = 0.9 Lk, and Tw = 0.1 Lk. In other words, the DNA which has been underwound finds it more favorable energetically to cross over itself repeatedly, than to alter its twist.

Some typical pictures of toroidal supercoils, where the coiling of the DNA has been stabilized by wrapping about proteins, will be shown in Chapter 7.

That concludes our survey of the physical properties of supercoiled DNA. We can now concern ourselves with its biological properties. Supercoiling is important in biology because it helps the DNA to unwind, so as to promote the synthesis of new RNA or DNA strands. As shown in Fig. 6.8, the various polymerase proteins that copy pre-existing DNA into new RNA or DNA must first unwind the DNA by one to two turns in the locations where they wish to act. This then enables them to 'read' the unpaired bases on one strand of the DNA, so that they can assemble a new strand according to the rules of Watson–Crick base-pairing. Sometimes it is not the polymerase that unwinds the DNA, but rather its 'helper' proteins. In any case, the DNA must be unwound or else accurate synthesis cannot proceed.

We said above that most of the missing turns in any negatively supercoiled DNA molecule are stored in the form of writhe Wr, whether by crossovers in an interwound supercoil, or by flat spirals in a toroidal supercoil. How, then, can supercoiling produce a reduction of twist Tw by one or two turns, as is needed for a polymerase protein to unwind DNA in various locations? Clearly, the DNA must be able to vibrate or fluctuate in solution, as a kind of Brownian movement, from shapes with high writhe to shapes with high twist. For example, an interwound supercoil might vibrate from the shape shown in Fig. 6.5(e) to any of the shapes shown in Fig. 6.5(d), (c), or (b), in order to generate twist. Similarly, a toroidal supercoil might vibrate from the shape shown in Fig. 6.6(b) to that shown in Fig. 6.6(a).

Unfortunately, we have few direct experimental data today, which might indicate how a DNA molecule fluctuates in solution over a large scale. We know, through probing for single-stranded regions using enzymes and chemicals, that negatively supercoiled DNA

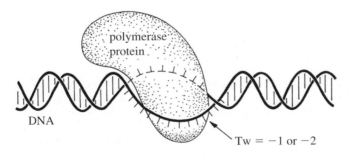

$$\text{Tw} = -1 \text{ or } -2$$

Figure 6.8 Untwisting of DNA by a typical polymerase protein, here shown schematically. Usually the twist of the DNA must be reduced by one or two turns in order for an RNA polymerase molecule to function.

vibrates much more efficiently than relaxed DNA to yield negative Tw; and we know also that many genes require negative supercoiling in order to be transcribed by RNA polymerase; but we do not know how DNA changes its shape over a large scale, to produce vibrations that lead to the generation of twist. Perhaps these involve changes in the local shape of the DNA from a right-handed supercoil to a plane curve, or from a plane curve to a left-handed supercoil, as described in Chapter 5. But all we have today are a great many lines of indirect evidence to suggest what might be going on. Furthermore, our indirect data are limited to observations about bacterial genes, because the genes in higher organisms are so poorly understood that one cannot draw any firm conclusions about how they work.

First, we have the observation that many different bacterial genes will synthesize more RNA, if the DNA in their 'promoter' region is appropriately curved due to its base sequence, than if this DNA is inappropriately curved or straight. Indeed, the most transcriptionally active genes in the bacterium *Escherichia coli* almost invariably have a region of curved DNA preceding the promoter. The structure of a typical bacterial gene is shown in Fig. 6.9(a). There we can see that an RNA polymerase protein starts to make messenger-RNA, just upstream of the long segment of DNA that codes for protein. The RNA copy of DNA then travels from the bacterial chromosome to

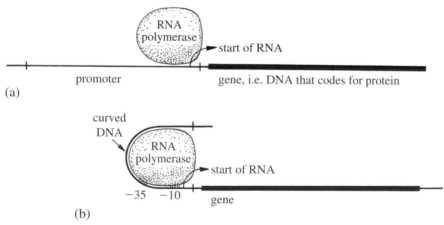

(a)

(b)

Figure 6.9 Highly schematic pictures of the structure of a typical bacterial gene, and its interaction with RNA polymerase. In (a), the polymerase protein binds just upstream of the DNA that codes for protein, and within the region of the 'promoter'. In (b), the presence of curved DNA within a promoter helps the polymerase to bind more tightly to the DNA, and hence to synthesize more RNA chains. Specific contacts between the protein and the DNA cluster in two regions that lie 35 and 10 base-pairs, respectively, upstream of the start-point for making RNA. The DNA unwinds at the −10 region (often of sequence TATA) to let the polymerase function as in Fig. 6.8.

the protein-making machinery, or ribosome, where the series of bases A, U, C, and G along its length specifies the synthesis of some protein with a certain number and ordering of amino acids. More protein is synthesized if there are more RNA molecules of a certain kind, and the number of RNA molecules depends in turn on how often RNA polymerase can initiate the synthesis of new RNA chains at any promoter.

Why should the curvature of DNA within a promoter affect how often RNA polymerase will start to make RNA? It makes no sense, if we consider the problem just in *one dimension* as shown in Fig. 6.9(a). It begins to make sense in *two dimensions*, when we consider that the curved DNA might wrap around the polymerase protein, as shown in Fig. 6.9(b), and so help it to bind at the promoter. In fact, there is considerable physical evidence that the *E. coli* RNA polymerase can bind at least up to 160 base-pairs, i.e. as much as a nucleosome, at a promoter. Most of this DNA is in front of the transcription start-site. When a polymerase approaches a promoter, it first probes for the identities of bases in two regions of the promoter called '−35' and '−10', as indicated in Fig. 6.9(b). Then it binds about 100 base-pairs of DNA upstream of the −35 region, and only then does it untwist the DNA in the promoter region to initiate transcription. Most importantly, binding of the DNA occurs best when it is negatively supercoiled; which suggests that it is wrapped onto the polymerase in a left-handed sense. Recent physical evidence for this wrapping of DNA around *E. coli* RNA polymerase seems compelling. For example, studies by Claudio Rivetti and colleagues using a combination of atomic force microscopy (see Chapter 9) and biochemical methods have shown that a typical promoter DNA wraps around the RNA polymerase protein by nearly 300° during the start of transcription, as shown in Fig. 6.10. Various protein-to-DNA contacts are necessary to stabilize that extensive curvature; but they represent only part of the full picture.

In this situation how could DNA supercoiling help the DNA in the −10 region to unwind? Now, when the DNA is free in solution the problem is relatively straightforward. Consider first the toroidal configuration of DNA in *three dimensions*. One turn of a left-handed, toroidal supercoil would yield Wr ≈ −1.0 turn if it were straightened out as in Fig. 6.6(b) and (a). Thus, due to thermal fluctuations in solution, this DNA will vibrate fairly often into an extended form with Tw ≈ −1.0 turn. In a free, highly supercoiled DNA this rapid interconversion of writhe and twist will result in transient strand separation at the most sensitive sequences, particular those containing TATA. However, when the DNA is wrapped around a protein such as RNA polymerase, it is constrained. Thus, if the

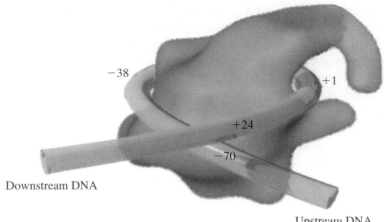

-38

+1

+24

-70

Downstream DNA

Upstream DNA

Figure 6.10 A model for the trajectory of promoter DNA around *E. coli* RNA polymerase, obtained by atomic force microscopy and biochemical methods. The DNA curves by about 300° in a negative (i.e. left-handed) supercoil; and the start-site for transcription is shown in blue. Courtesy of Claudio Rivetti.

wrapping of DNA about RNA polymerase could conceivably help the polymerase to unwind DNA near the start of the gene, i.e. in the −10 region, any change in shape of the bound supercoil comparable to the fluctuations of a free DNA molecule would be focussed. Alternatively, the energy of supercoiling could transmit a torque to the −10 region, by inducing a change in the polymerase protein itself.

Henri Buc and colleagues have shown that when RNA polymerase binds to a promoter without untwisting the DNA in the −10 region, one negative supercoil is constrained, presumably as writhe. However, the subsequent separation of the DNA strands around the transcription start-site increases the number of supercoils constrained to nearly 2: which shows that this step does not simply involve a direct conversion of writhe into twist, but that instead additional unwinding takes place. One possible solution to this paradox is to propose that the initial small untwisting in the −10 region could indeed be driven by the wrapped DNA, while the subsequent expansion of this untwisted region to encompass the transcription start-point could involve a concerted change in polymerase structure.

Furthermore, there is known to be a wide variety of proteins called 'activators' in bacteria, that seem to make genes work better in the following way: they bind to the DNA upstream of the polymerase, and curve it by up to 180°. It used to be thought that these activator proteins might simply stick to the polymerase, thereby helping it to bind the promoter as shown in Fig. 6.9(b). While this may still be true in part, more recent data have shown three things. First, the activator proteins do not have to occupy any precise location along the DNA relative to the polymerase, so long as they curve the DNA in a correct direction. Second, some proteins that curve the DNA can act as 'repressors' of gene activity in one location, by binding competitively to the same piece of DNA as that preferred by RNA polymerase; then, later they can act as 'activators' of gene activity, when moved to a new location upstream, further from the gene. Third, protein-free DNA can furnish similar activation of genes, if this DNA is appropriately curved on account of its base sequence. In other words, although the activators help the polymerase to wrap the DNA, the polymerase can do this by itself if the DNA is curved, or negatively supercoiled.

Note

1. See Appendix 1.

Further Reading

Asayama, M., Kato, H., Shibato, J., Shirai, M., and Ohyama, T. (2002) A curved DNA structure in the upstream region of light-responsive genes: its universality, binding factor, and function of cyanobacterial *psbA* transcription. *Nucleic Acids Research* **30**, 4658–66. A biological role for DNA curvature in the promoters of certain genes in cyanobacteria (ancient photo-synthetic bacteria).

Benhoff, B., Yang, H., Lawson, C.L., Parkinson, G., Liu, J., Blatter, E., Ebright, Y.W., Berman, H.M., and Ebright, R.H. (2002) Structural basis of transcription activation: the CAP-α CTD-DNA complex. *Science* **297**, 1562–5. Promoter DNA curves sharply about CAP protein, and also adheres nearby to the α subunit of *E. coli* RNA polymerase.

Berger, J.M., Gamblin, S.J., Harrison, S.C., and Wang, J.C. (1996) Structure and mechanism of DNA topoisomerase II. *Nature* **379**, 225–32. The X-ray crystal structure of topoisomerase II from yeast, which can break and rejoin the DNA so as to change Lk.

Boles, T.C., White, J.H., and Cozzarelli, N.R. (1990) The structure of plectonemically supercoiled DNA. *Journal of Molecular Biology* **213**, 931–51. Electron microscope pictures of interwound supercoils as in Fig. 6.7.

Bramhill, D. and Kornberg, A. (1988) Duplex opening by *dnaA* protein at novel sequences in initiation of replication at the origin of the *E. coli* chromosome. *Cell* **52**, 743–55. The transient unwinding of DNA as induced by its wrapping about proteins, near a bacterial origin of replication.

Cerf, C. and Stasiak, A. (2000) A topological invariant to predict the three-dimensional writhe of ideal configurations of knots and links. *Proceedings of the National Academy of Sciences, USA* **97**, 3795–8. A novel quantization of writhe Wr in certain knotted systems.

Coleman, B.C. and Swigon, D. (2000) Theory of supercoiled elastic rings with self-contact and its application to DNA plasmids. *Journal of Elasticity* **60**, 173–221. A detailed study of DNA supercoiling in terms of an elastic ring model of finite thickness.

Collis, C.M., Molloy, P.L., Both, G.W., and Drew, H.R. (1989) Influence of the sequence-dependent flexure of DNA on transcription in *E. coli*. *Nucleic Acids Research* **17**, 9447–68. The main part of the binding site for RNA polymerase can be curved DNA.

Crawford, L. and Waring, M. (1967) Supercoiling of polyoma virus DNA measured by its interaction with ethidium bromide. *Journal of Molecular Biology* **25**, 23–30. First demonstration that ethidium bromide can change the structure of supercoiled DNA, by altering the local twist of the double helix in many places.

Heggeler-Bordier, B., Wahli, W., Adrian, A., Stasiak, A., and Dubochet, J. (1992) The apical localization of transcribing RNA polymerases on supercoiled DNA prevents their rotation around the template. *EMBO Journal* **11**, 667–72. *E. coli* RNA polymerase induces a strong curve of 180° in the DNA, so strong that its binding site on DNA often becomes the curved end-loop of an interwound supercoil.

Kumar, A., Grimes, B., Fujita, N., Makino, K., Malloch, R., Hayward, R., and Ishihama, A. (1994) Role of the sigma-70 subunit of *E. coli* RNA polymerase in transcription activation. *Journal of Molecular Biology* **235**, 405–13. A careful study of the roles of -35 and -10 regions in bacterial promoters.

Lutter, L.C., Halvorsen, H.R., and Calladine, C.R. (1996) Toplogical measurement of protein-induced bend angles. *Journal of Molecular Biology* **261**, 620–33. A new method to determine with high accuracy, and in solution, the bending and twisting angles induced in DNA by a bound protein.

Ohyama, T. (2001) Intrinsic DNA bends: an organizer of local chromatin structure for transcription. *Bioessays* **23**, 708–15. A good review of DNA curvature, negative supercoiling, and their roles in RNA synthesis.

Schnos, M., Zahn, K., Inman, R.B., and Blattner, F.R. (1988) Initiation protein induced helix destabilisation at the origin: a prepriming step in DNA replication. *Cell* **52**, 385–95. The curvature of DNA around proteins at a viral origin of replication causes transient unwinding of the DNA, if it is negatively supercoiled in a test-tube.

Stump, D.M., Fraser, W.B., and Gates, K.E. (1998) The writhing of circular cross-sectional rods: undersea cables to DNA supercoils. *Proceedings of the Royal Society of London* **A 454**, 2123–56. A balanced-ply elastic model for DNA supercoiling.

Vinograd, J., Lebowitz, J., Radloff, R., Watson, R., and Laipis, P. (1965) The twisted circular form of polyoma viral DNA. *Proceedings of the National Academy of Sciences, USA* **53**, 1104–11. The earliest recorded work on DNA supercoiling.

Yan, H., Zhang, X., Shen, Z., and Seeman, N.C. (2002) A robust DNA mechanical device controlled by hybridization topology. *Nature* **415**, 62–5. Design of a nanoscale rotary device in which DNA topology can control structure.

Zaychikov, E., Denissova, L., Guckenberger, R., and Heumann, H. (1999) *E. coli* RNA polymerase translocation is accompanied by periodic bending of the DNA. *Nucleic Acids Research* **27**, 3645–52. Curvature of DNA about RNA polymerase while it is engaged in making RNA.

Bibliography

Amouyal, M. and Buc, H. (1987) Topological unwinding of strong and weak promoters by RNA polymerase: a comparison between the *lac* wild-type and UV5 sites of *E. coli*. *Journal of Molecular Biology* **195**, 795–808. They find a loss of linking number but no loss of base-pairs on the initial binding of RNA polymerase to DNA; which implies that this protein induces negative writhe before it unwinds the double helix completely.

Bracco, L., Kothlarz, D., Kolb, A., Diekmann, S., and Buc, H. (1989) Synthetic curved DNA sequences can act as transcriptional activators. *EMBO Journal* **8**, 4289–96. Curved DNA can modulate gene activity in the cell.

Drew, H.R. and Travers, A.A. (1985) DNA bending and its relation to nucleosome positioning. *Journal of Molecular Biology* **186**, 773–90. DNA sequences from the upstream region of a bacterial promoter are shown to direct the same kind of curvature for DNA wrapped around RNA polymerase, or wrapped around the chromosomal histone proteins, or wrapped into a tight, protein-free circle.

Fuller, F.B. (1971) The writhing number of a space curve. *Proceedings of the National Academy of Sciences, USA* **68**, 815–19. Development of the concept of writhe for both interwound and toroidal supercoils.

Gartenberg, M.R. and Crothers, D.M. (1991) Synthetic DNA bending sequences increases the rate of *in vitro* transcription initiation at the *Escherichia coli* lac promoter. *Journal of Molecular Biology* **219**, 217–30. How the phasing of bent DNA can increase transcription of genes in the cell.

Pemberton, I.K., Muskhelishvili, G., Travers, A.A., and Buckle, M. (2002) FIS modulates the kinetics of successive interactions of RNA polymerase with the core and upstream regions of the tyrT promoter. *Journal of Molecular Biology* **318**, 651–63. How DNA near the promoter wraps around the bacterial polymerase, and how the -10 and -35 regions interact with it.

Rivetti, C., Guthold, M., and Bustamente, C. (1999) Wrapping of DNA around the *E. coli* RNA polymerase open-promoter complex. *EMBO Journal* **18**, 4464–75. Direct visualization of DNA curvature about RNA polymerase by means of atomic force microscopy.

Exercises

6.1 For this exercise you will need a substantial length (say about 1.5 m) of flexible rod or cord. An electric extension cord, with circular cross-section about 6 mm diameter, is ideal; or else you may use rubber tubing (provided it is not *curved* when relaxed) or even a plaited rope of the kind which is sold at boat shops. For the sake of convenience, we shall refer to all of these models as 'cords'.

a   Use the cord to reproduce the transformation shown in going from Fig. 6.1(b) to (c). For each full turn of 360° wound into one end of the cord, how many times does it cross over itself?

b   Use the same cord to demonstrate that a left-handed toroidal supercoil, as in Fig. 6.2(a), transforms into a right-handed interwound coil, as in Fig. 6.2(b), when it is shaken out. To do this, first make a straight left-handed coil by wrapping the cord around a stick. (Hint: hold the stick and one end of the cord in your left hand, and wind the cord onto the stick using your right hand.) Then carefully remove the stick and bend the coil around to make a toroid; and finally hold the two ends together in one hand, and shake it out vertically. How many times does the right-handed interwound form cross over itself, for each turn of left-handed toroid around the stick?

6.2   For this exercise you should use the same cord as in Exercise 6.1, but now you must draw a black (or coloured) line (or stripe) along the entire length of the untwisted cord, in order to signify zero twist. To make a closed loop from the cord you can fasten the two ends together with sticky tape; or you can use a short dowel if you are working with rubber tubing; or you can hold the two ends together with your hand, as in Exercise 6.1.

Reproduce the transformations shown in Figs 6.4 and 6.5, by winding into the cord three full turns of either right-handed or left-handed twist, to give Lk = +3 or −3, respectively. Which of the shapes (b) to (e) is the most stable for your model? Count the number of crossovers, and then use Figs 6.4(e) and 6.5(e) as guides to decide upon the sign of Wr, by looking at the handedness of the crossovers. Also, follow the path of the black stripe in order to identify the sign and magnitude of twist Tw. Confirm that Lk = Tw + Wr.

Untwist the cord fully and repeat with, say, Lk = +5 or −5.

6.3   This is an exercise on determining the sense (+ or −) with which an individual crossover contributes towards the writhe Wr.

First, identify some particular crossover in a picture of a writhed cord. Put an arrow on the upper segment of the cord (i.e. the one nearer to you), and follow the cord round until you reach the lower segment at the same crossover. Mark an arrow on this segment too, in the same sense as the upper arrow, when going along the contour-length of the cord.

The two crossed segments should now both be marked with arrows, as shown in (a), below. Next, rotate this local picture of the crossover until the arrow of the upper segment points towards the top of the page. If the lower arrow now points from right-to-left, as in (b), then the crossover counts as +1 to Wr; but if it points from left-to-right, as in (c), then the crossover counts as −1 to Wr.

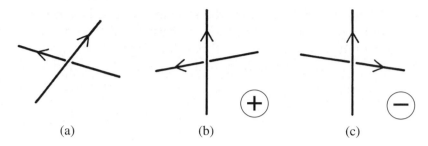

(a) (b) (c)

Use this double-arrow test to confirm that there is negative Wr in each of the coils shown in Fig. 6.2; and also to confirm the signs of writhe Wr assigned to the coils shown in Figs 6.4(e) and 6.5(e).

(You can, equivalently, associate a 'right-hand rule' with the situation shown in (b); and if you are familiar with rules of this kind in electricity and magnetism you might well find it more useful. Thus, if you straighten the thumb of your right hand and bend the fingers round, and then point the thumb along the upper segment, your fingers will indicate the direction of the lower segment, for positive Wr. But if you try this with (c), you will find that the lower segment goes the other way: so there we have a negative contribution to Wr.)

6.4 For this exercise you can use either a cord with a painted stripe, as in Exercise 6.2, or else a ribbon or strip of paper which has been colored on one side with a felt-tipped pen, as in Fig. 6.6.

a Wrap a portion of the cord twice, in a left-handed sense, around an ordinary soft-drink or beer can. Use your two hands to represent the two blocks to which the ends of the cord are attached, as in Fig. 6.6(c). Then remove the can from the wrapped cord, and pull the ends apart as in Fig. 6.6(a). How many turns of twist Tw have been created by the wrapping? Are these new turns right-handed or left-handed?

b The vast majority of DNA in our chromosomes wraps about protein spools of the kind shown in Fig. 1.5. There, the DNA wraps twice in a left-handed sense around each spool, just as in our ribbon-around-the-beer-can exercise. But when the DNA is removed from such a protein spool, it turns out by experiment that only one turn of twist is created for each two turns of wrapping. In other words, Tw = −1 instead of −2 as would be expected from the simple models of part (a) and Fig. 6.6(a). By how many turns would the DNA have to be pre-twisted, before being wrapped on the spool (or else during the process of wrapping), to account for this result? Why might the protein spools be designed to lessen the expected change in twist Tw, on wrapping or unwrapping the DNA? (See Germond, J.E., Hirt, B., Oudet, P., Gross-Bellard, M., and Chambon, P. (1975)

Proceedings of the National Academy of Sciences, USA **72**, 1843–7
for the first experiment of this kind ever performed.)

6.5 Suppose that DNA wraps around a cylindrical protein into n
complete turns of left-handed supercoil; for example, $n = 2$ in
Fig. 6.6(c). If we represent the stress-free, relaxed DNA as a ribbon,
we can easily find by direct experiment in such cases that Lk $= -n$,
where n is a positive, whole number of turns. Thus, Lk as so
defined is necessarily an integer; and the negative sign of Lk
corresponds to the left-handed sense of the wrapping.

Now we can further describe the path of the coiled DNA in terms
of a pitch angle α as shown previously in Fig. 5.4. The angle α is
likewise negative for left-handed wrapping. Furthermore, by combining results from Chapters 5 and 6, it can be shown that:

$$\text{Tw} = n \sin \alpha; \text{Wr} = \text{Lk} - \text{Tw} = -n(1 + \sin \alpha)$$

For $n = 1$, calculate Tw and Wr for $\alpha = 0°, -30°, -60°$, and $-90°$.

CHAPTER 7

The Assembly of DNA into Chromosomes

In Chapter 1 we gave a general description of the biology of a typical cell, and we explained how DNA plays a central role in that biology, by specifying the construction of protein molecules. Since then, we have considered DNA mainly as a simple, double-helical thread which undergoes transcription and replication, and which wraps itself around protein spools. Our task in the present chapter will be to describe the assembly of DNA into chromosomes. How does DNA fold into the highly compact chromosomes shown in Fig. 1.3, when a cell is about to divide? And how does it organize itself at other stages in the life of a cell, as in Fig. 1.2, when it has to make the RNA that codes for protein? Our answers to these questions will be, unfortunately, less secure than we would prefer. Only a few aspects of chromosome structure are known with confidence, while the rest require a lot of educated guesswork.

We begin by describing in some detail the proteins which make up any histone spool, and the DNA that coils around them. Next we explain how a string of successive histone spools can coil into a '300 Å fiber', if the conditions are right. Then we describe how these 300 Å fibers might fold into a series of loops along some protein 'scaffold'; how genes might work along the hypothetical loops; and how the scaffold complete with loops might coil once again into the form of the compact chromosome which we can see with a microscope on cell division. Finally, we explain how chromosomes are assembled from proteins and DNA, at least at a rudimentary level; and how they might be disassembled when genes are activated for the synthesis of RNA.

Now you may recall from Fig. 1.1 that the DNA in our chromosomes is compacted by a factor of about 10 000 in total, as compared with the length of a simple, double-helical thread. When DNA wraps around the histone spools, its overall length is reduced by a factor of about 6; so when it wraps into various other kinds of structure within a chromosome, its overall length must be reduced again by a further factor of 1500. Indeed, each individual chromosome contains a remarkably long length of DNA: typically 1 to 10 million base-pairs in yeast, or 50 to 400 million base-pairs in humans. If this DNA were not highly compacted in some definite way, then life as we know it could not exist. Throughout our presentation we shall consider, in a somewhat speculative way, how this degree of compaction might be accomplished.

Let us begin with a brief history of the subject. Chromosomes were studied for many years by geneticists, but not at a molecular level. Those scientists knew that chromosomes were made from a mixture of protein and DNA, but they did not know whether this mixture might possess any regular structure or organisation. In 1973, Dean Hewish and Leigh Burgoyne clarified the situation dramatically: they found that the majority of the DNA in a chromosome could be digested by a DNA-cutting enzyme into many small fragments of regular size, such as 200, 400, 600, 800, etc. base-pairs.

They were able to explain their result as follows. Suppose that chromosomes are made largely from a series of nearly identical units, each consisting of certain proteins in combination with 200 base-pairs of DNA. Now if the DNA-cutting enzyme were to act at every point where it found a weakness, perhaps in the regions of DNA which lie between units, then all of the long chromosomal DNA would be cut into short pieces of size 200 base-pairs. But if the cutting enzyme were to act at random, at only a limited number of points, then this long DNA would be reduced in size just to multiple units of 200, 400, 600, 800 base-pairs, etc., simply because not every point of weakness would be cut. Furthermore, if the individual units of 200 base-pairs were not of precisely determined size, but were to vary in size, for example from 180 to 220 base-pairs, then the lengths of the multiple units would vary also: from 360 to 440, 540 to 660, 720 to 880 base-pairs, etc. In such a case, these multiple units could hardly be seen as fragments of discrete size. But in fact, multiple units of discrete size are often seen up to DNA lengths as large as 2000 base-pairs in such experiments. Thus, one may conclude that certain proteins bind to this DNA in a rather precise fashion, once every 200 base-pairs, and thereby set its fundamental length.

It is now known that regular combinations of protein and DNA can be found once every 200 base-pairs along most of the length of DNA in any chromosome. These particles are called *nucleosomes*, and they are very important in biology. The fundamental length is not always exactly 200 base-pairs in every kind of animal or plant, or in every type of tissue within a particular kind of animal or plant. In fact, these fundamental 'spacings' are known to vary from as short as 160 base-pairs in certain chromosomes, to as long as 260 base-pairs in others. So one suspects that there may be some variety in the kinds or numbers of protein that make up any nucleosome. We have already shown some crude pictures of nucleosomes elsewhere in the book, for example in Figs 1.5 and 6.6(c), without giving these particles a name, or explaining how they were made.

For the sake of accuracy, one should note that the repeating pattern seen by Hewish and Burgoyne in 1973 had first been seen 3 years earlier by Robert Williamson. He was studying mouse cells in tissue culture, and he saw DNA fragments of size 200, 400, 600, 800 base-pairs, etc., as products of degradation from cells that were not growing so well. But he thought that these fragments might come from the incomplete synthesis of long DNA molecules, as shown by previous work, rather than from the degradation of intact chromosomes. Because of this, his work attracted little attention; yet progress in chromosome research could have proceeded more rapidly, if more people had realized its implications.

Following the correct interpretation of these data by Hewish and Burgoyne, other workers in many laboratories across the world quickly provided more detailed information about the newly found nucleosome. For example, early in 1974 Ada and Donald Olins visualized a preparation of nucleosomes at high magnification using an electron microscope (a technique to be described in Chapter 9), and saw a series of protein 'beads' along an extended DNA 'string'. Such a result was clearly consistent with the idea of a regular, repeating structure for DNA in chromosomes. The next significant advance came later in 1974, with a report by Roger Kornberg and Jean Thomas on the identity and approximate organisation of the proteins in a typical nucleosome. They studied the physical properties of all the major chromosomal proteins, which are known as 'histones'; and they found that the majority of histone types, known as H2A, H2B, H3, and H4, could associate with one another in a stable fashion, so as to form a large protein particle around which the DNA could wrap. This particle is now known as the 'histone octamer', because it contains eight proteins in total: there are two copies of each of the four distinct kinds listed above. At first it was thought that all of the DNA in any 200-base-pair nucleosome might wrap solely about this histone octamer, but later it was realized that the histone octamer as such binds only to about 145 base-pairs of DNA. A ninth protein molecule, known as histone H1, was found to bind to the remaining 50 or so base-pairs. These results are summarized in a

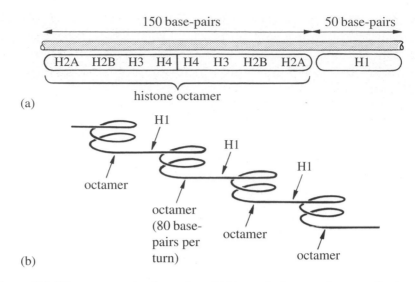

Figure 7.1 Histone proteins bound to DNA, and shown schematically in one dimension or three. The DNA is drawn as a long tube in (a), but as a long, curved string in (b).

simplified way in Fig. 7.1(a), which shows schematically the various histones strung out along the DNA in one dimension.

The three-dimensional structure of the histone octamer together with its DNA was not established firmly until 1977, when John Finch, Len Lutter and colleagues grew crystals of the histone octamer complete with DNA, that were suitable for analysis by X-ray diffraction methods. They also used electron microscopy to study the structure at low resolution, and enzyme-digestion methods to probe the structure in solution. By piecing together data gathered from all of these various methods (which will be described in Chapter 9), they were able to show that 145 base-pairs of DNA wrap almost twice around the histone octamer into a shallow, left-handed supercoil containing about 80 base-pairs per turn. Seven years later, in 1984, the same group with Tim Richmond succeeded in obtaining the structure of the histone octamer complete with its DNA to a resolution of 7 Å, solely by the method of X-ray diffraction. Finally in 1997, Richmond and colleagues analysed further the structure of the nucleosome core in better-ordered crystals at near-atomic resolution, thereby revealing many important details of how the DNA is wrapped on the surface of the histone octamer, as well as how the individual histone proteins fit together.

The old 1984 structure established that the diameter of the protein spool is about 60 Å, while the outer thickness of the DNA all around is about 20 Å, giving a diameter of about 100 Å for the particle overall. It also showed that the proteins H2A and H2B lie near both ends

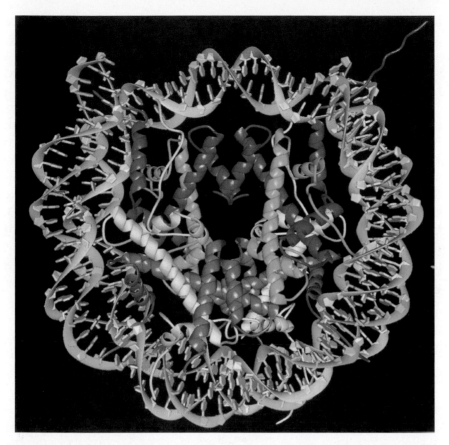

Figure 7.2 High-resolution X-ray diffraction structure of the complete nucleosome. The two-fold rotational symmetry of the protein about the vertical axis may be seen clearly. Colour-coding of proteins: H2A, orange; H2B, red; H3, blue and H4, green. The protein is mostly arranged as α-helices (see Chapter 8) and contacts between protein and DNA may be seen clearly. (The gaps between helices are filled by close-packed side-chains, not shown.) Courtesy of Tim Richmond.

of the DNA supercoil, while the proteins H3 and H4 lie near its center in a compact group. Hence the three-dimensional spool is a sort of wrapped-up version of the string of histones shown in Fig. 7.1(a). A model the new 1997 structure is shown in Fig. 7.2. Since the *x, y, z* coordinates in that new model are accurate to near-atomic resolution, many important aspects of nucleosome structure have been revealed that were not known before; for example: details of the protein fold and protein-to-protein contact; details of DNA roll, slide and twist; the locations of over 1000 water molecules; and the locations of cations such as magnesium or manganese.

The pictures shown in Figs 7.1 and 7.2 represent the simplest possible level of organisation for DNA and the proteins associated with it in a chromosome. Let us now think about the ways by which a string of nucleosomes could fold into some higher-order structure. The regularity of the spacing – 200 base-pairs per nucleosome – should be an advantage for the assembly of some sort of ordered fiber, as opposed to an irregularly shaped 'clump'.

The structure of a polynucleosome fiber is still not known for certain, but one widely cited model is that of John Finch and Aaron Klug, later extended by those workers and Fritz Thoma and Theo Koller. In 1979 those scientists reported that one could stabilize the structure of a polynucleosome fiber by soaking it overnight in fixative, prior to taking pictures by electron microscopy. They then saw structures in the microscope that look like the model shown in Fig. 7.3. Unfortunately, the electron microscope does not give the degree of resolution that one would like to have, and so some interpretation is necessary. In their model, individual nucleosomes wrap into a

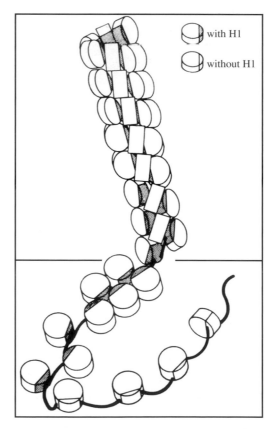

Figure 7.3 Histone octamers can be assembled onto long DNA like 'beads-on-a-string', and then these beads can wrap into some sort of more compact configuration, perhaps a flat spiral or '300 Å fiber' in the presence of histone H1. Adapted from F. Thoma *et al.* (1979).

compact spiral that advances by 110 Å per turn. Each turn of the structure contains about 5 to 6 nucleosomes, depending upon the solution conditions, and the DNA itself adopts a toroidal configuration. Histone H1 is thought to stabilize the folding of nucleosomes into this compact form: in the absence of histone H1, the nucleosomes are not 'frozen' by the fixative into some sort of spiral structure, but instead lie like 'beads-on-a-string' across the microscope support, as shown at the bottom of the picture. Whether or not histone H1 binds to the DNA depends on the chemical condition of the surrounding fluid, such as the amount of salt present.

The structure shown in Fig. 7.3 is known as a '300 Å fiber', because its outer diameter is typically 250 to 300 Å. The length of such a fiber is less than that of the constituent string of nucleosomes by a factor of about 6; so the length of the fiber is less than the length of a simple DNA thread by a factor of about 6 × 6, or approximately 40.

Histone H1 is located in the interior of this structure; but many other important aspects of the 300 Å fiber remain undetermined and are a subject of current research. Furthermore, due to the lack of precise structural definition in studies made by electron microscopy, various alternative models for the 300 Å fiber have been proposed: for example in the absence of fixative a highly irregular but coiled structure; or even an irregular zig-zag model, where the nucleosomes do not coil at all but follow a zig-zag path up the fiber axis.

Finally, we must emphasize that there are other abundant and important proteins in the cell nucleus as well as the histones, which serve to influence chromosome structure. In rank order after the histones, the next most abundant kinds of DNA-binding protein are the so-called 'high-mobility group' or 'HMG' proteins. These small proteins were noted by early investigators because they ran quickly through gels in electrophoresis experiments (see Chapter 9) when using an acetic acid–urea buffer. The three main types of HMG protein are known as HMGA, HMGB and HMGN. (They were known formerly as HMG-I,Y , HMG-1,2 and HMG-14,17 respectively.) Each class is about 1% to 10% as abundant as the histones in most tissues. All of these HMG proteins can associate with nucleosomes. HMGN is known to be associated with regions of DNA within actively transcribing genes, where the chromatin is partly unravelled; but the roles of HMGA and HMGB remain uncertain at present.

We have now summarized some of the more important small-scale features of chromosome structure. Let us therefore proceed to consider the large-scale features. Here the situation becomes very complicated. The problem is that, for most of the time, the DNA and histones are spread so uniformly in the cell nucleus that they show few distinguishing features, and so remain 'invisible' to both the light and the electron microscope. In other words, the 300 Å fibers seem to be disposed in the nucleus rather loosely, like a bunch of spaghetti in a bowl, as shown schematically in Fig. 1.2. It is only when a cell is on the point of dividing that these fibers condense or fold into structures that are sufficiently compact to be seen by use of a light or electron microscope. Such highly compact structures are known as 'metaphase'[1] chromosomes: some pictures of them were shown in Fig. 1.3. Clearly, the 300 Å fibers within metaphase chromosomes must be packed rather densely around each other into some sort of regular array, although we do not understand at present how this is accomplished.

Yet it seems likely that the 300 Å fibers in their dispersed or 'interphase'[1] state are not organized quite so loosely or randomly as spaghetti in a bowl. They must eventually undergo compaction when a cell divides at metaphase, and it is hard to see how they could do this if they were dispersed entirely at random during the intervening periods. Therefore, we should look for an additional level of structure between that of the 300 Å fiber and the condensed, metaphase chromosome.

The strongest evidence for such an intermediate level of organisation comes from the strange 'polytene' chromosomes which are found in the salivary gland of the fruit fly, and in a few other insects. The fly has four pairs of chromosomes, and by some trick of Nature, each chromosome in the salivary gland can make about 1000 copies of itself during interphase. These copies associate in a side-by-side, parallel fashion to create a highly ordered structure that can be seen by means of the light microscope. Some very clear, detailed pictures of fly polytene chromosomes were shown in Fig. 1.4. Each individual chromosome retains its dispersed form, but there are so many copies of it in the polytene chromosome, all in register with one another, that the assemblage almost becomes visible to the naked eye, its overall size being about 1/30 mm.

Furthermore, these giant constructions show a great deal of substructure, in the form of well-defined regions of dark and light, or 'bands' and 'interbands'. The bands contain 95% of the total DNA plus protein, while the interbands contain 5%. Thus, DNA and protein must be packed together much more tightly in the bands than in the interbands, because both regions are of about the same size. It would be good to know how the bands and interbands come about. They provide some of the clearest evidence concerning the structural organisation of a chromosome between the level of the 300 Å fiber and that of the final, compact form which is seen only when a cell is on the point of dividing. Many precise details of the fly polytene chromosomes are now known, as a result of careful studies by light and electron microscopy; but the structural basis for the band–interband organisation of DNA, which probably is determined somehow by the DNA base sequence itself, remains unclear.

In Fig. 6.3 we suggested that DNA might sometimes form 'loops' of size about 50 000 base-pairs, at somewhat regular intervals along the length of a chromosome. In such a model, each loop would contain an average of about 50 turns of 300 Å fiber, or 250 nucleosomes. (Incidentally, the thin curly line in Fig. 6.3 represents a string of nucleosomes, rather than the DNA itself.) It seems possible that the bands and interbands seen in polytene chromosomes are constructed from a series of such loops. Each loop would compact the DNA longitudinally, thereby providing for the dense packing of DNA and protein seen in the polytene 'bands'; while the intervals between loops would correspond to the 'interbands', where the packing of DNA and protein is much less dense. Of course, the individual loops themselves cannot be seen in the light microscope, because it takes 1000 nearly identical loops to make one band; but one can see these loops in other unusual sorts of chromosome, as we shall describe below.

Recall that the length of a DNA double-helical thread is reduced by a factor of about 40 when it wraps into a 300 Å fiber. Folding the 300 Å fiber into a series of loops will reduce its length by another factor of about 25, yielding a total compaction of about $40 \times 25 = 1000$, on going from free DNA to the dispersed, interphase form of a typical chromosome. So when a cell divides, it will only have to reduce the size of its chromosomes by a further factor of about 10, in order to reach the total compaction of 10 000-fold mentioned at the beginning of this chapter.

Now are these loops real, or are they just an attractive model, devized to explain phenomena which we cannot understand at present? There is much indirect evidence in favor of loops; for example the observation that DNA will fragment into pieces of somewhat regular size near 50 000 base-pairs, when chromosomes are degraded gently on a large scale. Fragments of size 300 000 base-pairs are also observed, as a turn or collection of six smaller loops. Giant loops of size 2 million base-pairs have even been suggested on the basis of indirect evidence, for the wrapping of DNA over very large distances in interphase nuclei.

Yet the most direct evidence in favor of such loops has long been known, and comes from the detailed study of so-called 'lampbrush' chromosomes, which are found in animals such as the frog or newt, in cells that are preparing to become egg cells. The loops are readily visible there for one simple reason: because genes along those loops are churning out huge amounts of RNA, in preparation for making a new frog egg or newt egg. Thus, those genes are covered almost

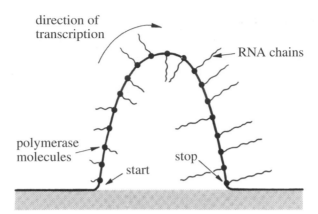

Figure 7.4 RNA polymerase molecules working their way around a loop of DNA in a lampbrush chromosome: schematic for both DNA and the base of the loop, or 'scaffold'. The RNA chains made by these polymerases grow longer as the polymerases travel further along the loop.

entirely by RNA polymerase molecules and their associated RNA chains, rather than by histone proteins. The DNA has lost almost all of its compaction due to wrapping about the histone proteins, and yet it remains relatively dense, owing to the great accumulation of protein and RNA along the length of any loop: so it is easy to see in the microscope.

As shown schematically in Fig. 7.4, the RNA polymerase molecules pack very densely along the length of the DNA in loops of a lampbrush chromosome, like cars in a queue at a traffic light. The RNA chains that emerge to either side are coated in protein (not shown in the diagram), and these chains grow longer as the polymerase molecules travel for greater distances around the loop. Each loop is anchored at its base in two places to certain unknown proteins (or other kinds of molecule) that provide a firm support or 'scaffold' for the flexible loop; and then there is an interval of some distance between loops, until another point of attachment to the scaffold is reached. These arrangements are rather similar to the 'band–interband' kinds of structure seen in fly polytene chromosomes, but in flies the loops become much more condensed, because they are covered in histone proteins.

Thus, in every case where we can actually see the fine structure of a chromosome by use of the light or electron microscope, we can see evidence for an intermediate level of structure between that of the 300 Å fiber and that of the folded metaphase chromosome. Furthermore, it is not unreasonable to suppose, as a working hypothesis, that this intermediate level of structure might consist simply of a series of large loops in the DNA, together with the intervals between loops.

But there are almost no certain data from biochemical studies today, to show which sequences in the DNA might attach themselves to the scaffold at the base of every loop, or to which proteins in such a scaffold the DNA might be attached.

A common procedure for investigating structures at this level, which has been followed by many workers, is to remove all the histones from the DNA by use of salt, detergent, or other reagents, and then to say that the proteins left represent a 'scaffold'. There are obvious dangers associated with this kind of approach, however, and the results remain controversial. For example, unless the DNA is attached to the scaffold more firmly than to the histones, the true attachments of DNA to any real scaffold might be lost during the treatment. Thus, in doing such an experiment, one might be left at the end with some sort of residual protein, rather than with any scaffold. A certain procedure devised cleverly by Uli Laemmli and colleagues uses only a relatively weak detergent to remove the histones, and does provide self-consistent evidence that the scaffold proteins which were left after treatment with such detergent, may have remained in their original places; but still the various procedures used are potentially destructive, as recent studies show. For example, Dean Jackson and Peter Cook have shown that the apparent size of 'loops' obtained by such methods is highly sensitive to the means of preparation of the sample. Furthermore, the loops as mapped by this kind of approach in fly polytene chromosomes do not show any correspondence to the band and interband structures that can be seen clearly by use of a light microscope. Finally, in one case it has been shown that a certain piece of DNA will attach itself to the scaffold only after, and not before, the addition of detergent or similar reagent. Still other studies show that topoisomerase II, the major proposed 'scaffold' protein which cannot be removed by detergent (probably because it binds covalently to the DNA while it alters the linking number Lk), does not play any sort of scaffolding or structural role when chromosomal loops are assembled in cell extracts.

Some interesting work on this difficult subject has been done in Siberia, by V.F. Semeshin, I.F. Zhimulev and colleagues. They have studied by electron microscopy the insertion of foreign bits of DNA into a fly polytene chromosome. For their foreign DNA, they used a special piece of DNA known as a 'P-element', that contains sequences which enable a fly enzyme to insert foreign DNA into fly DNA. They find that the P-element can either make a new band, or else split a previously existing band, depending upon its site of insertion in the chromosome. There are also other cases, not yet studied by electron microscopy, where the gene activity within a P-element can be 'insulated' from its position in a chromosome by attaching certain DNA sequences to both sides. Normally the activity of any gene is highly sensitive to its position in a chromosome; yet these experiments suggest that certain sequences in the DNA can set up a 'boundary', possibly in the form of a loop-attachment site, between different genes on a chromosome.

All of the compact metaphase chromosomes, the fly polytene chromosomes, and the frog lampbrush loops are highly specialized structures in biology, adapted to particular functions. Most of the time any cell nucleus resides in interphase; so in order to understand how chromosomes really work, we need to dissect the structure and function of that dispersed state. In the nucleus at interphase, the various long chromosomal DNA molecules are not all tangled up together, but instead remain segregated from each other in different spatial domains. Within any one chromosome, regions of transcriptionally active and inactive chromatin are also separated in space. Certain inactive regions (called 'heterochromatin') seem to be associated with the tough nuclear membrane; while certain active regions of chromatin, involved in transcription, seem to lie often on the two edges of any individual chromosomal domain.

An important advance in understanding how this interphase nuclear 'architecture' might be established comes from a recent discovery, by Kohwi-Shigematsu and colleagues, of a cage-like network within the nucleus composed of a protein called SatB1. That protein binds preferentially to regions of DNA which easily unpair into single strands. SatB1 anchors many genes to specific chromosomal sites, and also tethers the bases of chromosomal loops. It can further recruit enzymes which alter the structure of chromatin; and thus it has all the required properties of a protein which might regulate the function of an authentic chromosomal domain.

By this point you may be feeling quite frustrated at the general air of uncertainty in our presentation! Ever since we stopped talking about nucleosomes, and went on to talk about the 300 Å fiber, HMG proteins, loops and scaffolds, there has hardly been anything definite to learn. But before we leave this topic, we shall mention one further aspect of the looping behavior of DNA in chromosomes, which seems to have a lot to do with how genes work; it illustrates only too well how fluid is our knowledge of these important matters. In Fig. 7.4 we implied that the loops of DNA in a chromosome might be rigidly fixed structures, because the DNA in each loop seems to be held at its two ends by a protein rod or scaffold. But only some of the scientists working on chromosomes today think in that way. Others think that the loops are flexible structures, which allow the DNA to slide or thread itself through the base of a chromosome, in the same way that a piece of magnetic tape goes through the reading-head of a tape recorder.

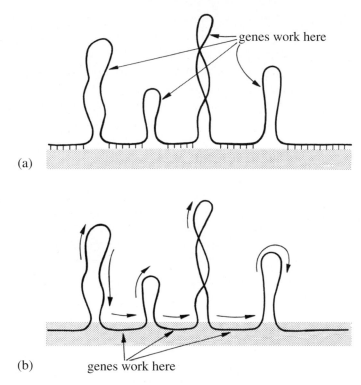

genes work here

(a)

(b)   genes work here

Figure 7.5 Two hypotheses for the location of gene activity in a typical chromo-some from higher organisms, and here shown schematically. We do not know if the hypothetical loops of DNA are rigidly fixed structures as in (a), or if they are flexible enough to let the DNA slide through their sites of attachment to a 'scaffold' as in (b).

   The two alternative models are sketched in Fig. 7.5. In Fig. 7.5(a), the loops are drawn as if they were fixed objects, and it is assumed that RNA polymerase travels along the genes which are contained in the outer parts of each loop. In other words, the polymerase and its associated proteins start making RNA near one end of the loop, and stop when (or before) they reach the other end. In Fig. 7.5(b), by contrast, the loops are able to slide through the base of the chromo-some, where it is supposed that all of the polymerases and their associated proteins are stored in a kind of 'active compartment'. The DNA then threads itself through this active region so as to come into contact with the polymerase, and thus to make RNA.
   At present, there is some evidence in support of each of these two theories. You must realize that biologists today are able to iso-late many of the proteins that make genes work (the 'transcription factors'), but they do not yet know where these proteins are located – whether on the base or the tip of a chromosomal loop. On the one

hand, it is known that a polymerase molecule can track around the outer parts of a loop in special cases (see Fig. 7.4), when the genes are making very large quantities of RNA. On the other hand, some loops within a lampbrush chromosome are thought to be able to change size, as shown schematically in Fig. 7.5(b). Also, in relation to normal chromosomes that are not making large quantities of RNA, it has been reported that one can cut away 90% of the DNA and protein from the outer parts, to leave just the central part or 'scaffold'; and in the process one keeps almost all of the enzymatic activities needed for making RNA.

After this excursion into looping and gene activity, let us now return to better-understood subjects. On a much larger scale than we have considered so far, each chromosome has two kinds of specialized structure, which correspond to specific sequences in the DNA. These are known as the 'centromere' (or 'central part') and the 'telomeres' (or 'end parts'), as shown schematically in Fig. 7.6. The centromere lies somewhere within the main part of each chromosome, and it is the feature which becomes attached to the tubular protein structures, known as 'microtubules', that assemble themselves when the cell is about to divide. These microtubules pull the duplicated chromosomes apart, thus providing one copy of each chromosome for each new cell. In the photograph of Fig. 1.3, the two DNA centromeres lie within the narrow, central X-shaped part of each duplicated chromosome. The telomeres, on the other hand, lie at either end of a long, linear chromosome, and their role remains uncertain. Obviously they can 'seal' the ends of any chromosome to prevent its joining to other chromosomes; but they may well do more than that. Some scientists think that the telomeres adhere to the nuclear membrane, so as to anchor the chromosomes in three-dimensional space. Other people think that the telomeres are needed to assist in the copying of a linear chromosome upon cell division,

Figure 7.6 Special functional regions of chromosomal DNA. A centromere need not necessarily be located at the very center of the chromosomal DNA, but the telomeres always occupy the ends. Telomeres are short, repetitive DNA sequences such as TTAGGG, repeated hundreds of times in a row (in fish, frogs, and humans). Centromeres are long, semi-repetitive sequences of repeat-length at least 1000 to 10 000 base-pairs, that may provide for multiple attachment sites of DNA to microtubules. Many of the chromosomal centromeric regions have been characterized and isolated from simple organisms such as yeast, but no functional centromere has yet been isolated from any higher organism, with certainty.

because the usual copying enzyme (DNA polymerase) cannot copy DNA all the way to the end of a linear molecule, on both strands of the double helix (owing to its requirement for a small 'primer', see Chapter 10); hence a special enzyme, known as 'telomerase', copies the telomeric ends.

Let us now consider in more detail the structure of a compact, metaphase chromosome. Many studies by light and electron microscopy have shown that such duplicated chromosomes, which are ready to be separated into two parts on cell division (where one copy goes to each daughter-cell), have a sort of spiral structure as shown in Fig. 7.7(a). You can see there the same overall shape as in Fig. 1.3, but now more detail is visible. A thick rod or sausage-shape coils round and round to make spiral arms that serve to reduce the overall size of the chromosome to assist in cell division. Of what, precisely, might this thick rod be made? One obvious possibility would be the scaffold-and-loop structure which we have already described (Fig. 7.7(b)). The loops are small in this picture by comparison with those shown earlier, because the two diagrams in Fig. 7.7(b), which provide both side and end-views of the loops, are drawn on the scale of the much larger Fig. 7.7(a). Yet it seems possible that the bulk of the spiral rod may consist mainly of these loops, while the protein scaffold constitutes perhaps a relatively narrow core.

The formation of compact coils from a scaffold-and-loop structure would account nicely for the final compaction by a factor of 10, that is needed to achieve an overall compaction of 10 000 between the length of the free DNA and the length of a metaphase chromosome. It seems possible that the cell could make (hypothetically)

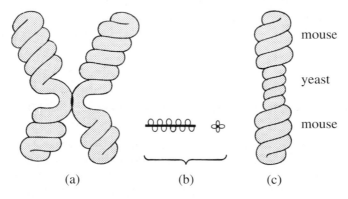

(a)   (b)   (c)

Figure 7.7 (a) Detailed fine-structure of the duplicated, metaphase chromosome as a wrapped-up spiral-rod or sausage-shape (cf. Fig. 1.3). (b) This same rod shown in detail as a hypothetical scaffold with loops, in two views. (c) A special mouse–yeast hybrid chromosome of two different diameters, revealing our ignorance of the factors that influence chromosome structure on a large scale.

some special protein that induces curvature in the scaffold-and-loop structure, as the cell gets ready to divide, thereby providing for the change from an interphase to a metaphase chromosome. Both left-handed and right-handed spirals – of the kind shown in Fig. 7.7(a) – have been seen by light and electron microscopy.

Scientists have recently found many different ways of transferring the DNA from one organism into the chromosomes of another. For example, the total DNA from a yeast chromosome of 9 million base-pairs can be inserted into the chromosome of a cell isolated from the mouse by a rare 'fusion' of the two kinds of cell in tissue culture. The resulting metaphase chromosome has a strange, dumb-bell shape (Fig. 7.7(c)): this picture is a simplified version, corresponding to one half of Fig. 7.7(a). The key feature is that the central part, which contains the yeast DNA, has only half the diameter of a normal mouse chromosome. One possible explanation is that the loops of yeast DNA are shorter than those of mouse DNA, thereby making the scaffold-and-loop structure smaller in diameter. Another possibility is that the protein scaffold in yeast curves into tighter spirals than does the scaffold in mouse. And perhaps both of these effects happen simultaneously. In any case this is a very deep result, because it reveals that there is a definite level of organisation in a metaphase chromosome, as specified by the sequence of DNA (or its pattern of methylation; see Chapter 11), which we do not currently understand.

Eventually, and probably in the present century, people will want to assemble full-length, authentic chromosomes in a test-tube, and use these in agriculture (and perhaps medicine) in order to do useful things that are not yet possible. Most present-day agricultural plants and farm animals are not wild species: instead, they are species that have been selectively bred for their food value by farmers over thousands of years. For example, the well-known 'Granny Smith' apple and the 'Parson Brown' orange were bred by people in the last century for improved quality of fruit. About twenty years ago, some scientists at the Calgene company in California made a tomato plant which produces tomatoes that will not rot, but will remain firm and red. This was done by adding to the tomato plant small amounts of foreign DNA which inhibit production of the 'rotting' enzyme. Since then, there have been many other examples of 'genetically modified' (GM) foods; for instance, cereals which incorporate a gene for insect resistance, or rice which is able to grow in salty water.

Much research is now in progress, aimed at developing methods for efficiently inserting large pieces of foreign DNA into cells, to make either 'transgenic' plants or animals for agriculture (for example, a transgenic pig that grows to maturity more rapidly than normal); or to deliver important genes to cells for 'gene therapy' in medicine, say as a new treatment for cancer or as a means to correct genetic defects. These new methods often use viruses or fat-DNA complexes known as 'liposomes' for their delivery into cells, and they will be discussed in Chapter 10. Other research is now in progress, to make functional mini-chromosomes that will grow normally and be inherited in human cells. It is not known whether any of these methods will involve histones and DNA in their eventual mode of action; but it seems worthwhile to study the folding of DNA about histones at least as a model system, in order to understand more clearly this kind of process, and how it might be carried out in the cell or in a test tube.

So far, scientists have made only the smallest start at understanding the assembly of DNA into chromosomes. They would like to be able to put into a test-tube moderate quantities of pure DNA, pure protein, and perhaps other substances, in order to assemble chromosomes in the laboratory. But that sort of thing is still a long way off. At present, histones and DNA can only be combined to make authentic nucleosomes in the test-tube by adding 'extracts' taken from living cells. These extracts are presumed to contain many important factors for the assembly of nucleosomes, and perhaps for the assembly of loops and scaffolds; but it is not yet known exactly what these factors are, or how they work. Some of the extracts can even assemble a nuclear membrane around the DNA. There is a lot of interesting work to be done here.

The most important early result in this general area was obtained by Ron Laskey and colleagues in 1977. They found that one could incubate DNA with an extract from frog eggs, in order to make authentic fragments of a chromosome. Eventually they and Juergen Kleinschmidt independently found two proteins in the frog egg, called 'N1' and 'nucleoplasmin', that bind to the histone pairs H3, H4 and H2A, H2B, respectively, and carry them onto the DNA. Why should 'carriers' such as these be needed to place the histones onto the DNA in the form of nucleosomes? Why should histones not be able to bind to DNA spontaneously?

The water in our cells contains various dissolved salts, at low concentration. In the test-tube, histones only place themselves on DNA in the form of nucleosomes if the salt concentration is first made much higher than it is in our cells, and is then reduced slowly. At physiological, cell-like salt concentrations, the positively charged histones tend to aggregate in clumps, rather than form nucleosomes. But the carrier proteins N1 and nucleoplasmin contain many negatively charged amino acids, such as aspartate and glutamate; and these bind tightly to the histones so as to prevent them from aggregating. In fact, a simple polymer of aspartate or glutamate can also assemble histones onto DNA. Nature uses complicated proteins such as N1 and nucleoplasmin instead for two reasons: first, to carry the histones from their site of synthesis on the ribosome, across the nuclear membrane to their site of assembly on the chromosome; and second, to pick out H3, H4 and H2A, H2B from all the different positively charged proteins in the cell, that might be able to bind to DNA non-specifically.

The discovery of these histone carriers was good progress, but then a stumbling block was reached. When the complexes of N1 with H3, H4, and nucleoplasmin with H2A, H2B were purified from the cell extract, it turned out that the histones would assemble onto DNA in the test-tube at a spacing of only 145 base-pairs, rather than at a spacing of 180 base-pairs as in the original extract from frog eggs. In other living cells, the spacing can vary from as low as 160 to as high as 260 base-pairs, depending on the proteins present in any nucleosome; but one never sees a very short spacing of 145 base-pairs in Nature.

Figure 7.8 shows some typical results for the assembly of histones onto DNA by an extract from living cells on the one hand, and by a pure system on the other. The procedure for determining the spacing of nucleosomes is to 'digest' the preparation by means of a DNA-cutting enzyme (like the one used by Hewish and Burgoyne), and then to measure the sizes of the fragments so obtained by using an electrophoretic gel according to a scheme which we shall describe in Chapter 9.

On the left-hand side of Fig. 7.8, one can see by 'reading' the gel that DNA may be assembled with histones from a frog-egg extract to make particles of size 180, 360, 540, 720, and 900 base-pairs; and so the size of the fundamental unit is 180 base-pairs. On the right-hand side of Fig. 7.8, DNA has been assembled with histones in a pure system (that is, without any cell extract), and it has made particles of size 145, 290, 435, 580, 725, and 870 base-pairs, after digestion with the enzyme; and so the size of the fundamental unit is 145 base-pairs. (The lowest band of size 145 base-pairs is obscured in this gel, but it can be seen in other experiments.)

One generally accepted explanation for the reduction in spacing from 180 to 145 base-pairs, on going from a cell extract to a pure system, is that we have lost certain important proteins when purifying the material. If these proteins normally associate with the histones as they assemble onto DNA in a cell, then in the cell extract there would be more proteins per particle than in the pure system, at least while they are being assembled; and hence each particle would extend for a greater length along the DNA. Which proteins might we have lost? The frog-egg extract is known not to contain any of the usual histone H1, so perhaps some other abundant protein (or proteins) might be important to the correct assembly of nucleosomes.

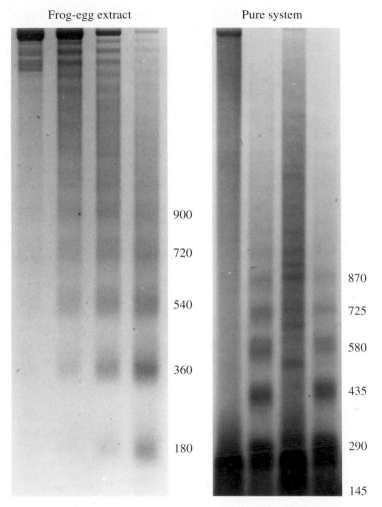

Frog-egg extract   Pure system

900

720

540

360

180

870

725

580

435

290

145

Figure 7.8 Analysis of nucleosome assembly by use of electrophoresis in gels, after digestion by enzymes. Histones are assembled onto DNA at a spacing of 145 base-pairs in a pure system, but at a more authentic spacing of 180 base-pairs in an extract from frog eggs. Left-hand gel by courtesy of David Tremethick.

The most likely candidate proteins for this task would be the high-mobility group proteins HMGA, HMGB and HMGN, which we mentioned earlier. These small proteins are fairly abundant in the cell nucleus, and they also bind to DNA with low specificity for the base sequence. Several experiments by David Tremethick have shown that either a moderately pure fraction of proteins taken from the frog-egg extract, and containing various HMG-like proteins, or else a completely pure preparation of phosphorylated proteins HMGN from human placenta, will increase the spacing of nucleosomes from 145 to 165 base-pairs. Other workers, for example, James Kadonaga and colleagues, have found that HMGN will increase the spacing of nucle-osomes in a fly-egg extract from 160 to 175 base-pairs per particle.

When histone H1 is added to a frog-egg or fly-egg extract, or to a partially-purified extract, the spacing of nucleosomes grows even larger: from about 180 to 220 base-pairs in a crude extract, or from 165 to 190 base-pairs in a partially-purified extract. Many inde-pendent experiments by various workers show that effect clearly. Most mature tissues in animals such as the fly, chicken, and human contain histone H1, and they also show spacings on the gel that cor-respond to a fundamental unit from 190 to 220 base-pairs long. Animals such as the sea urchin show a spacing of 260 base-pairs in some tissues; and perhaps those cells contain other special nucleo-some-assembly proteins that have not yet been identified.

A tentative summary of these results is shown schematically in Fig. 7.9. It seems likely that the assembly of DNA into nucleosomes proceeds by at least three steps. First, the N1 protein (or some other negatively charged carrier) binds two copies each of histones H3, H4 and assembles them onto DNA. Second, the nucleoplasmin pro-tein may bind H2A, H2B, and possibly an HMG-like protein, and add these proteins to DNA on either side of the already-assembled H3, H4 tetramer. The HMG-like protein may in some cases be related to HMGN. Without this HMG-like protein, during assembly there

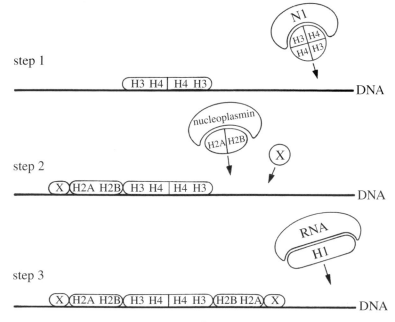

Figure 7.9 Three-step general scheme for the assembly of nucleosomes. The identity of the protein shown here as X is not yet known for certain; it is probably an HMG-like protein, or is perhaps related to HMGN, according to current work.

may be only eight proteins per particle, spanning 145 base-pairs of DNA; but with the HMG-like protein there are probably 10 proteins per particle during assembly, spanning 165 base-pairs. Finally, if histone H1 is present, it may add to the DNA between particles, to increase the spacing in the simplest case from 165 to 190 base-pairs, or in other cases from 180 to 220 base-pairs. The carrier for histone H1 in living cells is not known, but in the extract it seems to be some kind of RNA. No doubt the conjectured picture shown in Fig. 7.9 will be altered and refined in the future.

One important consequence of a stepwise assembly of DNA into nucleosomes, as in Fig. 7.9, is the following: each step in the procedure is reversible, so we can expect disassembly to proceed by the same series of steps but in a reverse direction. The relative rates of assembly *versus* disassembly within any step will therefore be of importance to specific biological processes, such as gene activation, that require the DNA to be at least partly unraveled from its folded state. We shall return to this subject later.

A large part of this book has been devoted to a study of DNA curvature. How does the curvature of DNA relate to its assembly into nucleosomes? In Figs 7.1 and 7.2 it was shown that DNA curves for almost two superhelical turns, each of about 80 base-pairs, around the histone proteins in any nucleosome. Thus, one might expect that *curved* or else very *flexible* DNA could assemble into nucleosomes more easily than DNA of mixed sequence, because it would take less energy for those special sequences to wrap into the required superhelical shape. Both of these factors are indeed important. Generally, even a small amount of curvature, as specified by the DNA base sequence (see Chapter 4), is sufficient to locate the path of DNA about the histone proteins. And also, in the absence of intrinsic curvature, the more flexible the DNA, the stronger is its affinity for histone proteins.

In addition, experiments using living cells have shown that the histone proteins often adopt highly ordered locations with respect to the DNA sequence in a cell nucleus, in a way that is influenced but not strictly determined by DNA flexibility. One can calculate the ease of curvature of any DNA sequence about a single set of histone proteins by Fourier algorithms similar to those described in Chapters 4 and 5. Yet it is difficult to calculate such things precisely over several successive sets of histone proteins, because successive nucleosomes may stick together, or fold into a 300 Å fiber, in a cooperative fashion that works against preferences in the DNA for curvature or flexibility. Recent experiments suggest however that the DNA sequence may specify the location of the first nucleosome in a long array, which then influences others nearby.

Several experiments have shown that the insertion of curved or else very flexible DNA near the start of a gene can actually repress the activity of that gene. In those cases it seems likely that the histone proteins bind to the DNA more tightly than usual, on account of curvature or flexibility; and so the specialized 'transcription factor' proteins which are used to initiate RNA synthesis cannot gain access to the DNA, in order to make the gene work. (The zinc-finger Zif268 protein described in Chapters 4 and 8 is one example of such a transcription factor.) Consistent with this idea, it is known that the insertion of non-curved or stiff DNA near the start of a gene (as for example long runs of sequence AA/TT or GG/CC) can actually activate that gene, by increasing the access of transcription factors to nearby DNA. Furthermore, when curved DNA is present on a large scale throughout an entire chromosome (as for example in certain birds and reptiles), then genes throughout the whole chromosome may become repressed, and the chromosome itself may remain so compactly folded and condensed, that it becomes visible at interphase.

Literally hundreds of experiments have been performed, where transcription factors and histone proteins (or HMG proteins) are added to the same piece of DNA, and the effects of one on the other measured. These studies are rather poor representations of what actually happens in the chromosome of a living cell, but they have turned up some interesting results. For example, certain transcription factors may bind to the same piece of DNA with widely different affinities, depending on which way the DNA curves about the histones, and hence which face of the DNA helix remains exposed. Those particular results seem to mimic well what happens in living cells, when transcription factors activate genes that are already wrapped in nucleosomes.

A broad conclusion that can be drawn from these studies is that the actions of DNA in a chromosome are very intricate and complex. If you try to study the workings of DNA solely in terms of abstractly defined genes and transcription factors, without paying attention to its three-dimensional structure, and to any proteins which are tightly associated with the DNA in a chromosome, then you will encounter many cellular phenomena which you will be unable to explain. In the past, cell-free studies on the mechanisms of transcription in higher organisms used free DNA, rather than DNA packaged into authentic pieces of a chromosome. But while testing the activities of such transcription factors, Kadonaga and colleagues found that the use of DNA bound to nucleosomal proteins, in a cell-free transcription assay, produced results that were more similar to those found in living cells, than by using as a template histone-free DNA.

A major precursor to the activation of any gene must be the unraveling of DNA from its tight folding about the histone proteins, at least transiently. Without such unravelling, RNA polymerase and other proteins required for 'reading' the DNA could never gain access to their required binding sites near the start of a gene. How, then, does the DNA become accessible to transcription factors and RNA polymerase, while at same time remaining associated with histones?

Our cells seem to have developed elaborate mechanisms for altering the local structure of chromatin, and for sliding or 'shuffling' nucleosomes along the DNA within active genes. In the test tube, nucleosomes possess an intrinsic but random ability to move along a piece of DNA. Yet in the living cell it seems that this ability is not used randomly, but instead is regulated actively. This implies that certain nucleosomes must carry some kind of 'flag' to indicate where such shuffling should take place.

The most likely mechanism for a 'flag' would be the chemical modification of histone proteins. Those histones H2A, H2B, H3 and H4 can be chemically modified in several different ways, the most important of which are acetylation, phosphorylation, and methylation. These three different modifications are performed by special enzymes known as *acetylases*, *kinases*, and *methylases* respectively. When a protein is acetylated, an acetyl group (CH_3CO) is added to one or more of its lysine amino acids, thereby removing the positive charge; see Fig. 7.10(a). When a protein is phosphorylated, a phosphate group (PO_3^-) is added, usually to one or more of its serine or threonine amino acids, thereby adding a negative charge; see Fig. 7.10(b). Finally, when a protein is methylated, either one, two or three methyl groups (CH_3) are added to its lysine amino acids,

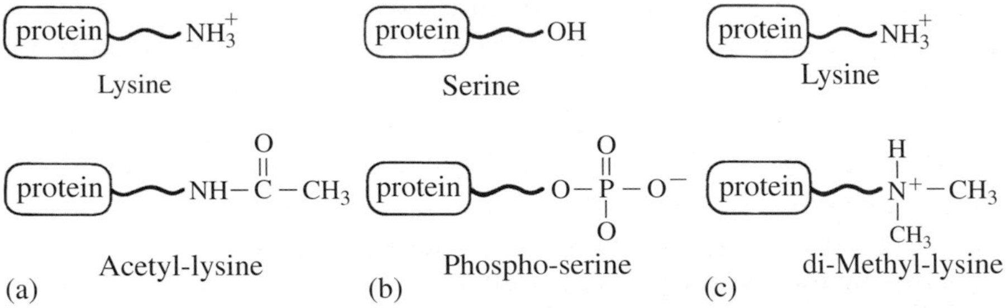

Figure 7.10 (a) Acetylation, (b) phosphorylation, and (c) methylation of typical amino acids in a protein. The first two modifications change the electric charge of the protein, in a way that may influence its structure and its ability to interact with DNA or with other proteins. In (c) two methyl groups have been attached; but a third might be added by replacing the H by a CH_3 group (or alternatively in the single-methyl form, just one of the three hydrogens could be replaced by a CH_3 group).

Table 7.1 The acetylation, phosphorylation or methylation of certain chromosomal proteins can affect their physical properties, and so influence their biological functions

Protein	Modification	Effect
H2A, H2B, H3, H4	Acetylation	Usually, easy sliding or shuffling on DNA
H3	Phosphorylation	Gene activation
H3	Methylation	Modify gene activity (see Chapter 11)
HMGB	Acetylation	Alters DNA binding
HMGN	Phosphorylation	Bind H2A, H2B more tightly
HMGA	Acetylation Phosphorylation Methylation	Not known
H1	Phosphorylation	Binds DNA less tightly
N1, nucleoplasmin	Phosphorylation	Bind histones more tightly

which increase their bulk but do not alter the electric charge; see Fig. 7.10(c). And arginine amino acids can also be modified by the addition of one methyl group.

The most important kinds of histone modification, in the context of chromosome dynamics, are listed in Table 7.1. These modifications are usually highly specific, and restricted mainly to the two exposed and flexible ends of each polypeptide histone chain. For example, acetylation, by reducing the net positive charge, reduces the strength of binding of those histones to the negatively-charged DNA. Indeed, histones carrying many acetyl groups have been known for a long time to be associated with chromatin that is being actively transcribed, or else easily available for transcription. Conversely, chromatin which is transcriptionally inactive or highly folded usually lacks acetyl groups. Similarly, phosphorylation of a particular serine residue (serine 10) in histone H3 is associated both with the condensation of metaphase chromosomes, and also with the activation of specific genes during interphase.

Histone methylation is generally associated with the recruitment of other proteins, which may then combine to form an inactive chromatin structure. Once formed, that inactive structure will remain stable to DNA replication and cell division; the underlying molecular basis for such remarkable stability will be described in Chapter 11. The most significant site of protein methylation seems to be located on lysine 9 of histone H3, where as tri-methyl-lysine it is recognized by another chromatin protein HP1, which has the ability to repress transcription. In another case, the methylation of lysine 4, also on histone H3, is associated with the *activation* of genes.

Histone modifications such as acetylation are intimately associated with increased nucleosome accessibility. Nucleosomes can be moved along the DNA or 'shuffled' by large remodelling complexes, which were first characterized by Carl Wu. This process can be helped by multiple acetylation of histones; and in certain cases it may result in the complete displacement of the histone octamer from a promoter region, thereby allowing both transcription factors and RNA polymerase to bind to the DNA.

In the nucleus, one type of RNA polymerase, responsible for the synthesis of messenger-RNA, is also heavily involved in changing the properties of nucleosomes. The largest protein subunit of RNA polymerase possesses a long 'tail' which consists of 30 to 50 repeats of a seven-amino-acid sequence (tyrosine-serine-proline-threonine-serine-proline-serine, or YSPTSPS). When RNA polymerase binds to any gene promoter, but before it starts to make RNA, the long tail-repeat may attract an 'activator' complex of proteins which is required for the proper initiation of RNA chains: see Fig. 7.11(a). Then during the initiation process itself, that same polymerase tail will become phosphorylated on one of its three serines (number 5 of the repeat); and will bind to a different, large 'elongator' complex of proteins which assists in 'capping' of RNA chains at their 5'-ends, to protect them from cellular nucleases: see Fig. 7.11(b). During subsequent stages of transcription, the tail becomes phosphorylated on serine number 2 of the repeat (instead of number 5), as shown schematically in Fig. 7.11(c). Then it may bind also to other large enzymatic complexes which help to 'splice' the messenger-RNA (i.e. remove non-coding parts before it leaves the nucleus), or 'polyadenylate' the messenger-RNA (i.e. add a string of A bases at its 3'-end to protect against nucleases). One of the protein subunits of that elongator complex is actually an acetylase enzyme, which can modify the lysine side-chains of histones within active genes.

So the cell acts very cleverly to couple transcription of DNA into RNA by any polymerase enzyme, and its associated 'activator' or 'elongator' complexes, with acetylation of histone proteins so as to establish an active state.

Now what exactly might happen, when RNA polymerase approaches some nucleosome in the chromatin which is blocking its path? Often the DNA which lies between that advancing polymerase and the bound nucleosome will become supercoiled in a right-handed sense (see Chapter 6); which will facilitate the unwrapping or 'loosening' of a left-handed, two-turn supercoil of DNA within the nucleosome itself. One of the histone H2A-H2B dimers

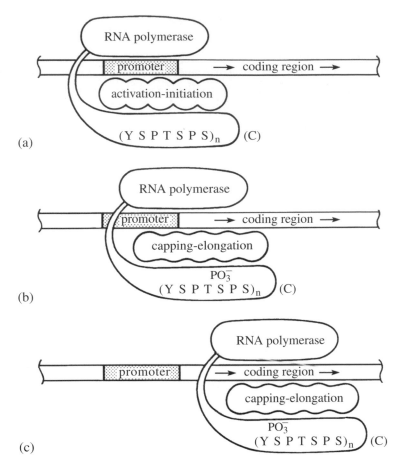

Figure 7.11 A highly schematic picture of RNA polymerase with its long 'tail' as it moves along the DNA. In (a), the tail has bound the 'activator' complex. Here, the seven-amino-acid repeating sequence is written in single-letter code: see the text. In (b), later on, one of the serines (S) in the tail repeat-region becomes phosphorylated; and a different, 'elongator' complex is bound instead. In (c), a different serine in the repeat has become phosphorylated; and now the complex bound to the tail includes a histone acetylase. Here, (C) indicates the carboxyl terminus of the protein.

may then be transiently released, as shown schematically in Fig. 7.12, thereby providing better access for the polymerase; although no one is sure how that might be accomplished. But another kind of RNA polymerase – polymerase III – appears to gain access to the DNA without displacement of histones from the octamer.

Over 20 years ago Brad Baer and Daniela Rhodes showed that some of the nucleosomes from actively growing cells lack one of those H2A-H2B dimers. Intriguingly, these particles were not only associated preferentially with RNA polymerase, but also were more accessible to nuclease cleavage than normal nucleosomes (i.e. they were unwrapped).

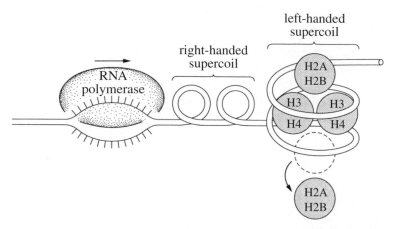

Figure 7.12 As the RNA polymerase advances (compare Fig. 4.2), right-handed supercoiling is imparted to the double-helical DNA ahead of it. This tends to loosen the left-handed supercoil of DNA around the histone octamer; and one of the H2A/H2B dimers may be expelled. Here, we have represented the histone octamer highly schematically as a bundle of four dimers.

Some nucleosome 'remodelling' enzymes seem to be attracted to regions of active chromatin; either by binding to acetylated histones, or else by binding to gene-activation proteins known as 'transcription factors' (see Chapter 8). Most 'remodelling' enzymes, like RNA polymerase, require an input of biochemical energy to move forward along any histone-bound DNA. Typically this comes from the conversion of adenosine tri-phosphate (ATP) to di-phosphate (ADP), by the replacement of a phosphate group with water and simultaneous release of energy (i.e. the three phosphates attached to an adenine base with a sugar ring are reduced to two; see Chapter 10). Moreover, most of those remodelling enzymes can also produce the transient release of H2A-H2B dimers, again just as for the action of RNA polymerase.

In summary, the acetylation of lysines in histones, plus the local supercoiling of DNA in a positive sense, as well as a transient release of H2A-H2B dimers, all combine to help RNA polymerase transcribe through any long array of nucleosomes without permanently displacing them.

Now we have covered many major aspects of current research into chromosome structure and gene activity. It is clear that the solution of some of the great problems in biology, such as the growth and development of higher organisms, will require a much deeper understanding of chromosomes and how they work.

Note

1. See Appendix 1.

Further Reading

Allshire, R.C., Cranston, G., Gosden, J.R., Maule, J.C., Hastie, N.D., and Fantes, A.P. (1987) A fission yeast chromosome can replicate autonomously in mouse cells. *Cell* **50**, 391–403. The yeast–mouse hybrid chromosome having two different diameters (shown schematically in Fig. 7.7) came from cell-line F1.1 of this paper.

Bednar, J., Horowitz, R.A., Dubochet, J., and Woodcock, C.L. (1995) Chromatin conformation and salt-induced compaction: three-dimensional structural information from cryoelectron microscopy. *Journal of Cell Biology* **131**, 1365–76. Proposal of an irregular zig-zag model for the packing of nucleosomes in a 300 Å fiber.

Cai, S., Han, H.J., and Kohwi-Shigematsu, T. (2003) Tissue-specific nuclear architecture and gene expression regulated by SATB1. *Nature Genetics* **34**, 42–51. Possible organisation of nuclear sub-structures by the SATB1 protein.

Callan, H.G. (1982) Lampbrush chromosomes. *Proceedings of the Royal Society of London* **B 214**, 417–48. Many beautiful pictures of scaffolds and loops.

Cook, P.R. (1989) The nucleoskeleton and the topology of transcription. *European Journal of Biochemistry* **185**, 487–501. Are genes located on the outer parts of loops, or at their bases?

Davey, C., Pennings, S., Meersseman, G., Wess, T.J., and Allan, J. (1995) Periodicity of strong nucleosome positioning sites around the chicken adult beta-globin gene may encode regularly spaced chromatin. *Proceedings of the National Academy of Sciences, USA* **92**, 11210–4. Histone octamers bind to natural DNA sequences with a variation in affinity of 300-fold; such wide variations in affinity may help to locate nucleosomes in living cells.

Disney, J.E., Johnson, K.R., Magnuson, N.S., Sylvester, S.R., and Reeves, R. (1989) High mobility group protein HMG-I localizes to GQ and C bands of human and mouse chromosomes. *Journal of Cell Biology* **109**, 1975–82. The HMG-I protein (i.e. HMGA) is found mainly in inactive regions of a chromosome.

Farr, C.J., Bayne, R., Kipling, D., Mills, W., Critcher, R., and Cooke, H.J. (1995) Generation of a human X-derived mini-chromosome using telomere-associated chromosome fragmentation. *EMBO Journal* **14**, 5444–54. Early steps towards making an authentic mini-chromosome for human cells.

Hirano, T. and Mitchison, T.J. (1993) Topoisomerase II does not play a scaffolding role in the organisation of mitotic chromosomes assembled in Xenopus egg extracts. *Journal of Cell Biology* **120**, 601–12. Study of the large-scale structure of chromosomes regarding scaffolds and loops.

Iyer, V. and Struhl, K. (1995) Poly (dA.dT), a ubiquitous promoter element that stimulates transcription *via* its intrinsic DNA structure. *EMBO Journal* **14**, 2570–9. Long runs of either AA/TT or else GG/CC will stimulate transcription from a nearby gene in yeast, by removing nucleosomes near the gene so as to let transcription factors bind more readily.

Kim, J.H., Lane, W.S., and Reinberg, D. (2002) Human 'elongator' facilitates RNA polymerase II transcription through chromatin. *Proceedings of the National Academy of Sciences USA* **99**, 1241–6. A large complex of proteins known as 'elongator' helps RNA polymerase to transcribe genes within chromatin, and also contains a histone acetylase enzyme.

Kurdistani, S.K. and Grunstein, M. (2003) Histone acetylation and deacetylation in yeast. *Nature Reviews in Molecular Cell Biology* **4**, 276–81. The roles of histone acetylation and de-acetylation in chromosome dynamics.

Lagarkova, M.A., Iarovaia, O.V., and Razin, S.V. (1995) Large-scale fragmentation of mammalian DNA in the course of apoptopsis proceeds *via* excision of chromosomal DNA loops and their oligomers. *Journal of Biological Chemistry* **270**, 20239–41. Indirect evidence for loops of size 50 000 or 300 000 base-pairs in chromosomes, by gentle fragmentation on a large scale.

Li, Q. and Wrange, O. (1995) Accessibility of a glucocorticoid response element in a nucleosome depends on its rotational positioning. *Molecular and Cellular Biology* **15**, 4375–84. Transcription factors may bind with widely different affinities to the same DNA sequence in chromatin, depending on how such DNA curves about the histone proteins.

Rattner, J.B. and Lin, C.C. (1985) Radial loops and helical coils coexist in metaphase chromosomes. *Cell* **42**, 291–6. Pictures of duplicated metaphase chromosomes as made from a spiral rod.

Richmond, T.J. and Davey, C.A. (2003) The structure of DNA in the nucleosome core. *Nature* **423**, 145–50. Detailed analysis of DNA curvature in terms of roll-slide-twist or other helical parameters for almost two turns of left-handed DNA supercoil.

Sandaltzopoulos, R., Blank, T., and Becker, P.B. (1994) Transcriptional repression by nucleosomes but not H1 in reconstituted preblastoderm *Drosophila* chromatin. *EMBO Journal* **13**, 373–9. Spacing by histone H1 from 180 to 220 base-pairs in a fly-egg extract, and studies of gene function there.

Sivolob, A.V. and Khrapunov, S.N. (1995) Translational positioning of nucleosomes on DNA: the role of sequence-dependent isotropic DNA bending stiffness. *Journal of Molecular Biology* **247**, 918–31. A good model for rotational and translational positioning of histone octamers on different DNA sequences, based on how easily various sequences will bend.

Suka, N., Shinohara, Y., Saitoh, Y., Saitoh, H., Ohtomo, K., Harata, M., Shpigelman, E., and Mizuno, S. (1993) W-heterochromatin of chicken: its unusual DNA components, late replication, and chromatin structure. *Genetica* **88**, 93–105. Curved DNA makes up a large part of the chicken W sex chromosome.

Sumner, A.T. (1990) *Chromosome Banding* Unwin Hyman, London. An authoritative account of the methods used to study metaphase chromosomes.

Tremethick, D. (1994) High mobility group proteins 14 and 17 can space nucleosomal particles deficient in histones H2A and H2B, creating a template that is transcriptionally active. *Journal of Biological Chemistry* **269**, 28436–42. HMG-14,17 (i.e. HMGN) can space histone octamers from 145 to 165 base-pairs, while making the chromatin more active towards transcription; and also can space histone hexamers from 125 to 145 base-pairs. (see Fig. 7.12)

Truss, M., Bartsch, J., Schelbert, A., Hache, R., and Beato, M. (1995) Hormone induces binding of receptors and transcription factors to a rearranged nucleosome on the MMTV promoter in vivo. *EMBO Journal* **14**, 1737–51. A careful study of nucleosome locations at high resolution in living cells, before and after gene activation by transcription factors.

Tsukiyama, T., Becker, P.B., and Wu, C. (1994) ATP-dependent nucleosome disruption at a heat-shock promoter mediated by binding of GAGA transcription factor. *Nature* **367**, 525–32. First direct demonstration of ATP-dependent nucleosome remodelling.

Urata, Y., Parmelee, S.J., Agard, D.A., and Sedat, J.W. (1995) A three-dimensional structural dissection of Drosophila polytene chromosomes. *Journal of Cell Biology* **131**, 279–95. A detailed study of the internal structures of fly polytene chromosomes, using light microscopy.

van Holde, K. (1988) *Chromatin* Springer-Verlag, New York. A major reference in the field, giving an overview of DNA and chromosomes in the 1980s.

Widom, J. (1996) Short-range order in two eukaryotic genomes: relation to chromosome structure. *Journal of Molecular Biology* **259**, 579–88. Periodicities of sequence near 10.2 base-pairs can be found in the total DNA of organisms such as bacteria, yeast or worm, especially for the dinucleotides AA/TT and GC/GC, as expected for the preferred bending of such DNA into tight curves.

Wolffe, A. (1995) *Chromatin* Academic Press, London. An important book citing much of the latest work on DNA and chromosomes.

Yoshida, M., Horinouchi, S., and Beppu, T. (1995) Trichostatin A and trapoxin: novel chemical probes for the role of histone acetylation in chromatin structure and function. *Bioessays* **17**, 423–30. Discovery of two chemicals that can inhibit the histone acetylase enzymes with high specificity.

Zinkowski, R.P., Meyne, J., and Brinkley, B.R. (1991) The centromere–kinetochore complex: a repeat subunit model. *Journal of Cell Biology* **113**, 1091–110. A centromere from a mammalian chromosome is shown to consist of many structural repeats, each of which attaches separately to microtubules on cell division.

Bibliography

Allfrey, V., Faulkner, R.M., and Mirsky, A.E. (1964) The acetylation and methylation of histones and their possible role in the regulation of RNA synthesis. *Proceedings of the National Academy of Sciences, USA* **51**, 786–94. First inference that the acetylation of histones might be correlated with transcription.

Arents, G., Burlingame, R.W., Wang, B.-C., Love, W.E., and Moudrianakis, E.N. (1991) The nucleosomal core histone octamer at 3.1 Å resolution: a tripartite protein assembly and a left-handed superhelix. *Proceedings of the National Academy of Sciences, USA* **88**, 10148–52. High resolution pictures of the histone proteins without the DNA.

Finch, J.T., Lutter, L.C., Rhodes, D., Brown, R.S., Rushton, B., Levitt, M., and Klug, A. (1977) Structure of nucleosome core particles of chromatin. *Nature* **269**, 29–36. First crystals of the nucleosome core particle, comprising the histone octamer plus 145 base-pairs of DNA, and their structural analysis at low resolution.

Hewish, D. and Burgoyne, L. (1973) Chromatin sub-structure: the digestion of chromatin DNA at regularly spaced sites by a nuclear deoxyribonuclease. *Biochemical and Biophysical Research Communications* **52**, 504–10. Discovery of a regular spacing of histone proteins along the DNA in chromatin.

Jackson, D.A., Dickinson, P., and Cook, P.R. (1990) The size of chromatin loops in HeLa cells. *EMBO Journal* **9**, 567–71. The 'loop sizes' obtained on histone depletion depend strongly on the exact procedures used.

Kireeva, M.L., Walter, W., Tchernajenko, V., Bondarenko, V., Kashlev, M., and Studitsky, V.M. (2002) Nucleosome remodeling induced by RNA polymerase II: loss of the H2A/H2B dimer during transcription. *Molecular Cell* **9**, 451–2. How transcribing polymerase displaces an H2A/H2B dimer from the nucleosome in its path.

Kleinschmidt, J.A., Fortkamp, E., Krohne, G., Zentgraf, H., and Franke, W.W. (1985) Co-existence of two different types of soluble histone complexes in nuclei of *Xenopus laevis* oocytes. *Journal of Biological Chemistry* **260**, 1166–76. Independent discovery of N1 and nucleoplasmin by these workers.

Kornberg, R.D. and Thomas, J.O. (1974) Chromatin structure: oligomers of the histones; Kornberg, R.D. (1974) Chromatin structure: a repeating unit of histones and DNA. *Science* **184**, 865–8; 868–71. Proposal for a histone octamer as the fundamental unit of protein–DNA assembly in chromosomes.

Laskey, R.A., Mills, A.D., and Morris, N.R. (1977). Assembly of SV40 chromatin in a cell-free system from *Xenopus* eggs. *Cell* **10**, 237–43. Discovery of a cell extract that will make authentic pieces of a chromosome.

Luger, K., Mader, A.W., Richmond, R.K., Sargent, D.F., and Richmond, T.J. (1997) Crystal structure of the nucleosome core particle at 2.8 Å resolution. *Nature* **389**, 251–60. The first near-atomic resolution crystal structure of the nucleosome core particle.

Olins, A.L. and Olins, D.E. (1974) Spheroid chromatin units (bodies). *Science* **183**, 330–1. First pictures of beads-on-a-string for histones and DNA in chromosomes.

Paranjape, S.M., Krumm, A., and Kadonaga, J.T. (1995) HMG-17 is a chromatin-specific transcriptional coactivator that increases the efficiency of transcription initiation. *Genes and Development* **9**, 1978–91. HMG-17 (i.e. HMGN) can increase the spacing of nucleosomes in a fly-egg extract from 160 to 175 base-pairs, while activating transcription, by increasing the rate at which new RNA chains are initiated.

Semeshin, V.F., Demakov, S.A., Perez-Alonso, M., and Zhimulev, I.F. (1990) Formation of interdiscs from DNA material of the P-element in *Drosophila* polytene chromosomes. *Genetika (USSR)* **26**, 448–56. The P-element DNA can make new bands or else split old ones.

Thoma, F., Koller, T., and Klug, A. (1979) Involvement of histone H1 in the organisation of the nucleosome and of the salt-dependent super-structures of chromatin. *Journal of Cell Biology* **83**, 403–27. Electron microscope study of long arrays of nucleosomes in a chemically fixed condition. The source of Fig. 7.3.

Tremethick, D.J. and Drew, H.R. (1993) High mobility group proteins 14 and 17 can space nucleosomes *in vitro*. *Journal of Biological Chemistry* **268**, 11389–93. Phosphorylated proteins HMG 14, 17 (i.e. HMGN) from human placenta can space nucleosomes from 145 to 165 base-pairs, just as for a spacing-fraction from the frog-egg extract that contains many HMG-like proteins, and depends on ATP (*via* a protein kinase) for its activity.

Williamson, R. (1970) Properties of rapidly labeled deoxyribonucleic acid fragments isolated from the cytoplasm of primary cultures of embryonic mouse liver cells. *Journal of Molecular Biology* **51**, 157–68. The 200-base-pair spaced pattern of chromatin, but not interpreted as such.

Exercises

7.1 A piece of double-helical DNA, 3010 base-pairs long, is bound in a test tube to a series of histone octamers in the manner of Figs 7.1, 7.2 and 7.3. Then an enzyme like the one used by Hewish and Burgoyne (1973) is added to a sample. Fine-mapping experiments show that the enzyme cuts only at certain locations along the length of the DNA molecule, which are known to be the empty spaces or 'linkers' between histone octamers. These locations, measured in terms of the distance from one end of the 3010-bp DNA, were found by experiment to be (in units of base-pairs): 220, 430, 670, 870, 1090, 1320, 1520, 1740, 1930, 2140, 2350, 2580, 2790.

    a Make a table of all possible DNA fragment lengths, for fragments containing either 1, 2, 3, or 4 histone octamers. This table should have four columns. Column 1 should list all fragment lengths of size $220 - 0 = 220$, $430 - 220 = 210$, etc. base-pairs, and should contain 14 numbers. Column 2 should list fragment lengths of size $430 - 0 = 430$, $670 - 220 = 450$, etc., and should contain 13 numbers. Column 3 should list fragments such as $670 - 0 = 670$, $870 - 220 = 650$, and contain 12 numbers. Finally, column 4 should list sizes of $870 - 0 = 870$, $1090 - 220 = 870$, etc., and contain 11 numbers.

    What is the mean fragment size in each of columns 1, 2, 3, 4, and its standard deviation?

    b What is the mean spacing of histone octamers in this particular sample? Divide each mean fragment size by the column number 1, 2, 3, or 4 to get an optimal value.

7.2   The model of the histone octamer and its associated DNA, that is shown in Fig. 7.2, may be represented crudely as a protein cylinder of diameter 60 Å and height 50 Å, around which are wrapped two complete turns of DNA having a diameter of 20 Å.

    a   Calculate the proportion of the total volume which is occupied by the DNA. For this rough calculation you may treat the DNA as two separate hoops, each with an inner diameter of 60 Å. Recall that the volume of a hoop is equal to the product of its cross-sectional area and its mid-circumference.

    b   Calculate the number of base-pairs in each hoop, assuming that the DNA has a length of 3.3 Å per base-pair.

7.3   The packing of nucleosomes into a usual model for the '300 Å fiber', as shown in Fig. 7.3, may be represented approximately by diagram (a), below. Here, to make the calculations especially simple, we have drawn the nucleosomes as closed rings of six, rather than as helical spirals with six nucleosomes per turn.

By considering each nucleosome as a cylinder of radius 50 Å and length 50 Å (for the protein plus DNA: see Exercise 7.2), estimate **a** the outer diameter of each ring, **b** the diameter of the inner hole, and **c** the height of the 300 Å fiber per ring of six nucleosomes.

A second possible packing scheme, also expressed in terms of closed rings, is shown in diagram (b) below. The nucleosomes are cylinders, just as before, but now their flat faces lie perpendicular to the axis of the 300 Å fiber, instead of lying parallel to this axis, as in diagram (a).

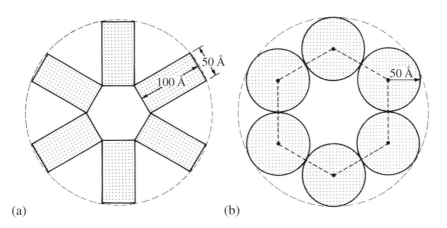

(a)   (b)

Compute the dimensions **a–c** also for model (b); and compare them with the corresponding figures for model (a). (See Widom, J. and Klug, A. (1985) *Cell* **43**, 207–13, for evidence about key dimensions, and in particular that the axial spacing **c** is ≈110 Å.)

7.4 The telomeres of human chromosomes are made from a long, multiple repeat of the sequence (5′) TTAGGG (3′), as stated in the caption of Fig. 7.6. Often, such a repeat projects beyond the end of the double-helical DNA at each end of the chromosome, to leave a short, single strand of 6 unpaired bases as shown below:

(5′) …TTAGGGTTAGGGTTAGGG (3′)
(3′) …AATCCCAATCCC (5′)

Such an arrangement is known as a 'sticky end' of DNA. There are known to be many enzymes in the cell that can join or 'fuse' two such single-stranded ends of DNA molecules to one another, if they detect sufficient Watson–Crick base-pairing.

  a Could two identical (5′) TTAGGG (3′) ends like the one shown above (with the second one found by rotating the first through 180° in the plane of the diagram) be joined by such an enzyme?

  b Suppose that human telomeres were made not from repeats of TTAGGG, but from repeats of some other sequence such as TAGCTA or CGATCG. Would such hypothetical telomeres prevent the ends of chromosomes from joining to one another by Watson–Crick pairing? (See Ijdo, J.W., Baldini, A., Ward, D.C., Reeders, S.T., and Wells, R.A. (1991) *Proceedings of the National Academy of Sciences, USA* **88**, 9051–5, for a rare instance of telomere fusion.)

7.5 It is often possible to prepare, in the laboratory, a 'soup' or extract of proteins from a living cell which is able to synthesize some particular RNA molecule when a DNA 'template' molecule, having a specific sequence, is added to it.

If you study the process of transcription solely in this cell extract, on 'naked', histone-free DNA, what important features of the chromosome structure – features that could affect the activity of a DNA template in a living cell – would be excluded from your study? By scanning the diagrams of this chapter, identify at least four different structures of DNA within a chromosome that might influence gene activity in this way. (See Felsenfeld, G. (1992) *Nature* **355**, 219–24, for a survey of such concerns.)

CHAPTER 8

Specific DNA–Protein Interactions

In the previous chapter we explained how the vast majority of chromosomal DNA is associated with packaging proteins, such as histones. Those abundant proteins bind to most parts of the lengthy chromosomal DNA with only a slight preference for base sequence; and in organisms whose cells have nuclei, they serve mainly to compact the long DNA by a factor of about 10 000, into the minute volume of a cell nucleus. Chromosome compaction can control the access of RNA polymerase enzyme to the start site of genes (i.e., the 'promoter' sequences) and thereby affect the transcription of DNA into RNA. However, the study of packaging proteins alone cannot teach us how specific genes might be controlled, during the growth or development of cells into a mature organism, or during cellular response to environmental stimuli, or during the routine 'housekeeping' activities of fully developed cells.

For that reason, scientists have also studied the many less abundant DNA-binding proteins that bind specifically just to a small part of the long chromosomal DNA, in the proximity of the start sites for genes, or in other important locations. Some of these proteins can bind to DNA with high specificity for the base sequence, and can thereby carry out many specific biological tasks, such as control of transcription, repair of damaged DNA or unwinding of DNA. A few of these tasks are shown in Fig. 8.1.

The first DNA-binding proteins studied by scientists were regulatory proteins from bacteria, where they act to control perhaps the simplest genetic systems found in Nature. Many of these bacterial proteins act as *repressors* of gene activity (see the upper part of the picture) if they bind tightly to a base-sequence of DNA which overlaps the 'promoter' sequence, where an RNA polymerase enzyme can also bind. They can thereby prevent the binding of RNA polymerase to a particular promoter, through direct competition for the same local segment of DNA. In general, such repressor proteins reduce the rate at which RNA is made from a promoter; and indeed such repression of RNA synthesis may be specific to just one or a few genes in an entire organism, if the repressor binds to only one or a few sites on an entire chromosome.

In bacteria, repressor proteins play an important role in reducing local rates of transcription; but in plants, animals and other organisms whose cells have nuclei – known collectively as eukaryotes[1] – the chromosome structure itself tends to repress transcription. Indeed, in nucleated organisms it is the activation of genes that seems to be the more important aspect of gene regulation. That process is managed by *'activators'* of transcription (see the middle part of Fig. 8.1) that bind specifically to DNA in the general vicinity of a binding site for RNA polymerase. The activator protein may then increase the rate at which RNA is made, by directly assisting the RNA polymerase enzyme and its auxiliary proteins to bind at the promoter sequence, through a network of protein-to-protein

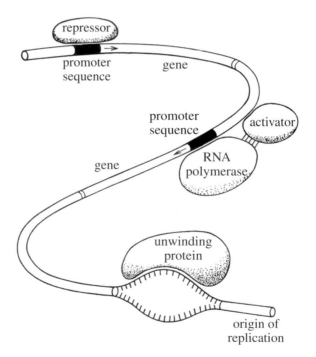

Figure 8.1 Proteins of various kinds may bind specifically to different DNA sequences, so as to carry out important biological tasks such as the repression or activation of individual genes, or unwinding the double helix in specific sites for copying or replication. A highly schematic picture, showing some of the DNA–protein interactions described in this chapter.

contacts; or else indirectly by helping to 'recruit' enzymes that can chemically modify the chromatin (see Chapter 7). For instance, certain transcription activators may direct histone acetylases to the general region of a specific gene. The resulting modification of histones may cause the chromatin to decompact near that promoter, and thereby make it more accessible to RNA polymerase and its auxiliary proteins.

'DNA looping' may represent a somewhat more complex example of how genes are regulated in three dimensions, and not just in one or two, by some linear or planar arrangement of DNA binding sites. In the latter case, two or more repressor or activator proteins may bind to the same piece of DNA, and then join together to create a small loop or coil, which can affect gene activity very strongly (either positively or negatively) on account of its stable structure.

We have mentioned so far repressors and activators and loops, but there are also many other kinds of regulatory protein that interact with the DNA, and so play important roles in the cell. Thus, a fourth kind of DNA-binding protein (see the lower part of Fig. 8.1) may help to unwind the DNA double helix near an 'origin of replication', where the duplication of old DNA into new DNA starts during every cell division. After such unwinding by these special kinds of protein, a DNA polymerase enzyme can read the bases on each separate strand by means of Watson–Crick pairing and then copy them; and so can make two new DNA double-helices from the two old strands. Also, proteins which bind to DNA may carry out many different biological tasks not shown here, such as the supercoiling of DNA, or the exchanging of DNA segments in 'recombination'. The cell also has elaborate assemblies of enzymes that recognize and repair DNA damage, which is a crucial aspect of cell viability.

In practice, all of the pictures shown in Fig. 8.1 are *greatly oversimplified*, since they omit so much important detail. For example, each specific interaction between protein and DNA must be highly complex from a chemical point of view, in order to distinguish one base sequence from others over the length of an entire chromosome. Also, each protein may be regulated in its ability to bind to specific sequences on the DNA, by a variety of factors including: (a) interaction with other proteins nearby; (b) any modification of the DNA such as base methylation (see Chapter 11); or (c) by binding to small molecules such as sugars and amino acids present in the cellular medium.

How can anyone hope to learn anything useful from such a cacophony of information, without making a big dictionary that would run to thousands of pages, and searching through the facts one at a time? The only hope for greater understanding seems to be to limit ourselves to studying a few, well characterized examples and to point out the salient principles.

Today, the best characterized information about specific protein–DNA interactions in a living cell lies within the detailed structures of many different protein–DNA complexes, at near-atomic resolution. Since 1990, hundreds of such large complexes between protein and DNA have been studied in detail by the X-ray diffraction of single crystals, and also by nuclear magnetic resonance spectroscopy of molecules in solution (see Chapter 9). Those structures now provide a highly useful and reliable set of data, by which we can understand in part the biological functions of DNA and protein in living cells.

One could perhaps organize a survey of such protein–DNA structures according to the scheme shown in Fig. 8.1, where 'repressors' fall into one class, 'activators' into another, and 'unwinding proteins' into another. Yet it seems that various proteins of widely different structure may lie within each of these biological classes. For example, repressors can recognize DNA by means of an α('alpha')-helix, by means of a β('beta')-sheet, or by other kinds of amino-acid 'motif'. Similarly, activators and unwinding proteins may recognize the DNA through a wide variety of protein structures. For example, in many eukaryotic regulators of transcription, short peptide loops are stabilized through the coordination of zinc metal: these are the

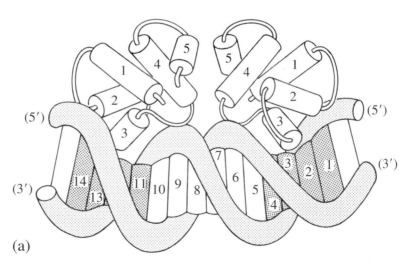

(a)

Figure 8.2 (a) A schematic drawing of the 434 repressor protein as it binds to two turns of DNA. The 434 protein consists of two identical halves, which are related by two-fold rotational symmetry. Within each half, a long peptide chain folds into five numbered α-helices, which are drawn as a series of short cylinders connected by flexible linkers. One of the α-helices, '3', lies deep within the major groove of DNA on each side of the complex, where its amino acids can contact directly the four base-pairs which are shaded.

'zinc fingers', and they occur in a very wide range of regulatory proteins of varied structure and function.

Hence, it seems more logical to organize our survey according to the general motif by which a protein recognizes DNA: whether it be by an α-helix, by two α-helices, by a β-sheet, or by a zinc-finger. Examples of each of these classes of 'reading head' will be described in detail below, while references to other important protein–DNA structures of a similar kind are listed in the bibliography. We shall also discuss some synthetic compounds which have been designed to recognize particular DNA sequences and to bind to them strongly.

Let us start our survey of protein–DNA interactions with the simplest kind of recognition, where an α-helix from a protein fits snugly into the major groove of a small piece of DNA, and makes direct contact with a few bases there. The best-known example of that kind of structure is shown in Fig. 8.2(a), where the 434 repressor

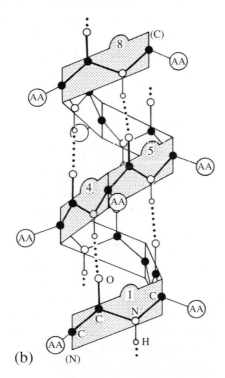

(b)

Figure 8.2 (b) A detailed explanation, at a much larger scale, of how a peptide chain folds into an α-helix. Each rigid peptide unit makes two hydrogen bonds (dotted lines) to other peptide units, three steps away along the chain, so as to fold the entire assembly into a regular right-handed spiral. Many different amino acids (AA) may be accommodated on this spiral structure, since the amino-acid side chains protrude out and away from the central peptide core. (The conventional numbering system for peptides starts at the N (amino) end and finishes at the C (carboxyl) terminus.)

protein – consisting of two identical molecules: a 'dimer' – is seen to bind to two successive turns of DNA. Within a small part of each turn, this protein inserts an α-helix, numbered '3' in the picture, into the major groove of the DNA, and is thereby able to recognize accurately the identity of the base-pairs there. A less detailed picture of the same thing was shown in Fig. 4.11, with regard to the bending of DNA in the very center of the complex, near base-pairs 5 to 10. Here we shall focus on specific recognition of DNA by the 434 protein at each end of the complex, near base-pairs 1 to 4 and 11 to 14, which are shaded in the picture.

Before proceeding to examine in detail the most important aspects of this protein–DNA complex, let us explain briefly some well-known features of protein structure, for the benefit of readers who are not already trained in biochemistry at a university level. In Chapter 1, we learned that every protein is manufactured as a long chain of peptides, with an amino acid attached to each peptide as prescribed by the DNA sequence, according to the three-base Genetic Code set out in Table 1.1. We also learned that there are 20 kinds of amino acid. We now need to recall that some of them are large and some are small; some carry an electric charge while others are neutral; and some are hydrophobic while others are hydrophilic.

Any long chain of peptides as found in most proteins will contain a broad mixture of amino acids, and it may be positively charged in one part of the chain but negatively charged in another; or hydrophobic in one part but hydrophilic in another. Different amino acids within the chain may then attract or repel one another in a complex way, so that a certain long sequence of amino acids can specify, for example, the entire detailed structure of the 434 protein shown in Fig. 8.2(a).

The process of folding any long polypeptide chain into its compact, final state must clearly be complex, as there will be a few, key interactions between the amino acid residues amongst an astronomically large number of possibilities. Still, one can understand certain key features, such as the stability and interaction of secondary structural elements, or *'motifs'*, like β-strands or α-helices, by examining the folded structures. For example, as shown in Fig 8.2(a), each of the two identical halves of the 434 protein consists of five small cylinders of various lengths, which are joined at their ends by flexible linkers to other cylinders of the same kind. Each of these cylinders actually contains a short length of peptide chain, that is wrapped into a single-stranded spiral or α-helix, as shown with more detail in Fig. 8.2(b).

Within each α-helix, successive rigid peptides coil into a right-handed spiral having 3.6 peptides per helical turn. Thus, you can see in Fig. 8.2(b) that after two turns of the coil, peptide 8 lies almost directly above peptide 1; and indeed it would lie *exactly* above it if the spiral were to contain 3.5 peptides per turn, instead of the actual 3.6. Each peptide is held weakly to its neighbors along the next turn of the chain by two hydrogen bonds, which run almost parallel to the helix axis and are shown here as dotted lines. Each of these bonds connects an oxygen atom (large open ball) in one peptide unit to a hydrogen atom (small open ball) in a neighboring peptide unit. Different amino acids, shown here as large balls labeled 'AA', may be attached along the outside of the central polypeptide core, something like large decorations on a small Christmas tree. This simple spiral structure for polypeptide chains was discovered in 1950 by the chemist Linus Pauling, when he was at home sick one day, and to occupy himself made models for proteins by drawing a polypeptide chain on a piece of paper and then rolling it up.

As mentioned above, the two identical parts of the 434 repressor protein are made from several such α-helices, which are numbered '1' to '5' in Fig. 8.2(a). The spaces between them are tightly filled by the amino acids that decorate their surfaces. Two of these α-helices, numbered '3', fit snugly into the major grooves of the DNA on either side of the complex. The amino acids which protrude from the α-helices '3' make direct contacts with base-pairs within the major groove of the DNA, in locations marked 1 to 4 or 11 to 14. The rest of the 434 protein holds these two recognition helices '3' the right distance apart and in the correct orientation, so that they fit well into both major grooves of the gently bent DNA, and can probe the edges of base-pairs there so as to determine their identities.

By binding tightly to DNA at just a few sequences in a cell, the 434 protein acts as a repressor of RNA synthesis for certain genes in a bacterial virus called 434. Its specific binding to DNA helps to decide whether the virus lyses (i.e. ruptures) and kills the bacterium which it infects, or just grows peacefully along with it. Thus, the biological action of 434 protein lies in its ability to recognize just one or a few DNA sequences from all others in a viral or bacterial chromosome.

How do those two α-helices '3' recognize a preferred DNA sequence, once an overall 'docking' of the protein onto DNA has been made? The sequence of base-pairs to which a 434 repressor binds most tightly is shown in Fig. 8.3(a). There we can see that each α-helix '3' binds to a base sequence ACAA in positions 1 to 4, or to its equivalent TTGT in positions 11 to 14. A more detailed view of the specific interaction is shown in Fig. 8.3(b), where we can see that four amino acids from helix '3', namely numbers 27, 28, 29, and 33, bind to each of the four base-pairs A1–T1', C2–G2', A3–T3', and A4–T4'.

Further details of these close contacts between protein and DNA are shown in Fig. 8.3(c) and (d). Figure 8.3(c) shows which amino

(a)

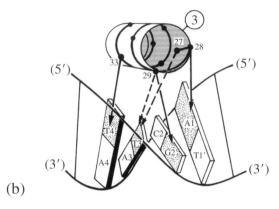

(b)

Figure 8.3 (a) A DNA base sequence which is recognized specifically by the 434 repressor protein. The four base-pairs on either end of this 14-base-pair sequence are shown in bold letters: they correspond to the four shaded base-pairs on each end of the DNA shown in Fig. 8.2(a), which are contacted directly by amino acids from α-helix '3'. (b) A more detailed view of how amino acids 27, 28, 29, and 33 from α-helix '3' contact different base-pairs within the major groove, at the sequence ACAA or TGTT shown in bold in (a). Hydrogen bonds between amino acids and base-pairs are drawn as continuous arrows, while hydrophobic contacts are drawn as dashed arrows.

acids connect to which base-pairs, while Fig. 8.3(d) shows which bases connect to which amino acids on the unrolled α-helix. For example, Thr (threonine) 27 connects to base T3' by a hydrophobic contact in the major groove; Gln (glutamine) 28 connects to base A1 by a hydrogen bond in the major groove; Gln 29 connects to bases G2' and T3' by a hydrogen bond to G2' and by a hydrophobic contact to T3'; while Gln 33 connects to base T4' by a hydrogen bond. All contacts made by hydrogen bonds, for example N–H to O or O–H to O, are shown schematically here as solid lines, while a hydrophobic contact (for example, CH_3 of the amino acid Thr to CH_3 of base T) is drawn as a dashed line. The detailed chemical formulae of the 20 amino acids may be found in any biochemistry text; and a few were shown in Chapter 4.

Other important contacts between the 434 protein and DNA are made between amino acids and DNA backbone phosphates, as shown also in Fig. 8.3(c). For example, Ser (serine) 30 makes a hydrogen bond to a phosphate just beyond base T4', while

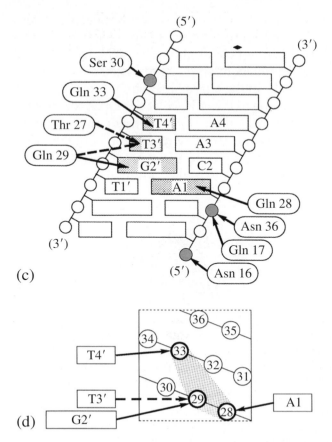

(c)

(d)

Figure 8.3 (c) Here the DNA helix has been 'unrolled' to display the major groove, and to show clearly which amino acids from the 434 protein contact which base-pairs or phosphates from the DNA. The small diamond at the top marks the center of two-fold rotational symmetry. (d) Here the protein α-helix '3' has been unrolled to show clearly which base-pairs from the DNA contact which amino acids on it. For example, base T4′ contacts amino acid 33, while base T3′ contacts amino acid 29.

Asn (asparagine) 16 and 36 and Gln 17 make hydrogen bonds to phosphates below base A1. Some of these specific contacts to phosphates come from amino acids in α-helix '2', while others come from the flexible linker which connects α-helices '3' and '4': see Fig. 8.2(a) for an overall three-dimensional view of such interactions.

Although the details shown in Fig. 8.3 are dauntingly complicated, we should note here that many simplifications have actually been made in drawing the pictures. In particular, we have omitted several water molecules that are found by X-ray crystallography at high resolution in the 'crevices' between the DNA and protein surfaces, and which provide many more, indirect hydrogen-bonded contacts between the two.

In summary, the specific interaction between 434 repressor protein and DNA seems to involve a sophisticated mixture of chemistry and three-dimensional geometry, which is known in general by the term 'stereochemistry'. Many different amino acids must first fit together to make a large protein, which is perfectly complementary in its shape to the surface of the DNA formed by base-pairs and phosphates; and then both surfaces must match closely so far as hydrogen bonds and hydrophobic contacts are concerned. Loss of just a few hydrogen bonds or hydrophobic contacts from an optimized protein–DNA complex will usually result in a large loss

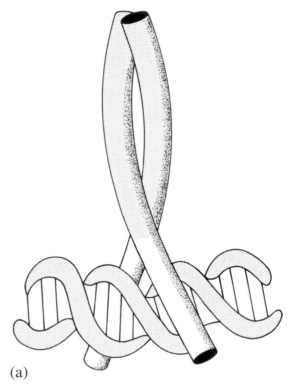

(a)

Figure 8.4 (a) A schematic drawing of the bZIP activator protein as bound to DNA. The bZIP protein consists of two related halves, not always identical. Each of these halves folds into a long α-helix that coils gently around the other half in what is known as a 'coiled coil' or 'leucine zipper' arrangement, here shown shaded. One end of each α-helix protrudes into the major groove of the DNA. (b) An unrolled view of a DNA sequence which is bound specifically by the bZIP protein, showing which amino acids from the α-helix contact which base-pairs or phosphates. The small diamond near the bottom indicates the center of two-fold rotational symmetry. (c) An unrolled view of the α-helix from bZIP which is bound within the major groove, showing which bases contact which amino acids. Amino acids encircled by a broken line make contact with phosphates.

of specificity for the chosen DNA sequence, in comparison with others to which the protein might dock.

Even before the protein and DNA come together, each 'polar group' (e.g., N–H, O–H, N or O) will make hydrogen bonds to surounding water molecules. Those hydrogen bonds to water will be mostly replaced in the protein–DNA complex by hydrogen bonds made directly between protein and DNA. It is essentially that concerted replacement of many different water molecules by protein or DNA, which causes the complex to become stable.

Let us consider next a second geometry by which a protein can bind specifically to DNA: by inserting *two* α-helices into the major groove of any DNA molecule, within just one double-helical turn. A well-known example is the complex between a short piece of DNA and a protein domain called 'bZIP', as shown schematically in Fig. 8.4(a). Two copies of this α-helical bZIP domain associate to form a dimer; and this is part of a larger protein known as GCN4, that is an activator of genes in yeast. In contrast to the 'bean' shape

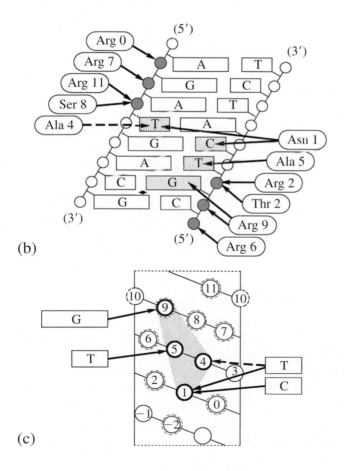

(b)

(c)

of the 434 repressor (Figs 4.11 and 8.2(a)), the bZIP protein domain has the shape of a letter 'Y', and its two arms are able to probe a single turn of the major groove. In this way it can engage just one double-helical turn of the DNA with two α-helices; while by contrast the bean-like 434 protein will engage two full double-helical turns with its two α-helices.

How are the two α-helices of bZIP held in the correct orientation and location by the rest of the bZIP protein, so as to recognize a specific sequence of DNA? Figure 8.4(a) shows that most of the bZIP protein forms a long pair of identical helices, which coil gently around one another in a left-handed sense. Those two α-helices adhere to each other very tightly, because their inner surfaces are covered by hydrophobic amino acids such as leucine and valine. In fact, the coiling of two α-helices as shown in Fig. 8.4(a) is sometimes called a 'leucine zipper', because many leucine amino acids form a kind of hydrophobic 'stripe' on each of the α-helices. In that picture, the helices are represented as smooth surfaces, but in fact the amino acids are more like knobbly protrusions on each α-helix, which fit into complementary holes between the knobs, on the partner helix. This 'knobs-into-holes' interaction then holds the two parts together like a zipper; and its existence was deduced by Francis Crick more than 50 years ago.

Why do the two α-helices from bZIP coil around one another in a left-handed sense? Why do they not lie side-by-side? Now we explained above in Fig. 8.2(b) that there are 3.6 amino acids for each turn of an α-helix. Suppose for a moment that this number were altered to 3.5. Then we should find that every seventh amino acid (where $2 \times 3.5 = 7.0$) would lie exactly in register with the first, in a view along the axis. In other words, amino acids numbered as 0, 7, 14, 21, ... would all lie on a long 'stripe' exactly *parallel* to the helix axis. If the amino acids along those stripes were all hydrophobic in nature for two separate α-helices, then the helices would stick to one another just like two pencils lying side-by-side on a table.

But it turns out that there are actually 3.6 peptides per helical turn, and not 3.5 as in our hypothetical case. The small difference between 3.6 and 3.5 means that peptides which are seven apart on the chain now form a gentle left-handed spiral on the surface of the α-helix, as shown in Fig. 8.2(b). Consequently, the hydrophobic stripes of those two helices will match only if the helices coil around one another in a gentle left-handed fashion.

Once they are each located within the major groove of DNA, over a short stretch of just one double-helical turn, how do the two protruding α-helices of bZIP recognize a specific base-pair sequence there? Figure 8.4(b) and (c) shows the details of specific contacts made between amino acids and DNA base-pairs, in the same manner as Fig. 8.3(c) and (d). Again, only one set of contacts is shown, because both the protein and the DNA are identical in the two halves: the entire arrangement has 'two-fold rotational symmetry'.

Each α-helix from the bZIP protein contacts four base-pairs of DNA, which are shown in Fig. 8.4(b): namely TGAC on one strand and its complement GTCA on the other. Thus, Ala (alanine) 4 makes a hydrophobic contact with T, while Asn (asparagine) 1 makes a hydrogen bond with T and C, and so on. You can also see where positively charged amino acids from the protein, such as Arg (arginine) 0, 7, and 11 on the left and Arg 2, 6 and 9 on the right, make contacts with negatively charged DNA phosphates along both strands. The mechanism of recognition for bZIP is thus similar to that shown above for 434 repressor, even though the relative locations and orientations of the two α-helices that are used for recognition are different in the two cases.

We will now broaden the scope of our survey, to include other examples where some protein does not deploy an α-helix to recognize base-pairs, but uses instead some other kind of structure or *motif*. For example, the met repressor protein shown in Fig. 8.5(a) inserts a β-sheet into the major groove of DNA, so as to recognize the base-pairs there. This met repressor protein is used by a bacterium to regulate the amount of the amino acid methionine which is made. It binds to the promoter of a gene which controls methionine synthesis, and represses that gene if a certain amount of methionine-derived chemical (called S-adenosyl-methionine) is present in the cell. From close inspection of Fig. 8.5(a), we can see that the met repressor consists of two identical peptide chains that interweave to build a single structural unit. There are α-helices labeled 'A' and 'B', which help to hold a two-stranded β-sheet (long arrows) deep within the major groove. Amino acids which protrude from the β-sheet may then contact DNA base-pairs, so that the met repressor protein can recognize a specific base sequence.

What is the detailed structure of a β-sheet, by way of comparison with that of the α-helix shown in Fig. 8.2(b)? Within any β-sheet, as drawn in Fig. 8.5(b), two or more polypeptide strands, here shown crimped up-and-down, lie side-by-side and close together in a plane. Each of the polypeptides is held weakly to its closest neighbor on a nearby strand by two hydrogen bonds (dotted lines), which connect an oxygen from one peptide unit (large open ball) to a hydrogen from a peptide unit on the other strand (small open ball). Moving along the strand, one finds that different amino acids (AA) protrude alternately above and below the plane of the β-sheet. The picture of Fig. 8.5(b) also shows two more strands which

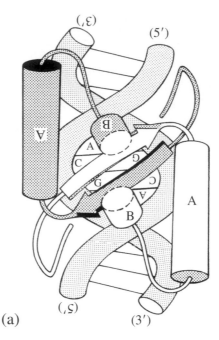

(a)

Figure 8.5 (a) A schematic drawing of the met repressor protein as bound specifically to DNA. The protein consists of two identical halves, each of which contains several α-helices (A and B) and part of a two-stranded β-sheet (broad arrows), which lies within the major groove. Other parts of the protein are not shown, for the sake of clarity. The identical halves are colored differently, and the upside-down lettering on some components corresponds to the two-fold rotational symmetry of the assembly. (b) A more detailed view of how a peptide chain folds into a β-sheet. Two peptide chains lie side-by-side (either parallel, or antiparallel as shown here), so that each peptide unit may make two hydrogen bonds (dotted lines) to nearby peptide units on other chains. Each peptide chain in a β-sheet can be drawn also as a broad arrow, as shown in the lower part of the diagram. (c) An unrolled view of the DNA sequence which is bound specifically to met repressor protein, showing which amino acids from the β-sheet contact which base-pairs or phosphates. The twofold symmetry of these contacts is also indicated by the upside-down lettering at the top, which is identical to the rightside-up lettering at the bottom.

extend the same β-sheet to the right, but which are represented in less detail by arrow-like strips, of the kind shown in Fig. 8.5(a).

When the met repressor protein binds to DNA, both strands of its β-sheet fit snugly into the major groove, so as to recognize base-pairs there. Close contacts between amino acids and base-pairs are shown in Fig. 8.5(c). The met repressor protein recognizes a conserved eight-base-pair sequence of DNA, although specific bonds are made by amino acids to only four of the eight individual bases in each part. Thus, adjacent G and A bases form hydrogen bonds with Lys (lysine) 23 and Thr (threonine) 25 from different strands of the β-sheet; while the outer two bases A and C in AGAC do not seem to contact any amino acids directly. Several DNA phosphates are

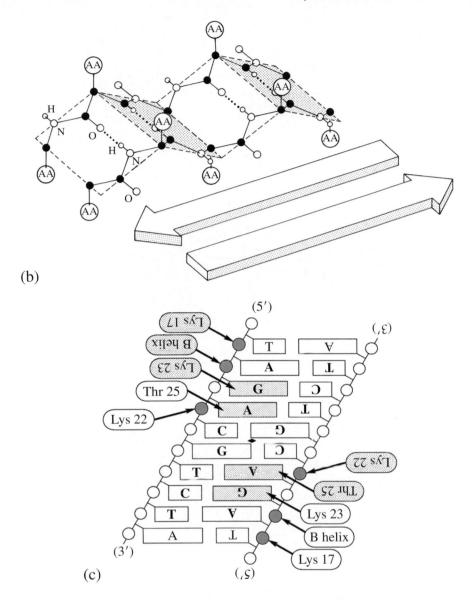

(b)

(c)

bound by various amino acids, namely Lys 17, 22, and a few peptides from α-helix B.

Thus, the specific recognition of DNA by the met repressor protein resembles closely the recognition of DNA by both 434 and bZIP
proteins; except that in the case of met repressor, those amino acids
which contact the base-pairs directly are held in place by a β-sheet
rather than by an α-helix. The met repressor protein has two-fold
rotational symmetry, and it recognizes a two-fold symmetric base
sequence in the DNA. The same is true for the 434 repressor and
bZIP proteins, and the DNA sequences which they recognize. But in
addition to the interaction with DNA, two met repressor proteins

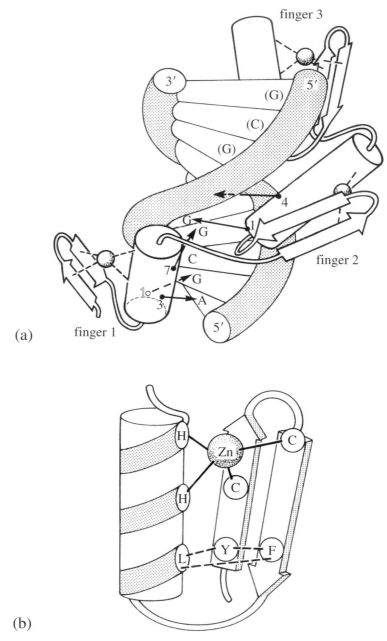

Figure 8.6 (a) A schematic drawing of the Zif268 protein as bound specifically to DNA, showing how its three zinc-finger modules coil in a right-handed sense, so as to follow the DNA major groove over almost a full turn. Each zinc-finger places an α-helix in the major groove at a steep angle of attack, so as to contact base-pairs there. Some bases are identified by letters, which are enclosed in parentheses on the minor-groove side. (b) A more detailed view of how a peptide chain folds into a zinc-finger. Part of the peptide chain wraps into an α-helix, while the other part wraps into a β-sheet. The zinc atom provides a link between them.

can also bind to each other if they to bind to adjacent sites on the DNA. This enhances the strength of DNA binding; and later we shall see several other cases where analogous protein-to-protein interactions occur on a repeated DNA target-sequence.

We mentioned in Chapter 4 the protein TBP (or 'TATA-binding-protein'), which unwinds and sharply bends the TATA promoter sequence in DNA: see Fig. 4.14. That special protein also uses a β-sheet to recognize DNA, as for the met repressor, except there the sheet fits into a greatly widened minor groove at the low-twist TATA sequence, rather than into a normal-width major groove as for met repressor. We shall return to TBP when we discuss how it inter-acts with another protein which is also bound to the DNA nearby.

The 'zinc-finger' family of proteins is another example of alter-native structures used by proteins to recognize DNA. One example of such a structure was shown in Fig. 4.13, for the zinc-finger pro-tein Zif268. There a series of zinc-fingers were pictured schemati-cally as binding to a series of three-base-pair recognition sites on the DNA, by means of specific contacts between arginine amino acids and guanine bases in the major groove. The 'modular' arrange-ment permits small proteins to select long recognition sequences in DNA, despite using a small protein *motif.*

A more detailed drawing of the same specific complex between Zif268 and DNA is shown in Fig. 8.6(a). There we can see how three

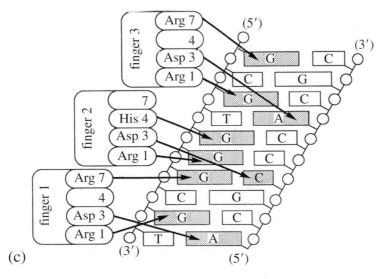

(c)

Figure 8.6 (c) An unrolled view of the DNA sequence which is bound to Zif268, showing which amino acids from fingers 1, 2 and 3 contact which bases. Note that Asp 3 of finger 2 and Arg 7 of finger 1 contact different bases of the same base-pair; in this way, the binding sites of neighboring fingers overlap (as they do in other cases of zinc-fingers).

successive zinc-fingers coil through space in a right-handed sense, so as to follow the path of the major groove. Each zinc-finger inserts an α-helix into the groove at the same, fairly steep, 'angle of attack', and then joins to the next finger along the chain through a flexible peptide linker. Zif268 is thought to function as an activator of transcription, because it is made by the cell in response to certain growth factors, which induce the cell to grow rapidly and divide.

Let us now examine the structure of a zinc-finger in even more detail. Within any finger, part of the peptide chain wraps into an α-helix and part into a β-sheet, as shown in Fig. 8.6(b). The α-helix is bound to the β-sheet by hydrophobic contacts between amino acids leucine, phenylalanine and tyrosine – here shown as discs labelled in the single-letter code as L, F and Y, respectively – and also by a zinc ion that binds to two histidine (H) amino acids from the α-helix, and two cysteine (C) amino acids from the β-sheet. The binding of zinc stabilizes the folding of the peptide so as to present an unique recognition surface to the DNA. The zinc-finger in fact is very small in comparison with other proteins, such as the 434 repressor shown in Fig. 8.2. Zinc-fingers may be found in a wide variety of DNA- and RNA-binding proteins from animals or plants, but only rarely in DNA-binding proteins from bacteria – perhaps because the ancestors of those simple organisms could not protect the zinc-cysteine bond from the damaging effects of a high-oxygen environment.

How does a zinc-finger recognize base sequences within the major groove of DNA? Specific examples of such recognition are shown in Fig. 8.6(c) for the three zinc-fingers of Zif268. Both fingers 1 and 3 recognize bases along one strand of sequence GCG, by means of hydrogen bonds between Arg (arginine) 1 and 7 and atoms on the major-groove edge of guanine G: see Fig. 4.12 for more detail. The base C in each triplet GCG is not recognized directly by either finger.

Finger 2 of the same protein recognizes bases of a sequence TGG along the same strand, rather than GCG. This finger lacks an amino acid Arg at position 7, and so it cannot easily recognize G as the first base; but finger 2 now contains an amino acid His (histidine) at position 4, that recognizes the edge of the second base as G by a hydrogen bond. All three fingers have Asp at position 3, that contacts an A or C on the complementary strand. Because of this Asp 3 contact, the binding sites of finger 1 and finger 2 overlap. All three fingers together recognize a ten-base-pair sequence GCGTGGGCGT, with more strength and specificity than for any one of these zinc-fingers considered separately.

Zinc fingers for their size would seem to be very efficient at recognizing a long series of base-pairs in the DNA, because they can follow the right-handed, twisted path of the major groove. The end-to-end tethering of several modular zinc-finger units enables them to recognize target DNA sequences that are not necessarily symmetric – in contrast to the 434 repressor, met repressor and bZIP examples, where the 'reading heads' are held together by two-fold rotationally symmetric structures.

A final group of sequence-specific DNA-binding proteins from animals are the 'receptors' for small 'signalling-molecules' such as hormones. Once those proteins have bound a specific hormone molecule, they can stimulate the transcription of a target gene in the close vicinity of their specific DNA binding site. The cells of animals may contain many different types of signalling receptor, each of which will recognize some specific 'ligand' – a small molecule such as androgen, estrogen, vitamin D or vitamin A. One part of the receptor protein is specific for recognition of the small ligand,while another part is specific for recognition of the DNA. Crystal structures show how those receptor proteins use an α-helix, stabilized and oriented by zinc binding, to make specific base-pair contacts in the major groove.

Surprisingly, many of these hormone-receptor proteins bind to the *same* DNA sequence, yet they control different genes! In this case, gene-specificity arises from the recognition of the unique *spacing* and *orientation* between two separate, but nearly identical DNA sequences.

Let us first consider a well-known example: the estrogen receptor (Fig. 8.7(a)). Overall, the estrogen receptor is somewhat similar to the 434 repressor protein, in that it binds to two identical 'half-sites' of the DNA target, which run in *opposite directions*. Also, it consists of two identical protein molecules connected together in a head-to-head fashion. The entire estrogen receptor protein and its recognition half-sites are two-fold rotationally symmetric, just as for the 434 repressor; but its half-sites are each six base-pairs long, and they are separated by any three base-pairs.

Next, let us consider the vitamin D receptor (Fig. 8.7(c)). Again, two identical protein molecules recognize two identical DNA half-sites of six base-pairs, separated by three base-pairs; but in this case those half-sites run in the *same* direction, and so the protein-to-protein contact is of a head-to-tail kind.

Finally, let us consider the thyroid receptor (Fig. 8.7(b)). Here the DNA recognition half-sites are almost the same as for the estrogen and vitamin D receptors; except now they are separated by *four* base-pairs, and they run in the same direction. Again, two protein molecules contact one another so that the 'reading heads' adopt the right relationship. Two other examples, shown in Fig. 8.7(d) and

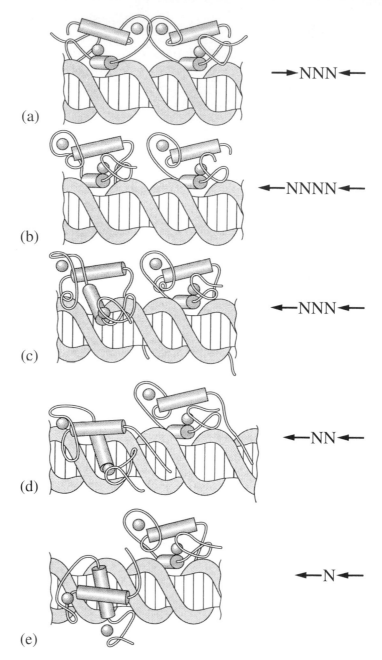

(a)

$\rightarrow$NNN$\leftarrow$

(b)

$\leftarrow$NNNN$\leftarrow$

(c)

$\leftarrow$NNN$\leftarrow$

(d)

$\leftarrow$NN$\leftarrow$

(e)

$\leftarrow$N$\leftarrow$

Figure 8.7 Examples of how protein-to-protein interactions are used to recognize the spatial pattern of DNA recognition sites. The steroid receptors, as represented by the estrogen receptor, bind to 'inverted' repeats, but other receptors bind to 'tandem' repeats. Receptor proteins for (a) estrogen, (b) thyroid, (c) vitamin D_3, (d) retinoic acid, (e) 9-cis-retinoic acid all bind to repeats of the same hexameric sequence, 'AGGTCA'. However, they do not confuse these sites, as each requires a unique pattern of protein-to-protein contacts. The schematics on the right show the recognition sites as directional arrows, separated by different numbers of base-pairs. These pictures show only the DNA-binding parts of the receptors; the hormone-binding parts are not shown.

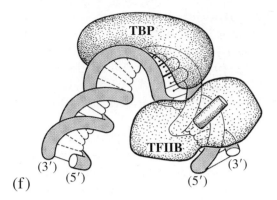

Figure 8.7 (f) The symmetry of the TATA–TBP complex is broken by interactions with a second protein, TFIIB, which contacts both TBP and upstream DNA. The TATA element has a rough two-fold rotational symmetry, and the TBP could in principle align with it in either of two directions; however, binding of TFIIB orients the complex by protein-to-protein and protein-to-DNA contacts. This complex signposts the direction for the start of transcription.

(e), remain similar to the vitamin D and thyroid receptors, except the half-sites are separated by two or one base-pairs, respectively.

To conclude, let us now return briefly to our discussion of the binding of TBP to an eight-base-pair TATA-containing sequence. We drew a picture of TBP in Fig. 4.14 as if the entire arrangement – both protein and DNA – had two-fold rotational symmetry. But to be precise, although the TBP protein has approximate two-fold rotational symmetry, the DNA sequence to which it binds need not have such symmetry; for example, its sequence might be TATAAAAT. Now TBP usually has a 'partner protein' on the promoter DNA for any specific gene, called TFIIB. That partner protein binds specifically to just one end of TBP, and simultaneously to a G–C-rich DNA sequence that precedes the TATA *motif*. Such a unique, well-directed arrangement ensures that the transcription of the promoter-associated DNA takes place in the correct direction.

In summary, we have seen how Nature has devised many distinct means of recognizing a specific DNA sequence, using a variety of protein structural elements. It turns out that typically two-thirds of the atomic contacts in protein-DNA complexes arise from a close fit of similar surfaces, while roughly one-third arise from hydrogen-bonded contacts, that may form either directly between amino acids and the DNA, or else through intervening water molecules.

The direct recognition of a DNA sequence by hydrogen bonds to either amino acids or water is often called 'direct readout'. By contrast, 'indirect readout' results from some subtle recognition of a DNA-base sequence, through the ability of the DNA to change its shape, for example by curving or twisting (as described in Chapter 4) in order to fit the protein better. Since the greater proportion of protein-to-DNA contacts are hydrophobic, we can see that 'indirect readout' could make an important contribution to the specificity of binding of any protein to DNA.

Now certain common modes of DNA recognition can be used by diverse proteins, say where either an α-helix or a β-sheet may be buried in the major groove to make an optimal match of the two surfaces. Might these rough similarities suggest some kind of 'recognition code', for predicting what specific sequence of DNA will be recognized by a particular sequence of amino acids? The simple pattern of recognition of three-base-pair DNA sequences by individual zinc-fingers in Zif268 suggests a special recognition code for the zinc-finger class of proteins. Indeed, Fig. 8.6(c) shows that either three or four successive bases of DNA may be recognized individually by amino acids at positions 7, 4, 3 and 1 on the zinc-finger α-helix, that dips into the major groove. Thus, an arginine (Arg) at position 7 recognizes a G base at the 5' end of four base-pairs GXXX, while a histidine (His) at position 4 recognizes a G in the next base-pair, XGXX. Also, an Arg at position 1, on the very tip of the α-helix, recognizes a G in the third base-pair of XXGX, while an aspartic acid (Asp) at position 3 'reaches over' and recognizes either an A or a C from the opposite strand in the fourth base-pair, XXXA or XXXC.

Based on these promising ideas, several attempts have been made to design novel zinc-finger proteins, and to deliver them to therapeutic targets. For example, Andrew Jamieson and colleagues have made many random mutations in finger 1 of the protein Zif268, and have found that two different sequences, GTG or TCG, can replace the original GCG if Arg 1 and Arg 7 are changed to other amino acids. In a more extensive study, Yen Choo and Aaron Klug have made many random mutations so as to select for a series of three different zinc-fingers, that will bind specifically to the sequence GCAGAAGCC, which is present in certain cancer cells but not much in normal cells. They first selected for different zinc-fingers which would bind preferentially to each of the three short sequences GCA, GAA and GCC; then they combined three chosen zinc-fingers into a long protein molecule, which can indeed recognize all nine base-pairs together. Carl Pabo and colleagues have also used similar methods to find coding patterns that match the sequence of DNA to the four most relevant amino acids (1, 3, 4 and 7) of the fingers. Some of their results are shown in Fig. 8.8. For example, the base G in each of the four targeted positions may be recognized by Arg, His, or Lys at α-helical positions 7, 4 and 1.

Could we also design variants of the other proteins described above, such as 434 repressor or the hormone receptors, so that by replacing just a few amino acids, they might pick out slightly different sequences of DNA? In the case of zinc fingers, it seems that the recognition α-helix always lies in much the same orientation relative to the major groove of the DNA, which makes possible the simple scheme of Fig. 8.8. But the situation for those other proteins is far more complicated, because their α-helices can adopt a wide range of inclinations when they dock onto the major groove of DNA. Any conceivable recognition scheme would therefore have to specify the overall geometry by which the critical α-helix is located within the major groove. Finally, as another severe difficulty, it has been found that the interactions of amino acids with base-pairs are not in general of a one-to-one kind, as they are in Fig. 8.8; instead they are context-dependent in ways that are presently difficult to predict.

Since the task of designing novel proteins seems so daunting, many workers have chosen instead to design novel *chemical*

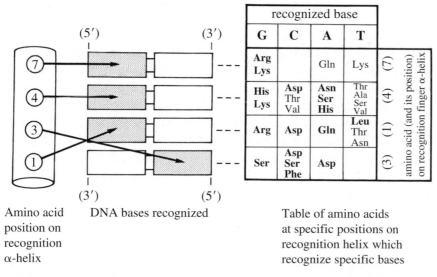

Amino acid position on recognition α-helix

DNA bases recognized

Table of amino acids at specific positions on recognition helix which recognize specific bases

Figure 8.8 A picture-table which summarizes the frequently observed contacts between DNA and designed zinc-fingers, as determined by Choo and Klug and by Miller and Pabo. The pattern of contacts between amino acids and bases is shown in the simple schematic on the left. Each finger can potentially contact four bases, and the binding sites of consecutive fingers may overlap, as they do for instance in Fig. 8.6(c). The table on the right indicates which bases are recognized by which amino acids in the four recognition positions. Bold type denotes contacts seen in crystal structures. (See Table 1.1 for the three-letter code for amino acids.)

compounds which recognize short DNA sequences specifically. These attempts often start from the X-ray crystal structures of complexes between short pieces of DNA and various antibiotics, or chemical compounds such as ethidium or porphyrin. In another important kind of study, scientists have tried to design novel peptides, or peptide-like compounds, that will bind specifically to base-pairs within either the major or the minor groove of DNA. There are many natural examples of this sort, which show useful biological activities, including the antibiotics actinomycin, daunorubicin, chromomycin and bleomycin – some of which are still used for the treatment of cancer.

As the best known example, Peter Dervan and colleagues have synthesized a series of distamycin-related compounds, which place an amide chain deeply into the minor groove of DNA, so as to recognize base-pairs there. The native distamycin recognizes a short base sequence such as AAAA or AATT, by means of hydrogen bonds from N–H groups on the amides to nitrogens N or oxygens O on the minor-groove edges of A–T base-pairs . But Dervan and his chemists have gone one step further, and have made poly-amides which place *two* chains in the minor groove, lying side-by-side, as shown schematically in Fig. 8.9(a). Those synthetic poly-amides may contain separate ring units of pyrrole (Py), imidazole (Im) and hydroxy-imidazole (Hp). Pyrrole is the five-membered ring compound found in the red pigment of blood, while imidazole is part of the side chain of the amino acid histidine. Two such poly-amide chains can be linked to form a kind of hairpin, so that the Py, Im and Hp stack as pairs. Those amide units may bind to the minor-groove oxygen and nitrogen atoms by means of hydrogen bonds; and it turns out that Py and Im, side-by-side in the minor groove, will recognize uniquely a C–G base-pair, as shown schematically in Fig. 8.9(b) and (c). Similarly, Py and Hp side-by-side will recognize uniquely an A–T base-pair, as also shown. Those poly-amides can rotate around the bonds that link their amide units together; and so they can coil helically in space to maintain close register with base-pairs in the helical minor groove.

To conclude our survey, we have seen here how different proteins can recognize specific base sequences of DNA by many different mechanisms, most of which are not predicted by current theory. Most of those proteins seem first to 'dock' in a rough fashion onto the sugar–phosphate chains of DNA, so as to insert an α-helix or β-sheet into the major groove; then in a second step, the amino acids which protrude into the groove may 'probe' for the identities of base-pairs, by means of hydrogen bonds or hydrophobic contacts. Such precise interactions cannot easily be predicted by theory,

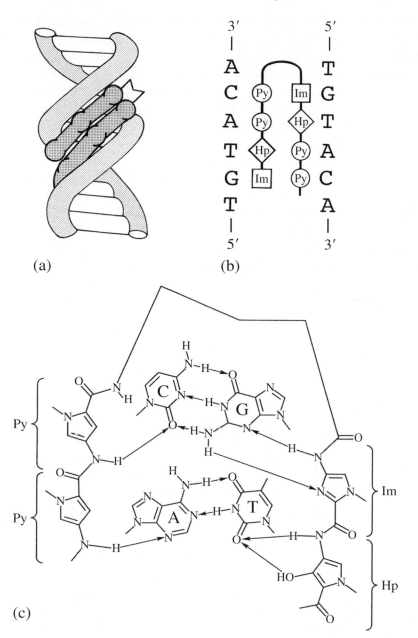

(a)   (b)   (c)

Figure 8.9 Recognition of specific base sequences by side-by-side poly-amide molecules in the minor groove. (a) General arrangement, showing two chains linked at one end. (b) Spread-out schematic view of the minor groove of a particular sequence, with the two amide chains in register with their recognition partners. This shows the unique recognition code. (c) Detailed layout of the hydrogen bonding for the recognition of C–G and A–T base pairs. Here, the poly-amides are shown on either side of the base-pairs: the poly-amides have been separated and spread out on the page, while the base-pairs have been rotated onto the same plane. The scales of the poly-amides and the bases are not the same. Hydrogen bonds are shown by arrows, in the direction of donation.

because they involve intricate stereochemistry. Moreover, the DNA helix is flexible in a way that depends on its base-sequence; which makes prediction even more difficult.

Nevertheless, some workers have made progress in this difficult field, by selecting novel DNA-binding sites for zinc-finger proteins, through random mutation of amino acids in a single zinc-finger, which then binds with altered specificity to the DNA. Other workers have made novel chemicals which bind to DNA of specific base sequences; and those small molecules may also be useful in biology and medicine. Studies of the specific interaction between proteins and DNA are still at an intermediate stage; and no doubt more progress will be made in the next ten to twenty years, which should result in some useful inventions.

The cellular biology of specific protein–DNA interactions remains far beyond the scope of our limited text; but a few points about such biology may be made briefly. First, one might imagine that every gene in a complex organism would have its own particular regulatory protein, say a repressor or activator, which could bind specifically to some particular DNA sequence near the start-site for transcription of the gene. Given 30 000 genes in the human genome, such a one-to-one scheme would require 30 000 unique regulatory proteins! Yet Nature often does things differently from ways that one might naively imagine.

Thus, in most kinds of bacteria, several related genes may be clustered together, using only *one* protein to control their overall expression. A gene-cluster of that kind, known as an 'operon', may contain most or all of the necessary genes for some particular biochemical process: say all of the enzymes needed for synthesis of the amino-acid methionine. The bacterial methionine operon is in fact controlled by just one protein: the *met* repressor that we described earlier in this chapter.

Now in animals or plants, several related genes may still cluster together; yet due to the great complexity of a large genome, one protein is usually not enough to control transcription with sufficient specificity. And so we find that most genes in animals or plants are controlled by two or more proteins known as 'transcription factors', which can act jointly in various combinations so as to achieve a much higher gene-specificity than for any single protein acting separately.

Furthermore, the cells of animals or plants contain very complex DNA-packaging devices in the form of chromosomes, which also regulate genes in specific ways. Thus, the histone proteins which condense very long DNA into chromosomes may be modified chemically by enzymes known as 'acetylases' or 'methylases', so as to reduce the tightness of packing around specific genes, and thereby activate transcription in particular locations (see Chapters 7 and 11). The genes of animals or plants, therefore, would seem to be controlled by a complex hierarchy of molecular interactions; whereas the genes of bacteria may often be controlled more simply.

Further Reading

Aggarwal, A.K., Rodgers, D.W., Drottar, M., Ptashne, M., and Harrison, S.C. (1988) Recognition of a DNA operator by the repressor of phage 434: a view at high resolution. *Science* **242**, 899–907. An early example of specific protein-DNA binding, for the 434 repressor bound to its preferred base sequence. Source of Figs 8.2(a) and 8.3.

Arora, P.S., Ansari, A.Z., Best, T.P., Ptashne, M., and Dervan, P.B. (2002) Design of artificial transcriptional activators with rigid poly-L-proline linkers. *Journal of American Chemical Society* **124**, 13067–71. How polyamide based compounds may be used to control gene expression.

Fersht, A.R. (1999) *Structure and Mechanism in Protein Science*. W.H. Freeman, New York. Elucidates the general principles of specific associations within and between biological molecules, including interactions that establish the fold of a protein, and the protein-to-protein interactions that affect their stability and specificity.

Gowers, D.M. and Halford, S.E. (2003) Protein motion from non-specific to specific DNA by three-dimensional routes aided by supercoiling. *EMBO Journal* **22**, 1410–18. How proteins can find target sites on a length of DNA by sliding or hopping. These processes influence the rates at which sites can be found within the cell.

Judson, H.F. (1979) *The Eighth Day of Creation*. Simon & Schuster, New York. Chapter 2(a) provides an excellent description of Pauling's early studies of proteins, and of his discovery of the simple α-helix structure.

Lewis, M., Chang, G., Horton, N., Kercher, M., Pace, H., Schumacher, M., Brennan, R., and Lu, P. (1996) Crystal structure of the lactose operon repressor and its complexes with DNA and inducer. *Science* **271**, 1247–55. A well-known repressor protein bound to DNA; two such repressor proteins often stick together, to form a tight DNA loop.

Li, T., Stark, M., Johnson, A., and Wolberger, C. (1995) Crystal structure of the MATa1/MATα2 homeodomain heterodimer bound to DNA. *Science* **270**, 262–7. A DNA-binding protein from yeast, which curves the DNA around itself into a large loop. The structure also illustrates how protein-to-protein interactions are an important aspect of specificity and control of gene expression.

Lilley, D.M.J. and White, M.F. (2001) The junction-resolving enzymes. *Molecular Cell Biology* **2**, 433–43. A summary of proteins which recognize 'four-way junctions' in DNA to facilitate homologous recombination.

Marmorstein, R. and Fitzgerald, M.X. (2003) Modulation of DNA-binding domains for sequence-specific DNA recognition. *Gene* **304**, 1–12. Factors that influence DNA sequence recognition, including protein-to-protein interactions in multi-component assemblies of DNA-binding proteins.

Nolte, R.T., Conlin, R.M., Harrison, S.C., and Brown, R.S. (1998). Differing roles for zinc fingers in DNA recognition: structure of a six-finger transcription factor IIIA complex. *Proceedings of the National Academy of Sciences, USA* **95**, 2938–43. Example of different modes of DNA binding by the zinc-finger family.

Ogata, K., Sato, K., and Tahirov, T. (2003) Eukaryotic transcriptional regulatory complexes: cooperativity from near and afar. *Current Opinion in Structural Biology* **13**, 40–8. Principles of combinatorial control of gene regulation in eukaryotes; experimental evidence for DNA looping in a multi-component regulatory complex, using atomic force microscopy – a technique described in Chapter 9.

Perutz, M.F. (1992) *Protein Structure: New Approaches to Disease and Therapy.* W.H. Freeman, New York. An authoritative survey of protein structures, and their relevance to drug design and medicine.

Rice, P.A., Yang, S., Mizuuchi, K., and Nash, H.A. (1996) Crystal structure of an IHF-DNA complex: a protein-induced DNA U-turn. *Cell* **87**, 1295–306. The remarkable structure of the IHF protein bound to DNA, which includes 180 degrees of curvature over 35 base pairs, and proline intercalation at two large kinks.

Web-based Resources

Atlas of amino-acid/base interactions by Janet Thorton and colleagues:
http://www.biochem.ucl.ac.uk/bsm/sidechains
A summary of DNA-binding protein structural families
http://www.biochem.ucl.ac.uk/bsm/prot_dna/prot_dna_cover.html
A repository of crystallographic and nuclear magnetic resonance spectroscopy structures of nucleic acids and protein complexes
http://www.ndbserver.rutgers.edu

Bibliography

Choo, Y. and Klug, A. (1997) Physical basis of a protein-DNA recognition code. *Current Opinion in Structural Biology* **7**, 117–25. Describes studies on the engineering of specificity of DNA binding protein, with emphasis on the zinc-fingers for base triplets in DNA, when amino acids are altered in key positions. A source for Fig. 8.6.

Dervan, P.B. (2001) Molecular recognition of DNA by small molecules. *Bioorganic Medicinal Chemistry* **9**, 2215–35. An account of the development of the poly-amide compounds, based on observations of small molecules binding to DNA. The source of Fig. 8.9.

Jamieson, A.C., Kim, S.-H., and Wells, J.A. (1994) *In vitro* selection of zinc fingers with altered DNA-binding specificity. *Biochemistry* **33**, 5689–95.

Changes of amino acids within the zinc-fingers of Zif268 allow them to recognize different DNA sequences.

Keller, W., König, P., and Richmond, T.J. (1995) Crystal structure of a bZIP/DNA complex at 2.2 Å resolution: determinants of DNA specific recognition. *Journal of Molecular Biology* **254**, 657–67. Crystal structure of the bZIP protein bound to DNA. Source of Fig. 8.4.

Kielkopf, C.L., White, S., Szewczyk, J.W., Turner, J.M., Baird, E.E., Dervan, P.B., and Rees, D.C. (1998) A structural basis for recognition of A-T and T-A base pairs in the minor groove of B-DNA. *Science* **282**, 111. Crystal structure showing how a poly-amide can encode recognition of all four possible base-pairs (A–T, T–A, G–C and C–G) through minor groove contacts. Source of Fig 8.8.

Luscombe, N.M., Laskowski, R.A., and Thornton, J.M. (2001) Amino acid-base interactions: a three-dimensionsal analysis of protein-DNA interactions at an atomic level. *Nucleic Acids Research* **29**, 2860–74. A review of amino acid-to-base contacts in a wide variety of protein-DNA complexes.

Miller, J.C and Pabo, C.O. (2001) Rearrangement of side-chains in a Zif268 mutant highlights the complexities of zinc finger-DNA recognition. *Journal of Molecular Biology* **313**, 309–15. Explores detailed issues concerning the design of zinc-finger proteins to recognize specific DNA targets. A source for Fig. 8.6.

Nair, S.K. and Burley, S.K. (2003) X-ray structures of Myc-Max and Mad-Max recognizing DNA. Molecular basis of regulation by proto-oncogenic transcription factors. *Cell* **112**, 193–205. Illustrates how DNA is recognized by one class of zipper proteins, and shows how the Myc-Max heterodimer could favor loop formation in the DNA.

Otwinowski, Z., Schevitz, R.W., Zhang, R.G., Lawson, C.L., Joachimiak, A., Marmorstein, R.Q., Luisi, B.F., and Sigler, P.B. (1988) Crystal structure of trp repressor/operator complex at atomic resolution. *Nature* **335**, 321–9. Most of the contacts between protein and DNA bases are mediated indirectly by water in this structure.

Pabo, C.O. and Nekludova, L. (2000) Geometric analysis and comparison of protein-DNA interfaces: why is there no simple code for recognition? *Journal of Molecular Biology* **301**, 597–624. Demonstrates geometrical requirements for favorable hydrogen-bonding interactions between the DNA bases and amino-acid side chains, and how they depend on the orientation of α-helices in the groove.

Pavletich, N.P. and Pabo, C.O. (1991) Zinc finger-DNA recognition: crystal structure of a Zif268-DNA complex at 2.1 Å. *Science* **252**, 809–17. A chain of three zinc-fingers that bind to DNA in a regular way, for the protein Zif268. A source of Fig. 8.6.

Rastinejad, F. and Khorasanizadeh, S. (2001) Nuclear-receptor interactions on DNA-response elements. *Trends in Biochemical Scences* **26**, 384–90. Summarizes crystal structures of different receptor proteins on different DNA half-sites, and recognition of symmetry through protein-to-protein interactions. A source of Fig. 8.7.

Schaffer, P.L. and Gewirth, D.T. (2002) Structural basis of VDR-DNA inter-actions on direct repeat response elements. *EMBO Journal* **21**, 2242–52. A source of Fig. 8.7.

Somers, W.S. and Phillips, S.E.V. (1992) Crystal structure of the *met* repressor-operator complex at 2.8 Å resolution reveals DNA recognition by β-strands. *Nature* **359**, 387–93. The first detailed example of a β-sheet structure which binds specifically to DNA. Source of Fig. 8.5(a) and (c).

Suzuki, M. and Yagi, N. (1996) An in-the-groove view of DNA structures in complexes with proteins. *Journal of Molecular Biology* **255**, 677–87. Studies of the detailed fit between transcription factor proteins and DNA.

Tsai, F.T., Littlefield, O., Kosa, P.F., Cox, J.M., Schepartz, A., and Sigler, P.B. (1998) Polarity of transcription on PolII and archael promoters: where is the 'one-way sign' and how is it read? *Cold Spring Harbor Symposium in Quantitative Biology* **63**, 53–61. A source of Fig. 8.7.

CHAPTER 9

Methods Used to Study the Structure of DNA

Our goal in this book has been to explain, as simply as possible, how DNA works in biology. For that reason, we have tried not to dwell too much on the methods which are used by scientists to study DNA: instead we have tried to give an integrated picture of DNA as obtained by many different methods of analysis. We have emphasized on many occasions that DNA is a very tiny object; yet our pictures of DNA have been drawn in terms of images which may be perceived by the reader on a 'household' scale.

A student who wants to understand any subject in depth will want to know exactly how the evidence has been obtained, from which the overall conclusions have been reached. And the historian of whom we spoke in Chapter 1 was puzzled not so much by the fact that the DNA in the cells of our bodies is so exceedingly small, but by the problem of how one can *find out* anything about an object so small. In this chapter, therefore, we shall explain some of the techniques which scientists today use to study the structure of DNA.

The most important method, at least from a historical point of view, has been the analysis of DNA structure by *X-ray diffraction*. This is the tool which was used to discover the basic double-helical form of DNA in 1953. Ten years earlier, in the 1940s, studies of pneumococcal bacteria by Oswald Avery and colleagues had shown that a pure preparation of DNA could cause a harmless form of the bacterium to become infectious, and so impart pneumonia to mice. (We know now that the DNA used by Avery contained a gene for making a strong shell or coat around the bacterium, but this was not known at the time.) By the 1950s, enough evidence had piled up to convince even physical scientists that DNA might constitute the invisible 'gene' of which biologists had spoken for more than 40 years. Therefore, some physicists and chemists began to investigate the structure of DNA by various methods, including X-ray diffraction, to see whether its physical structure might shed any light on how DNA could act as the genetic material.

A talented early X-ray worker was Rosalind Franklin. She pulled fine fibers of DNA from natural sources, and found that when those fibers were exposed to X-rays, they could give either of two distinct X-ray diffraction photographs. She called these two patterns 'A' and 'B'. The 'A' form was seen when she kept the fibers relatively dry, whereas the 'B' form was seen when she kept the fibers wet. Her 'B' form photograph – which was much the simpler of the two – was interpreted by James Watson and Francis Crick in the spring of 1953 in terms of a right-handed double helix containing A–T and G–C base-pairs. Robert Langridge subsequently confirmed the essential points of the 'B' form model in 1960; and Watson Fuller produced a similar model for the 'A' form in 1965, refining a model first proposed by Franklin and Raymond Gosling late in 1953. These 'A' and 'B' form models were shown as part of Fig. 2.7.

Later work by Struther Arnott and colleagues showed that DNA of a regularly repeating sequence, such as A_n/T_n (that is, all A on one strand and all T on the other) or $(AT)_n/(AT)_n$ (that is, the alternating sequence ATATAT on both strands) could be extremely polymorphic. Thus, each fiber preparation could produce as many as three or four different kinds of X-ray pattern, depending on its salt and water content during exposure to X-rays. For example, a fiber of the sequence $(AT)_n/(AT)_n$ can produce a total of four different X-ray patterns under different conditions, which are known as 'A', 'B', 'C', and 'D'. Other base sequences can produce related X-ray patterns, for example 'B''', 'C''', and 'C''''', which are clearly variants of the basic forms; while still others show X-ray patterns such as 'E' that are plainly distinct. Thus by 1980, much evidence had accumulated that the structure of DNA might be more complex than Watson and Crick could ever have anticipated in the 1950s. Yet although X-ray pictures of a fiber sample can show well enough that the forms 'A' to 'E' are distinct, they do not yield enough information to determine the three-dimensional structures of those different forms at sufficient resolution to see the individual atoms clearly.

Fortunately, by 1980, chemists such as Keiichi Itakura, Shoji Tanaka, and Jacques van Boom had learned how to synthesize DNA chemically in large amounts, and how to purify it so that one could grow crystals of particular, short base sequences. Crystals will not grow unless the preparation is pure; that is, unless the short fragments of DNA (or oligomers,[1] as they are known) have identical sequence and are all of the same length. The first structure of DNA to be solved by X-ray analysis of a crystal, as distinct from a fiber, was that of the sequence ATAT by M.A. Viswamitra in 1978. It proved to be disappointing, for the molecule did not form a complete double helix, perhaps because the TA step unwinds easily. But the next few X-ray structures were to produce astonishing results: the sequences CGCG and CGCGCG, as analysed independently by Andrew Wang, J. Crawford, Horace Drew, and their colleagues in 1979–80, both crystallized as left-handed double helices. Biologists had assumed for over 20 years that DNA could only be right-handed; and then it was discovered that DNA could be left-handed as well! Earlier solution studies by Fritz Pohl and Tom Jovin, using circular dichroism methods (see below), had suggested that alternating C–G sequences such as CGCG might be either right-handed or left-handed, depending on the salt concentration; but only a few crystallographers and other specialists had taken them seriously.

Since 1980, many important structures of DNA oligomers, and of their complexes with antibiotics or proteins, have been analysed in single crystals by X-ray diffraction methods. These studies form the essential background for all of the previous chapters in this book. They are now far too numerous to be cited here, but we list some references to this large body of work at the end of the chapter. Here we shall explain the typical steps of an X-ray analysis with reference to a particular DNA molecule, or protein–DNA complex of interest.

First, the crystallographer must decide what sequence of DNA to study, and then prepare large amounts of the material in pure form, usually by the method of chemical synthesis. In the case of a protein–DNA complex, one has to prepare also large amounts of the protein in chemically pure form, usually by cloning a gene for the protein into bacteria, and then expressing the protein in large amounts. Next, he or she must grow crystals of this substance that are suitable for X-ray diffraction studies. Growing crystals is a very chancy business! Some crystals of DNA, suitable for X-ray analysis, are shown in Fig. 9.1. They are each about 1 mm long. These particular crystals contain a complex of an eight-base-pair double helix of sequence AGCATGCT, together with the antibiotic nogalamycin to which it binds tightly. The antibiotic is orange; the DNA itself has no color. The crystals shown in this figure were grown slowly in a cold room for 2 weeks before they reached the required size of 1 mm.

Next, one of the crystals is carefully mounted in a wet, sealed capillary tube and placed in an X-ray beam. If the crystal is well-ordered in structure, an X-ray photograph such as that shown in Fig. 9.2 will be obtained. Any large crystal of DNA is made from millions of identical DNA molecules, all of which are close-packed into some sort of regular array. The geometrical locations of spots in Fig. 9.2 tell us

Figure 9.1 Crystals of DNA of sequence AGCATGCT in combination with the antibiotic nogalamycin. Each crystal is about 1 mm long. Courtesy of Maxine McCall and Louise Lockley.

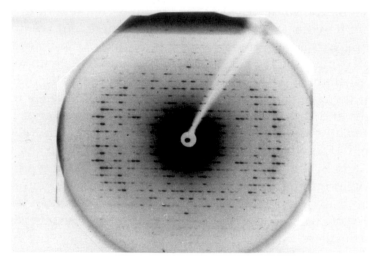

Figure 9.2 Typical X-ray diffraction photograph of a DNA crystal. The DNA molecule in the crystal which produced this picture contained 12 base-pairs of sequence CGCGAATTCGCG on each strand.

what sort of repeating array the molecules have formed. DNA can pack into a crystal having any one of 65 different kinds of three-dimensional symmetry; and so the first task is to determine which kind of symmetry is present in any particular crystal. The next job is to measure the relative intensities of the spots, for each of many photographs taken with different orientations of the crystal in the X-ray beam. In the early days this was a time-consuming task, but now it

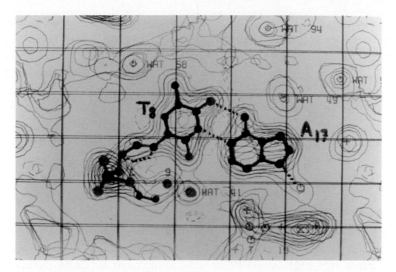

Figure 9.3 Assignment of atomic structure to part of an electron-density map as produced by X-ray diffraction methods. This part of the map shows an A–T base-pair as in Fig. 2.11(a). Water molecules are labeled 'WAT'. From Dickerson and Drew (1981) *Journal of Molecular Biology* **149**, 761–86.

can be done routinely in a short time by automated methods, and by use of a powerful X-ray beam. The final task, which is the most difficult, is to translate the relative intensities of spots into a model of the atomic structure, for DNA and any antibiotic or protein in the crystal. Each non-hydrogen atom (for example, carbon, nitrogen, oxygen, or phosphorus) can be located to an accuracy of about 0.1 to 0.2 Å in three-dimensional space if this last job is done properly; and so even the fine details of a structure can be found.

Figure 9.3 shows one small part of a completed DNA structure as determined in this way. There we can see an adenine–thymine base-pair, surrounded by many ordered water molecules. The locations of carbon, nitrogen, and oxygen atoms are identified by successive contours of increasing electron density. In fact, the X-ray scattering power of any atom is proportional to the square of its electron number; so carbon scatters X-rays as $(6)^2 = 36$, nitrogen as $(7)^2 = 49$, and oxygen as $(8)^2 = 64$. That is why hydrogen atoms cannot usually be located, because they scatter X-rays only weakly as $(1)^2 = 1$. And that is also why heavy atoms such as bromine, iodine, or platinum can be used to help solve X-ray structures, because platinum, for example, scatters as $(78)^2 = 6084$, or much more strongly than the other light atoms.

From the final assignment of atomic positions in a crystal, and after many cycles of refinement by a computer, one can obtain a highly accurate model of the whole DNA molecule (plus protein or antibiotic, as appropriate) in three dimensions. Such three-dimensional models are usually regarded as being broadly representative of the structure in solution, on average: for if the structure in the crystal were too different from that in solution, the molecule would never have crystallized! These three-dimensional models form the whole underpinning for the science of molecular biology; and that is why we have explained how they are derived. In addition, the principles of symmetry and molecular structure which one learns during the course of an X-ray analysis are useful in understanding other, wider aspects of biology which do not have anything to do with crystals. For example, one cannot understand filaments of any sort, whether they take the form of flagella, muscles, microtubules, or DNA, without knowing something about symmetry. As J.D. Bernal once wrote, 'generalized crystallography is the key to molecular biology'.

We have not explained here how scientists actually convert the relative intensities of spots in a set of X-ray patterns to a detailed three-dimensional atomic structure. Our reason for this is that the mathematics used in the process are extremely difficult; all the student needs to know is that the final structure is built up from the superposition of many waves, and that the relative intensity of each spot on the film defines the height of one particular wave. One should also know that the mathematics are highly statistical in nature: each X-ray particle (or photon) behaves unpredictably when going through the crystal, and can emerge at any spot. Thus, from a single X-ray scattering event you cannot learn anything, but by averaging over many events, you can obtain a consistent probability of the photon's arriving at any spot, which is then proportional to its intensity.

In addition to X-ray diffraction methods for analysing atomic structure, one can also carry out *electron microscopy* experiments. Here one first lays a molecule of DNA onto a 'grid', and applies a heavy-metal stain such as uranium or platinum in order to help visualize the DNA in the electron beam. Then one surrounds the grid by a vacuum and shoots electrons through the sample. The electrons are focused, and clean pictures of DNA such as those shown in Figs 5.5 and 6.7 may be obtained. Electron microscopy experiments are easy to perform, but the pictures are lacking in atomic detail. Furthermore, the DNA – or whatever – can easily be distorted from its natural shape in the course of preparation for electron microscopy, since it must be removed from the fluid which normally surrounds it and be placed in a vacuum. The images produced by electron microscopy generally show the molecule of interest in just two dimensions, unless special care is taken to tilt the grid and shoot successive pictures from different perspectives.

Both X-ray structure analysis and electron microscopy are *direct* techniques for determining the structure of DNA. In the end, you can simply look at a three-dimensional model of DNA as determined by X-ray diffraction, or at a picture of DNA on a grid as determined by electron microscopy, and be confident that the thing you are looking at corresponds to physical reality. However, not all scientists practise these two methods, because: (a) the necessary equipment is expensive; (b) a scientist must be highly trained in order to carry out such analyses; and (c) it is often difficult to prepare a suitable crystal, or indeed a sample for electron microscopy, of a biologically interesting substance.

For those reasons, many scientists today use a variety of *indirect* methods for finding out about the structure of DNA. Most of the indirect methods are less reliable than the direct methods described above, but they are generally cheaper and simpler to perform. We shall explain about several different kinds of indirect method here. First there are the *spectral methods* such as nuclear magnetic resonance, Raman spectroscopy and circular dichroism. Then there are the *enzymatic methods* such as 'footprinting' with a DNA-cutting enzyme. Finally there are the *electrophoretic methods*, where DNA is passed through a gel in the presence of an electric field, and thereby separated according to its size, shape, and electric charge.

In the technique of *nuclear magnetic resonance* or NMR (which of late has become very expensive to perform, even more so than X-ray diffraction), a concentrated sample of DNA is brought into the presence of a strong magnetic field, so that the magnetic moments of all of the hydrogen atoms in the DNA align themselves with this major field. Next, the sample is exposed to a low-energy electromagnetic field over a range of radio frequencies; and individual protons within the nuclei of the hydrogen atoms of the DNA may absorb energy at some particular frequency, and thereby align their magnetic moments *against* the main field. The amount of energy required to flip the magnetic moment of a hydrogen atom against the main field is very sensitive to its location in the molecule, how it is chemically bonded to other atoms, and what atoms are located near it in three-dimensional space. Figure 9.4 shows part of the NMR spectrum for a 12-base-pair molecule of sequence CGCGAATTCGCG at several different temperatures. Because both ends of the molecule are related by symmetry (in other words, CGCGAA can pair to TTCGCG), there are only six peaks rather than 12 in the spectrum. Each of those peaks represents the magnetic alignment of a single kind of hydrogen atom in millions of identical DNA molecules. These particular hydrogens lie in the center of Watson–Crick base-pairs as N–H···O or N–H···N hydrogen bonds (recall Fig. 2.11). There are six peaks, at slightly

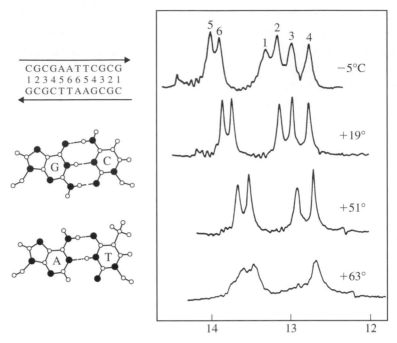

Figure 9.4 NMR spectra of a DNA molecule of sequence CGCGAATTCGCG, at four different temperatures. The arrows alongside the sequence show the (5') to (3') directions, and the horizontal scale under the spectra indicates radio frequency. Courtesy of Dinshaw Patel.

different frequencies of radio absorption, because there are six slightly different kinds of base-pair in different chemical environments. When the temperature of the sample is increased from −5°C to +51°C, the base-pairs on either end of the molecule (numbers 1 and 2) begin to fall apart: then the NMR peaks for their hydrogen atoms are lost, as they exchange with water. The main virtue of NMR methods is that they provide information about the dynamic structure of DNA in solution, which is not available from X-ray or electron microscopy studies. Many of the applications of NMR to DNA in solution were pioneered by Dinshaw Patel in the 1970s, and in recent years hundreds of scientists have entered the field.

In the 1980s, Kurt Wüthrich and colleagues applied a new technique in NMR spectroscopy, first developed by Richard Ernst, that allows you to measure the transfer of magnetic alignment from one hydrogen atom to any other in a chemical molecule, and then use that information to tell you something about its chemical or three-dimensional structure. The two hydrogens must be close together in space for such transfer of magnetism to occur. This technique allows one to measure approximately all of the interatomic distances between different hydrogens in biological molecules, and so to determine their structure in solution. The method has worked well for proteins, but in practice has proved only qualitatively useful for DNA, because there are so few hydrogen atoms on the bases and sugar. But such 'transfer' methods are useful in studying complexes of DNA with antibiotics or with proteins, where they show which hydrogens on the DNA are close to which hydrogens on the antibiotic or protein. NMR methods are limited to molecules having no more than a few thousand atoms, because of the increasing complexity of the spectra for many atoms, and because very large molecules do not turn over (or tumble) rapidly enough in solution to produce a clean spectrum.

Raman spectroscopy measures the vibrational frequencies of individual bonds in the DNA, and hence it is a sensitive measure of chemical bonding and structure. *Circular dichroism spectroscopy* measures the absorption of polarized ultraviolet light by DNA, and shows whether the molecule absorbs more left-handed or right-handed polarized light. Both of these methods were used more in the past than today – for example, in 1972 by Pohl and Jovin to find evidence for left-handed DNA; so they will not be discussed further here.

A variety of *enzymes* and *chemicals* can be used for the analysis of DNA sequence and structure. Some of these will break the DNA into bits at certain short series of nucleotides such as GAATTC, or else at certain single nucleotides such as A, G, C, or T. Others will cut the DNA in practically any location, except where an antibiotic or protein has bound itself to the molecule. Still others will cut the DNA only in places where they detect an unwound single strand, instead of a double helix. What all of these methods have in common is that they use *electrophoresis in gels* to separate the fragments of DNA according to their size. So we must say something about the motion of DNA through gels, before we can explain how enzymatic or chemical methods can be used to probe the structure.

A gel is nothing more than a three-dimensional array of tiny, randomly oriented fibers, like the fibers in a grass mat that you wipe your feet on before going into the house. Most of the spaces in a gel are filled with water. For example, when you make 'jello' as a dessert at children's parties, you simply boil a small amount of gelatin powder in a large volume of water, and let it cool; then the final gel will be no more than 10% gelatin and 90% water. It is easy to see why a typical gel should be highly porous to small molecules such as DNA or protein: they can move easily through the gel by passing through the water spaces between the gel fibers.

Some small molecules can move only slowly through a gel by diffusion, if they are uncharged. But DNA and protein both carry a net electric charge, and so they can move quickly through a gel in the

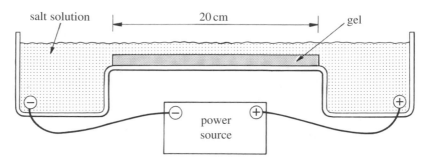

Figure 9.5 A typical gel-running apparatus. Other set-ups may be as long as 50 cm, and may be vertical as well as horizontal.

presence of an electric field. An ordinary gel as used in DNA work can be poured between two glass plates, or else as a thin slab onto a flat surface of size typically 20 cm × 20 cm, as shown in Fig. 9.5. (More recently, gel-like polymers have even been put inside narrow capillary tubes.) Positively and negatively charged electrodes can be placed at its ends in a suitable salt solution in order to impart the desired voltage gradient.

We need not concern ourselves yet with details of the gel-running experiment, such as how to choose the correct density of gel; or indeed how to describe the motion of DNA through a gel by use of mathematics. For present purposes, only two things really matter: one is that short molecules of DNA can travel through a gel more rapidly than long ones, and the other is that every kind of DNA molecule runs through the gel at a very definite, size-related speed.

Suppose that we have a pure sample of linear[1] DNA of some given length and sequence. If we load this sample into a small 'well' at one end of the gel, and turn on the voltage for a few hours, we find that the DNA migrates as a tight band towards the other end, without significant broadening or diffusion. There are two commonly used ways to locate the DNA in a gel. One is to stain the gel with a dye such as ethidium bromide (see Chapter 2), which fluoresces strongly under ultraviolet light when it is bound between two base-pairs. The other is to incorporate one or more radioactive phosphorus atoms into the DNA at its 5′- or 3′-end, or perhaps throughout the length of the molecule. Then the radioactive band of DNA will darken an ordinary photographic film after only a few hours. A third, less commonly used way to locate either DNA or protein in a gel, is to stain the gel with silver metal: then any DNA or protein within the gel binds to the silver metal and so turns the gel brown locally.

Suppose next that we take the same sample of DNA, but treat it with a 'restriction enzyme' called 'Eco RI', that cuts wherever it can find the particular sequence of nucleotides GAATTC. If there are *n*

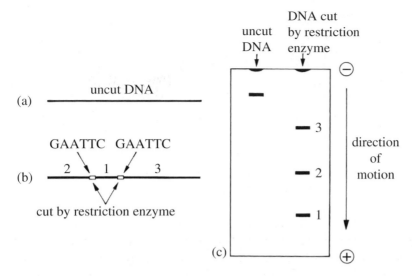

Figure 9.6 Running of DNA fragments through a gel, before and after cleavage with a restriction enzyme.

sequences of the kind GAATTC along its length, our DNA sample will now run through the gel as a series of $n + 1$ distinct bands, as shown schematically in Fig. 9.6. This scheme enables us to find out how many sequences GAATTC are contained in the DNA, and something about their location, since the smaller fragments will run further down the gel than the larger ones.

Finally, suppose we treat the same double-stranded DNA with some chemical that cuts only at the nucleotide A on any strand. Then if the products of this chemical reaction are run through a 'denaturing' gel that contains a high concentration of urea in order to separate the two strands, they will produce a 'ladder' of single-stranded fragments that mark the relative locations of all bases A in the DNA sequence, as shown on the left-hand side of Fig. 9.7. In order to locate bases G, C, and T, one can use other chemical reactions that are specific for these bases, and run these DNA fragments through the gel as well (Fig. 9.7). You can select which strand of the double helix you want to see in the gel photograph, by attaching a radioactive phosphorus to either one strand or the other; fragments from the 'other' strand will be non-radioactive and therefore invisible.

Another commonly-used method is to attach fluorescent dyes of four different colors to the four sequencing reactions A, G, C, and T. A dye will be attached only at one end of the molecule, and to terminal bases which may be distinguished by their distinct colours. All four kinds of reaction for any sequence can then be run in the same gel lane – or in the same gel-filled capillary tube – so as to improve

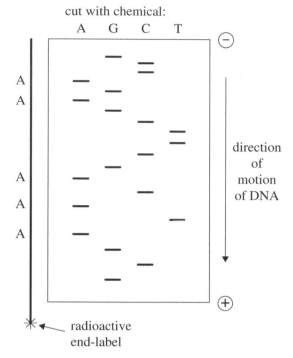

cut with chemical:

A G C T

direction
of
motion
of DNA

radioactive
end-label

Figure 9.7 A gel for determining the sequence of a DNA molecule, which is shown in part on the left.

efficiency. We shall describe this four-colour sequencing method further in Chapter 10.

Modern sequencing methods allow one to determine the complete sequence of a DNA molecule as long as 800 to 1000 nucleotides, because single strands of length 50 to 800 nucleotides will run at slightly different speeds through a gel, or through a polymer-filled capillary, for most sequencing analyses that are done today. Strands of length greater than about 800 nucleotides tend to run at more identical speeds, and so individual lengths cannot easily be resolved from one another (say 800 from 799 or 801). In order to determine the complete sequence of a very long DNA, say from a chromosome or a virus, you have to break it into many different pieces of size about 800 base-pairs or less, and then sequence the pieces individually. By another commonly used method, one can start reading a sequence reaction on long DNA at roughly 800-base-pair intervals or less; in that case, a polymerase enzyme is used to build up many DNA sequencing strands for application to a gel or capillary. The strand to be sequenced is recognized specifically through Watson–Crick base-pairing to a series of small oligonucleotides or 'primers'.

These are not the only ways of determining the sequence of a long DNA molecule, but they are representative of the other methods.

The techniques described here have become so routine that scientists today have already completed large-scale sequencing for the complete genomes of many different animals, plants or micro-organisms – including the well-known Human Genome Project. The medical and scientific implications of these data will be discussed in Chapter 10.

Most of the enzymes and chemicals discussed so far can cut the DNA in very precise locations, according to its base sequence. Certain other enzymes and chemicals, for example DNAase I, can cut the DNA in practically any location, with only a mild specificity for the base sequence. How might such a generalized DNA-cutting activity be useful?

Suppose we have isolated from an animal or plant some important protein that affects gene activity, by binding to an unknown DNA sequence along the length of a chromosome. How can we determine where it prefers to bind to the DNA? Usually the protein will bind so tightly to the DNA that it blocks the cutting activity of an enzyme such as DNAase I; so we can locate the bound protein by looking to see where the cutting activity of DNAase I is reduced in the presence of protein. This technique is known as 'footprinting', because the regions of reduced cutting by DNAase I look like 'footprints' of the protein on the DNA when we study a gel photograph.

One example of such an experiment is shown in Fig. 9.8. There we are looking to see where a small antibiotic called 'echinomycin' binds along the DNA. Our detailed procedure is as follows: we label a DNA molecule of 200 base pairs at either of its two 3'-ends with radioactive phosphorus atoms in separate experiments; then we add the antibiotic to each of these DNA preparations, and wait for a few minutes until the antibiotic has located its preferred binding sites; finally, we add DNAase I for a certain length of time, until some cutting has taken place at every nucleotide. When the two kinds of DNA sample are run on a urea-containing gel, we obtain the patterns shown in Fig. 9.8.

The left- and right-hand sides of this figure show the results of DNAase I cutting along either of the two strands of the double helix. The first three gel lanes on either side show how DNAase I cuts the free DNA, in the absence of echinomycin. The bands there show evidence for some cutting at every nucleotide; yet these bands are of greatly varying intensity. It seems that DNAase I prefers to cut more at some base sequences than at others. Many studies have shown that DNAase I binds across the minor groove of DNA, and only cuts well if this groove is of a correct size, and if the bond to be cut is positioned properly relative to the active site of the enzyme. Each of these structural features depends on the base sequence of the DNA, and so we see a rather complex pattern of cutting even in the absence of the antibiotic.

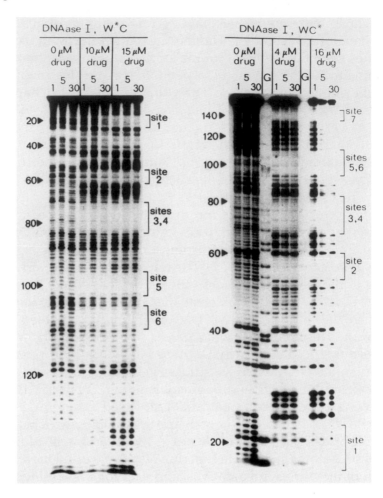

Figure 9.8 Use of a gel for 'footprinting' of the antibiotic echinomycin onto a DNA molecule of 200 base-pairs. The left- and right-hand sides of the figure show cutting by DNAase I on either of the two strands of the double helix, at different drug concentrations. Courtesy of Loretta Low and Michael Waring.

The effect of adding the antibiotic to DNA can be seen in the remaining six gel lanes on either side: different sets of lanes contain different concentrations of echinomycin. The lanes labeled 'G' are markers for guanine. One can easily identify the sites of binding of echinomycin to DNA, by looking for short regions within any lane of the gel where the bands are relatively faint, as compared with their intensities for free DNA. The antibiotic binds tightly to seven different locations along this 200-base-pair DNA, and blocks DNAase I cutting for 5 to 6 bonds at each of its binding sites. 'Site 1' is located near one end of the molecule, far from the radioactive end-label used on the left-hand side of the figure; so there it runs slowly, near the top of the gel. But 'site 1' lies close to the radioactive end-label used on the right-hand side; so there it runs rapidly, near the bottom of the gel. The opposite holds true for 'site 7', which runs off the bottom of the gel on the left, but near the top of the gel on the right. If we were to compare the locations of these echino-mycin binding sites with the base sequence of the DNA used in this experiment, we would find that each binding site is centered on a short sequence of the kind CG. After this experiment was published, it was found by X-ray diffraction methods that echinomycin binds at a step CG in its crystalline complex with DNA.

Near 'site 2' on the left-hand side, or 'site 1' on the right-hand side of Fig. 9.8, several bands adjacent to each echinomycin-binding-site actually become more intense in the presence of the antibiotic than in its absence. It turns out that these are AT-rich regions of high propeller twist, where DNAase I cuts only poorly in the free DNA. Once echinomycin binds next to such DNA, it flattens the propeller twist and lets DNAase I cut more rapidly. Other antibiotics such as ethidium bromide are also thought to flatten the propeller twist, so as to let DNAase I cut more rapidly.

Footprinting studies like the one shown can now be carried out even on single-copy DNA sequences within a whole chromosome, to see where different proteins bind along the DNA in living cells.

As a final example of useful enzymatic methods, let us look at the cutting activity of an enzyme called 'S1 nuclease'. This enzyme cuts DNA only where it can detect an unwound single-strand, as opposed to a double helix. Figure 9.9(a) shows two particular 20-nucleotide strands of DNA, called '20-Watson' and '20-Crick' respectively. When these two strands are mixed in equal amounts, they form a 20-base-pair double helix with Watson–Crick pairs as shown. But if the two strands are kept separate, each may fold back on itself to form a 'hairpin loop', with six base-pairs in the stem and eight unpaired bases in the loop. These hairpin loops are shown schematically in Fig. 9.9(b).

When the 20-base-pair double helix from part (a) is treated with S1 nuclease, and run through a denaturing gel, the cutting pattern shown on the right-hand side of Fig. 9.9(c) is obtained. In this experiment, the 3'-end of 20-Watson has a radioactive label, while the strand 20-Crick remains unlabeled and therefore 'invisible'. It seems that S1 nuclease can cut only at the very ends of a double helix, where the two single strands of DNA are not so firmly connected to one another. Yet when you treat the hairpin 20-Watson from Fig. 9.9(b) with S1 nuclease, you see the pattern shown on the left-hand side of Fig. 9.9(c). Once again the lowest part of the double helix is cut at its fraying end, but in addition you can see extensive

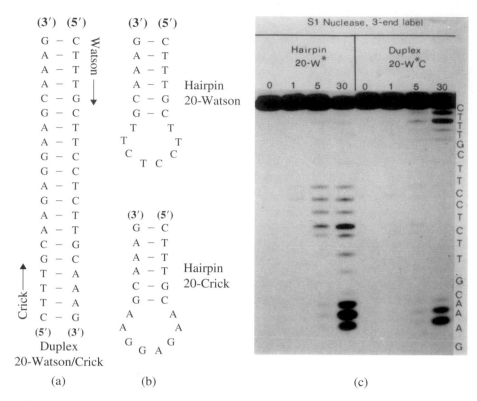

(a)   (b)   (c)

Figure 9.9 Investigation of hairpin-loops by means of cleavage with single-strand specific S1 nuclease, followed by gel electrophoresis of the cleavage products.

cutting of the strand within its loop of eight unpaired bases, halfway up the gel. This result shows conclusively that S1 nuclease recognizes the structure of the DNA rather than its sequence, for the same sequence TTCCTCTT is not cut when it is part of a double helix on the right.

In summary, one can cut the DNA by using a wide variety of enzymes and chemicals, in order to probe its structure. Some of these recognize the base sequence, while others recognize the double-helical structure, or lack thereof. In all such cases, the standard procedure is to separate the fragments of DNA so obtained by means of electrophoresis in gels, where each double helix or single strand runs according to its size.

Finally, although it is not widely known, it seems that the gels themselves can be used to find out something about DNA structure. It is not strictly true that molecules of DNA run in gels according to their *size*; rather, they run according to their *size, shape*, and *net electric charge*. For example, if a piece of negatively charged DNA is bound to a positively charged protein, it will have a reduced overall negative electric charge, and so it will run more slowly through a gel than free DNA of the same length. That is easy enough to understand; but how can a molecule of DNA run in gels according to its shape? How can the gel fibers sense the shape of the DNA, in addition to its chain length or size? This result also follows in a trivial way, once you understand the theory of DNA motion through gels. We shall explain briefly how it works, without recourse to the difficult mathematics that are required to account precisely for the motion of DNA through gels.

Actually, there are two kinds of gel electrophoresis, that use *either constant* or *regularly alternating* electric fields. All of the examples we have given so far have been for the method of constant field. In this case, a sample of DNA is applied to one end of a gel near a negatively charged electrode (see Fig. 9.5), and this sample then migrates through the gel for a distance of 20 cm (or 50 cm in some cases) towards a positively charged electrode. The potential difference between the two electrodes is typically 100 to 1000 V, or enough to generate a current of 10 to 100 mA for a typical salt solution, at neutral pH. Most of the current between the two electrodes is carried by positively and negatively charged ions in the salt solution, such as sodium (Na^+) and chloride (Cl^-), which move through the gel in opposite directions owing to their different charges. The DNA, of course, is negatively charged, so it moves in the same direction as the chloride ions; that is, it moves towards the positively charged electrode. It is necessary to keep a solution of salt in the pores of the gel, because DNA in distilled water will not move at all; it remains bound to sodium ions that pull it in one direction, while it wants to go in the other. In the presence of excess salt, DNA can move towards the positively charged electrode by exchanging sodium ions continuously as it moves forward. (The most commonly-used salt solutions for gels actually include ions such as $Tris^+$ and $borate^-$ or $acetate^-$, which maintain a firmer control on the pH than do Na^+ and Cl^-.)

When this experiment is carried out in a vessel filled with simple salt solution, in the absence of any gel, all DNA molecules move toward the positively charged electrode with about the same velocity, regardless of their size or shape. Thus, a single nucleotide will move at the same speed as several thousand base-pairs of double helix. The explanation for this is that the larger DNA molecules contain more phosphates, and hence they have more electric charge to pull themselves forward in the electric field; but they also experience more viscous drag from their contacts with water molecules, because they are bigger. The two effects cancel, and so DNA moves through a simple salt solution at a speed independent of its size.

The purpose of a gel is to exert an even greater friction or drag on the larger DNA molecules, so as to cause them to move more slowly than the smaller ones. Gels can be made from fibers of agarose or polyacrylamide, at concentrations of 0.5 to 20 gram of solid per 100 ml of fluid. Single nucleotides of DNA move through a gel at the same speed as they move free in solution, but large DNA molecules are retarded in proportion to their size by contacts with gel fibers. The gel method at constant field is capable of separating by size DNA molecules as long as 50 000 base-pairs, with a resolution of better than 1% of the size of the DNA. It is also capable of separating by size a wide variety of protein molecules, with a similar resolution.

There is thus no mystery as to how a gel works, in principle. But how exactly does a gel sense the size and shape of the DNA (or a protein), and retard its motion accordingly? To answer this question, we must first see what a gel looks like on a molecular scale. A typical agarose gel, as visualized by electron microscopy, is shown in Fig. 9.10. It contains many fibers of diameter about 100 Å that cross over one another like the strands of a grass mat. The most important property of an agarose gel is that its fibers are arranged *randomly in space*. Suppose that we could somehow make a gel from *regularly* arranged fibers or points, as shown schematically in Fig. 9.11(a). Then all of the passages through the gel would be of the same size, and hence a plot of particle velocity *versus* size would show a sharp cut-off where the particle (whether protein or DNA) becomes too large to go through any passage. But, as we have said, real gels are constructed from fibers that are arranged randomly in space, as shown in Fig. 9.11(b). Thus some passages through the gel are small, while others are large, and the speed of the particle will decrease gradually with increasing size. Scientists usually adjust

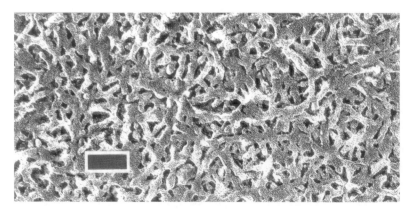

Figure 9.10 Electron micrograph of a portion of a 2% agarose gel, 1 mm × 0.5 mm overall: the small black rectangle is 1000 Å × 500 Å. Individual gel fibers are about 100 Å wide. Courtesy of Sue Whytock and John Finch.

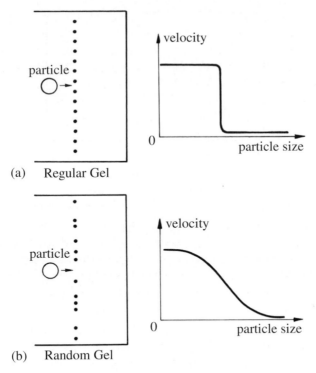

Figure 9.11 Sorting effect of (a) regular or (b) random-sieve gel.

the concentration of their gel so that the size of the molecule they wish to isolate lies about in the middle of the smooth curve shown in the diagram.

How does the DNA pick its way through a series of randomly sized passages in a gel? Obviously, there can be no regular pattern to its motion, and so its path is usually described through the use of statistics. There are several ways of thinking about this. One way is shown in Fig. 9.12(a), where a 'disc' must move from left to right across a plane field of randomly distributed point obstacles. In the free spaces it goes quickly, just like DNA free in solution; but where it overlaps one or more points it goes slowly, perhaps at only 5% or 10% of the speed of DNA in solution. The bigger the disc, the greater the fraction of its path where it contacts point obstacles: hence, big discs go more slowly through the field of obstacles than do small discs.

Another way of looking at the same thing is shown in Fig. 9.12(b). There we see cars traveling along a road, either singly or in cara-vans. The road is smooth for most of its length, permitting travel at 100 km/hr, but it is rough in parts, permitting travel at just 10 km/hr. A single car will average a speed of 92 km/hr along this road, if the rough patches amount to just 1% of the total road surface: recall that

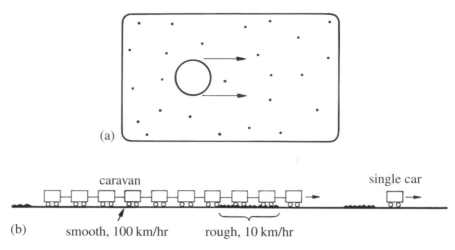

Figure 9.12 Two analogies for gel electrophoresis of DNA at constant field: (a) a disc passes through a two-dimensional 'gel' of randomly-located point obstacles; (b) a long caravan is impeded more than a single car by rough patches in a road.

average speed = (total distance)/(total time). But the long caravan will proceed at about 10 km/hr, because it always contacts a rough spot somewhere along its length. Both parts of Fig. 9.12 describe a process of non-uniform motion, where the DNA starts and stops a lot while going through a gel. It does not proceed steadily, but its speed is uniform when averaged over a long period of time – which is itself short in comparison with the total time of testing.

The probability that a DNA molecule will make contact with a gel fiber is given by statistical theory for several different hypothetical cases. For a gel made of long fibers, the DNA is slowed according to its surface area; while for a gel whose obstacles are points, such as the junctions between fibers, the DNA is slowed according to its volume. Many workers now agree that DNA is likely to be slowed in a gel according to its volume, as if the gel behaved like a set of points arranged randomly in space; but such fine points of gel theory are still not certain.

Many experiments show that *curved* DNA moves more slowly through gels than does *straight* DNA of the same size or length. From the discussion presented in Chapter 5, it seems obvious that curved DNA will occupy a larger volume, or have a larger effective surface area, than straight DNA of the same size, because it can coil around itself to include a lot of empty space into which the gel fibers cannot enter. Thus, from careful measurements by gel electrophoresis, one can estimate the increased volume or surface area of curved DNA compared with straight DNA. But one cannot learn anything definite about local structural parameters such as roll, slide, or twist (see Chapter 3), because the gel does not sense such parameters directly.

Other experiments show that *supercoiled* DNA moves through gels with a velocity that depends on its linking number Lk. Some pictures of supercoiled DNA from bacteria were shown in Fig. 6.7. Circular DNA molecules with Lk near 0 move more slowly through gels than do molecules with a large value of Lk, either positive or negative, because the shape of the DNA becomes more compact as it gets more supercoiled. To be precise, it is thought that the cross-sectional diameter of an interwound supercoil decreases as Lk departs from zero. For example, in Fig. 6.5 the circle (a) with Lk = 0 has a larger cross-sectional diameter (or 'fatness') than circles such as (c) to (e) with Lk = −3. Given a mixture of supercoiled DNA molecules of identical size but different linking numbers Lk, one can determine the true linking number Lk of any one of them by running the total mixture through an agarose gel, and then counting the number of discrete bands from the slowest with Lk = 0 at the top of the gel to, say, Lk = −10 or −20 near the bottom (see Exercise 9.6).

In general, the method of gel electrophoresis as described above can separate DNA molecules of different length out to about 50 000 base-pairs. Above this length the DNA is never out of contact with gel fibers: it is like a very long caravan in Fig. 9.12(b), always impeded by rough patches on the road. Thus, all DNA molecules beyond a certain length travel at the same speed; and so the gel is of little use in separating them.

To separate very large DNA molecules according to their size, one can carry out gel electrophoresis in the presence of a regularly alternating electric field. Suppose that the voltage applied to a gel is reversed at regular intervals of time: how will that affect the motion of DNA through the gel? For example, the voltage from left to right in a gel may be set at +100 V for 10 s, then at −100 V for 5 s, then at +100 V again for 10 s, and so on. The DNA will move first in one direction for 10 s, then in the other direction for 5 s, then in the first direction again for 10 s. How will its net motion be altered? After an interval of 15 s, will it go simply (10 − 5)/15 = 1/3 as far as it would have gone at a constant 100 V? If the DNA is small, say less than 1000 base-pairs, then periodic reversal of the voltage has precisely this effect: the DNA goes 1/3 as far as it would have gone at constant voltage. But if the DNA is very large, say 100 000 to 10 million base-pairs, then periodic reversal of the field has a dramatic and unexpected effect: the large DNA cannot 'turn around' in the gel as quickly as small DNA, when the voltage is reversed, and so the large DNA proceeds much more slowly than expected. In other

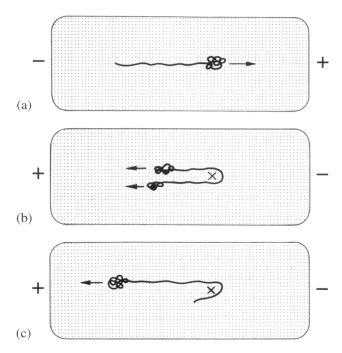

Figure 9.13 Movement of a very long DNA molecule through a gel, in alternating-field electrophoresis: (a) forward motion; (b) 'tie on coathook' situation; (c) reverse motion.

words, there is a 'time delay' for the large DNA to change direction in the gel; and consequently it does not go as far through the gel as expected. In the absence of this effect, all of the very large DNA molecules would go through a gel at the same speed.

Several approximate models for the behavior of long DNA in gels with an alternating electric field have been described by Carlos Bustamante, Bruno Zimm, and others. One possible mechanism of this time-delay is shown schematically in Fig. 9.13. During the first 10 s at +100 V, the DNA proceeds from left to right at the expected speed. Studies by light microscopy show that the long DNA is sperm-shaped, with most of its mass concentrated into a 'head', while a small part follows as a 'tail'. The large head of DNA must force its passage between gel fibers, like an icebreaker or bulldozer, because there are few if any pre-existing passages in the gel large enough to accommodate it. This kind of motion of DNA through the gel does not depend much on DNA size. But when the voltage is reversed, to −100 V, the DNA must reorganize itself to create a new 'head' in the opposing direction. For a brief interval, the DNA is suspended over the gel fibers in motionless equilibrium, like a tie over a coat-hook. After this brief interval or 'time-delay', the DNA falls from its unstable position and forms a 'head' that can proceed in the reverse direction. The time required for the DNA to 'fall off' the coat-hook depends quite strongly on DNA size. The same time-delay will occur when the voltage is switched back to +100 V. This interesting and unexpected behavior of large DNA in gels was discovered by David Schwartz and Charles Cantor in 1984. It enables one to separate in size, by means of gel electrophoresis, DNA to a length of 5 or 10 million base-pairs. Thus, one can separate in a gel all the chromosomal DNA molecules from a simple organism such as yeast, where the chromosomes do not exceed 10 million base-pairs. But one cannot yet separate by this method the chromosomal DNA molecules from a human being, which are on the order of 50 to 400 million base-pairs in length.

The key to the separation of large DNA molecules in gels, by this method, is to adjust the times for the forward and reverse pulses of voltage to be of the same order as the time to 'turn around', for any given length of DNA. To a first approximation, the time required for a DNA molecule to 'turn around' in the gel increases as the 2/3 power of its size. This is not hard to understand in principle: short ties fall from coat-hooks much more frequently than long ties.

One can also understand why DNA larger than 5 to 10 million base-pairs cannot be separated in size: when a tie gets very long, say the distance from your floor to your ceiling, then the time required for it to fall from a coat-hook, by random vibration, will not depend very much on its length. Often one can improve the separation of such DNA in gels by fiddling with the voltages, or by reversing the voltage at an angle slightly less than 180° (say 120°); but the fundamental difficulty of separating 50 to 400 million-base-pair DNA in a gel remains. Whoever can solve that problem will be a great hero to molecular biologists, because he or she will for the first time make possible the detailed study and manipulation of human chromosomes. There seem to be two possible paths to a solution of this problem: (a) to change the structure of the gel so that its 'coat-hook' properties apply to larger DNA; or (b) to reduce the contour length of the DNA uniformly, perhaps by wrapping it around proteins, to make the DNA seem smaller relative to the structure of the gel. Here is an interesting, if extremely difficult, problem for a clever student.

So far we have described methods to study the properties of large numbers of DNA molecules in solution (or in a gel or a crystal). Any properties so measured will represent the average over many different individual values. For example, there will typically be 10^{12} DNA molecules within any gel sample described above, all of which will move through the gel along slightly different paths and at slightly different rates!

Yet scientists have recently developed other novel methods, which now permit the manipulation and detection of single DNA molecules in solution under near-physiological conditions. These single-molecule methods may be used, for example, to measure the local force needed to separate two strands of DNA; to study supercoiling of DNA when it is twisted; or to measure the forces required for packaging of DNA into the shell or 'capsid' of a bacterial virus.

Manipulation of single DNA molecules may be achieved with the help of micron-sized latex beads and micropipettes: the DNA is first attached to a latex bead, by a 'sticky tag' located at its very end, which then adheres to the surface of the bead. *Detection* may be achieved with the help of light microscopes and sensitive video-recorders, which can image the small beads (but not the DNA) and allow the path of the DNA to be deduced from the separation of the beads.

Several methods have been devised for applying small forces to the DNA. The earliest was to attach a double-stranded DNA molecule to the tip of a tiny glass needle, which then acts as an elastic cantilever beam as shown in Fig. 9.14 (a). Hence the force required to separate two strands of DNA may be deduced from the small elastic deflection of the glass needle, as it moves to the right in the picture. Another method is to attach double-stranded DNA to a tiny magnetic bead that is controlled by a magnet which can be rotated, as shown in Fig. 9.14(b). In that case, both tension as well as rotational torque may be applied to single molecules of DNA, allowing supercoiling to be studied.

Finally, the DNA may be attached to a tiny bead made from some material with high refractive index, such as latex. Then the bead can be held in place by intense laser light which has been focussed through a microscope lens: this is called an 'optical tweezer' or 'optical trap'. When a force is applied to the bead by tension in its attached DNA, the bead will be pulled elastically away from the laser light-focus, as if it were held there by an imaginary spring. Figure 9.14 (c) shows how this light-based method may be used to study the packaging of DNA into the shell or capsid of a bacterial virus.

Already scientists have measured these microscopic forces with sufficient accuracy to distinguish the separation of the two strands in A–T *versus* G–C rich DNA, and to study forces involved in binding DNA to protein. We expect that refinement of these techniques will have many other fruitful applications to DNA and DNA-protein interactions in the future.

Finally, we shall describe another technique that has been used in recent years to study the conformation of individual molecules

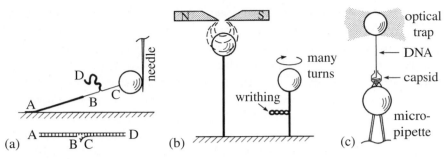

Figure 9.14 (a) Unzipping a single molecule of DNA. Double-stranded DNA with a break in one strand (lower diagram) was attached at end A to a microscope slide, but at end C to a small latex bead, which in turn was attached to a tiny glass needle. When that needle is moved to the right, the DNA becomes separated into two strands between B and C. The force is measured *via* the tip-deflection of the needle. (b) DNA attached to a magnetic bead can be twisted by means of a rotating magnetic field and made to writhe, if the tension imposed is sufficiently small – typically a fraction of 1 pico-newton (10^{-12} newton). On the left is the relaxed DNA, while on the right, after many turns, a portion of the DNA has writhed, as in Fig. 6.1(c). The length of DNA is about 10 000 base-pairs. (c) Packaging of DNA into a viral capsid. The capsid is attached to a bead held by a microscopic pipette, while one end of the DNA is attached to a bead held in an optical trap. The setup can be arranged to provide a constant force; and the rate at which the DNA is pulled by the tiny motor at the neck of the capsid can be measured. The length of the viral DNA is about 20 000 base-pairs.

attached to mica: Atomic Force Microscopy (AFM). In principle, this method involves dragging a hard, sharp probe over a flat specimen in a 'raster' of closely-spaced parallel straight lines in the x, y plane, and measuring the height z to which the probe must be raised in order to clear the object. Then a computer can process the measured $z(x, y)$ into a colour-coded contour map of the surface. In practice, the sharp probe – usually of silicon nitride – is mounted at the end of a fixed, elastic cantilever (of length about 50 μm); and the specimen is moved under it in three dimensions by piezo-drivers controlled by a computer. The specimen may be up to about 0.5 mm square, but normally much smaller areas are scanned.

AFM was first used to assay the variation of adhesive forces over the surface planes of metal or mineral crystals – hence the name; and it could also produce topographical contour maps of the surfaces, showing patterns of individual atoms. Assays of surface forces were achieved by measuring – with one of several possible methods – the minute deflection of the tip of the calibrated cantilever. But even if those forces are not of interest in a particular application, the deflection of the cantilever is still needed, as part of the control-loop for the z-direction movement of the specimen.

When AFM is used to visualize, say, the path of an individual DNA molecule attached to a mica surface, a number of special problems, which are not present with 'hard' specimens, have to be addressed. Thus, there is a tendency for the passing probe to dislodge the specimen, unless it has been stuck down by judicious application of magnesium to the mica base – where it forms a di-valent 'bridge' between the negatively charged DNA and the negatively charged mica. And in order to avoid damage to the 'soft' specimen, the sharp probe is usually oscillated vertically as it traverses, in so-called 'tapping mode', so that it taps across the specimen, something like a blind person tapping the ground with a stick. This procedure reduces lateral forces between the probe and the specimen, since the probe is in contact with the specimen for only a small fraction of the time.

The resolution that can be achieved with AFM on soft specimens such as protein and DNA typically can be as good as 1 Å vertically and 5 Å horizontally. A study of the wrapping of DNA around RNA polymerase using AFM and biochemical methods was mentioned in Chapter 6.

Other typical applications include the study of persistence length (a measure of elastic flexural stiffness) by using statistical analysis on the observed profiles of hundreds of DNA specimens; and investigation of the mode of action of restriction enzymes that break the DNA into pieces. An advantage of AFM is that it is equally effective on wet or dry specimens.

Note

1. See Appendix 1.

Further Reading

Brower-Toland, B.D., Smith, C.L., Yeh, R.C., Lis, J.T., Peterson, C.L., and Wang, M.D. (2002) Mechanical disruption of individual nucleosomes reveals a reversible multistage release of DNA. *Proceedings of the National Academy of Sciences, USA* **99**, 1960–5. The use of 'optical tweezers' to study the unravelling of single-molecule DNA from nucleosomes.

Berge, T., Jenkins, N.S., Hopkirk, R.B., Waring, M.J., Edwardson, J.M., and Henderson, R.M. (2002) Structural perturbations in DNA caused by bis-intercalation of ditercalinium visualised by atomic force microscopy. *Nucleic Acids Research* **30**, 2980–6. Use of atomic force microscopy to study supercoiled DNA-drug complexes.

Bustamante, C., Bryant, Z., and Smith, S.B. (2003) Ten years of tension: single-molecule DNA mechanics. *Nature* **421**, 423–7. An overview of the application of single-molecule methods to study the mechanical properties of DNA.

Calladine, C.R., Collis, C.M., Drew, H.R., and Mott, M.R. (1991) A study of electrophoretic mobility of DNA in agarose and polyacrylamide gels. *Journal of Molecular Biology* **221**, 981–1005. Application of the ideas of A.G. Ogston to DNA gel-running. Source of the photographs in Fig. 5.5.

Diekmann, S. and Wang, J.C. (1985) On the sequence determinants and flexibility of the kinetoplast DNA fragment with abnormal gel electrophoretic mobilities. *Journal of Molecular Biology* **186**, 1–11. The kinetoplast DNA curves naturally in a plane, giving Wr = 0; but it gives non-zero Wr when distorted by supercoiling, due to its curvature of about 1.3° per base-pair.

Drew, H.R. (1984) Structural specificities of five commonly-used DNA nucleases. *Journal of Molecular Biology* **176**, 535–7. Source of the S1 nuclease gel shown in Fig. 9.9.

Hagerman, P.J. (1985) Sequence-dependence of the curvature of DNA: a test of the phasing hypothesis. *Biochemistry* **24**, 7033–7. First proof that the slow gel-running of certain DNA sequences is due to their curvature; and a deduction that these sequences proceed through the gel as superhelices of definite shape.

James, T.L. (ed.) (1995) Nuclear magnetic resonance and nucleic acids. *Methods in Enzymology* **261**, 1–640. A comprehensive series of articles, describing current progress on the use of NMR to study DNA and RNA.

Keller, W. (1975) Determination of the number of superhelical turns in simian virus 40 DNA by gel electrophoresis. *Proceedings of the National Academy of Sciences, USA* **72**, 4876–80. Gel-running of supercoiled DNA in the absence or presence of ethidium bromide, an intercalating dye.

Kozulic, B. (1995) Models of gel electrophoresis. *Analytical Biochemistry* **231**, 1–12. An excellent, knowledgeable and critical review of various models for DNA gel motion.

Leuba, S.H., Karymov, M.A., Tomschik, M., Ramjit, R., Smith, P., and Zlatanova, J. (2003) Assembly of single chromatin fibers depends on the tension in the DNA molecule: magnetic tweezers study. *Proceedings of the National Academy of Sciences, USA* **100**, 495–500.

Low, C.M.L., Drew, H.R., and Waring, M.J. (1984) Sequence-specific binding of echinomycin to DNA: evidence for conformational changes affecting flanking sequences. *Nucleic Acids Research* **12**, 4865–79. Source of the footprinting gel shown in Fig. 9.8.

Marini, J.C., Levene, S.D., Crothers, D.M., and Englund, P.T. (1982) Bent helical structure in kinetoplast DNA. *Proceedings of the National Academy of Sciences, USA* **79**, 7664–8. First indication that the slow gel-running of certain DNA sequences might be because they are curved.

Ogston, A.G. (1958) The spaces in a uniform random suspension of fibers. *Transactions of the Faraday Society* **54**, 1754–7. The classic work on the application of random statistics to gel-running phenomena, on which almost all later studies are based.

Rivetti, C., Guthold, M., and Bustamante, C. (1999) Wrapping of DNA around the *E. coli* RNA polymerase open-promoter complex. *EMBO Journal* **18**, 4464–75. Use of atomic force microscopy to map protein/DNA conformations.

Rodbard, D. and Chrambach, A. (1971) Estimation of molecular radius, free mobility and valence using polyacrylamide gel electrophoresis. *Analytical Biochemistry* **40**, 95–134. Application of Ogston's work to protein gel-running.

Sayre, A. (1975) *Rosalind Franklin and DNA*. W.W. Norton and Company, New York. The story of the discovery of a double-helical structure for DNA, told from the perspective of Rosalind Franklin, who collected the critical X-ray data from fibers.

Smith, S.B., Heller, C., and Bustamante, C. (1991) Model and computer simulations of the motion of DNA molecules during pulsed-field gel electrophoresis. *Biochemistry* **30**, 5264–74. A good theoretical study of alternating-field electrophoresis.

Whytock, S. and Finch, J.T. (1991) Substructure of agarose gels as prepared for electrophoresis. *Biopolymers* **31**, 1025–8. Source of Fig. 9.10.

Zimm, B.H. and Lumpkin, O. (1993) Reptation of a polymer chain in an irregular matrix: diffusion and electrophoresis. *Macromolecules* **26**, 226–34. A reptation theory for DNA motion through gels, that includes friction between DNA and the gel fibers.

Bibliography

Avery, O.T., MacLeod, C.M., and McCarty, M. (1944) Induction of transformation by a desoxyribonucleic acid fraction isolated from *Pneumococcus* Type III. *Journal of Experimental Medicine* **79**, 137–58. The classic work implicating DNA as the substance of genes.

Bernal, J.D. (1967) *The Origin of Life*. Weidenfeld & Nicolson, London. A good discussion of pre-biotic evolution and structures in biology, by one of the founders of the science.

Crawford, J.L., Kolpak, F.J., Wang, A.H.-J., Quigley, G.J., van Boom, J.H., van der Marel, G., and Rich, A. (1980) The tetramer d(CpGpCpG) crystallises as a left-handed double helix. *Proceedings of the National Academy of Sciences, USA* **77**, 4016–20. The sequence CGCG in a low-salt crystal form.

Dekker, N.H., Rybenkov, V.V., Duguet, M., Crisona, N.J., Cozzarelli, N., Bensimon, D., and Croquette, V. (2002) The mechanism of type IA topoisomerases. *Proceedings of the National Academy of Sciences, USA* **99**, 12126–31. Use of an attached magnetic bead to produce writhing in supercoiled DNA under tension. Source of Fig. 9.14(b).

Drew, H.R., Takano, T., Tanaka, S., Itakura, K., and Dickerson, R.E. (1980) High-salt d(CpGpCpG), a left-handed Z' DNA double helix. *Nature* **286**, 567–73. The sequence CGCG also goes left-handed in a high-salt crystal form.

Essevaz-Roulet, B., Bockelmann, U., and Heslot, F. (1997) Mechanical separation of the complementary strands of DNA. *Proceedings of the National Academy of Sciences, USA* **94**, 11935–40. Use of a flexible glass needle to measure the force required to separate the strands of double-helical DNA. Source of Fig. 9.14(a).

Franklin, R.E. and Gosling, R.G. (1953) Structure of sodium thymonucleate fibres: importance of water content. *Acta Crystallographica* **6**, 673–7. First identification of the 'B' and 'A' types of DNA in fibers.

Fuller, W., Wilkins, M.H.F., Wilson, H.R., Hamilton, L.D., and Arnott, S. (1965) The molecular configuration of deoxyribonucleic acid: X-ray diffraction study of the 'A' form. *Journal of Molecular Biology* **12**, 60–80. A refined model for the 'A' form of DNA.

Itakura, K. and Riggs, A.D. (1980) Chemical DNA synthesis and recombinant DNA studies. *Science* **209**, 1401–5. An early, important survey of the advances made in biology by the chemical synthesis of DNA.

Kozulic, B. (1994) On the 'door-corridor' model of gel electrophoresis. *Applied and Theoretical Electrophoresis* **4**, 117–59. An improved two-phase model for the motion of macromolecules through gels, which includes deformation of the gel by a migrating molecule.

Langridge, R., Marvin, D.A., Seeds, W.E., Wilson, H.R., Hooper, C.W., Wilkins, M.H.F., and Hamilton, L.D. (1960) The molecular configuration of deoxyribonucleic acid: molecular models and their Fourier transforms. *Journal of Molecular Biology* **2**, 38–64. A refined model for the 'B' form of DNA.

Leslie, A.G.W., Arnott, S., Chandrasekaran, R., and Ratliff, R.L. (1980) Polymorphism of DNA double helices. *Journal of Molecular Biology* **143**, 49–72. The wide variety of double-helical forms 'A, B, C, D, E' and their variants as seen by fiber X-ray diffraction.

Patel, D.J., Pardi, A., and Itakura, K. (1982) DNA conformation, dynamics and interactions in solution. *Science* **216**, 581–90. A review of some of the earliest work by NMR on large synthetic DNA. Source of parts of Fig. 9.4.

Pohl, F.M. and Jovin, T.M. (1972) Salt-induced cooperative conformational change of a synthetic DNA: equilibrium and kinetic studies with poly (dG-dC). *Journal of Molecular Biology* **67**, 375–96. The left-to-right-handed transition of DNA in solution.

Schwarz, D.C. and Cantor, C.R. (1984) Separation of yeast chromosomal-sized DNAs by pulsed-field gradient gel electrophoresis. *Cell* **37**, 67–75. The discovery that alternating electric fields can separate very long DNA in gels according to size.

Smith, D.E., Tans, S.J., Smith, S.B., Grimes, S., Anderson, D.L., and Bustamante, C. (2001) The bacteriophage ϕ29 portal motor can package DNA against a large internal force. *Nature* **413**, 748–52. A study demonstrating how optical tweezers may be used to measure the force required to compact DNA into a bacteriophage capsid. Source of Fig. 9.14(c).

Viswamitra, M.A., Kennard, O., Jones, P.G., Sheldrick, G.M., Salisbury, S., Falvello, L., and Shakked, Z. (1978) DNA double-helical fragment at atomic resolution. *Nature* **273**, 687–8. The crystal structure of ATAT, which 'melts' at the central TA step.

Wang, A.H.-J., Quigley, G.J., Kolpak, F., Crawford, J.L., van Boom, J.H., van der Marel, G., and Rich, A. (1979) Molecular structure of a left-handed double-helical DNA fragment at atomic resolution. *Nature* **282**, 680–6. First visualisation of left-handed DNA in a sequence CGCGCG.

Wüthrich, K. (1995) *NMR in Structural Biology*. World Scientific Publishing, Singapore. Methods for analyzing structures of biological molecules in solution.

Exercises

9.1   In an electron-density map such as that shown in Fig. 9.3, which of the atoms should scatter X-rays most strongly?

  a   Rank in order of scattering power, from highest to lowest: carbon, phosphorus, oxygen, hydrogen, nitrogen.

  b   Which part of the electron-density map in Fig. 9.3 was produced by the strongest scattering of X-rays? Use Figs 2.8(b) and 2.11(a) to identify the atom types.

  c   Why are the hydrogen atoms in the Watson–Crick base-pairs not seen in Fig. 9.3?

9.2   Referring to Fig. 9.4, how many NMR peaks would you expect to find in the spectral region shown, and at –5°C, for double-helical DNA molecules which are specified by the following single-strand sequences?

  a   (5′) CGCAATTGCG (3′)

  b   (5′) AGCATGCATGCT (3′)

  c   (5′) AGCATGCGCG (3′)

   In each case, first construct the two-stranded version of the molecule, and then look for two-fold symmetry that may make some peaks equivalent to others.

9.3   Suppose we have a DNA molecule of length 1000 base-pairs, which contains three cutting sites, GAATTC, for a particular restriction enzyme; and suppose that these sites are located at 100, 350, and 550 base-pairs, respectively, from one end.

  a   On complete digestion of the DNA by the restriction enzyme, how many kinds of DNA fragment will be produced, and of what size?

  b   Which fragment will run fastest in an ordinary electrophoretic gel, and which will run slowest?

9.4a   Suppose that you have the DNA from a virus, consisting of 100 000 base-pairs, and you wish to determine its complete nucleotide sequence. Find, approximately, the smallest number of fragments that you could sequence individually, by use of present-day gel technology, in order to complete this task. Don't count the many overlapping fragments which would be needed to align partial sequences.

b   A typical human cell contains 6×10^9 base-pairs of DNA, located on 23 pairs of chromosomes. What is the smallest number of fragments which you could sequence individually, in order to carry out the gigantic task of sequencing the complete DNA from a human cell? Again, don't count overlaps.

9.5   Figure 9.12(a) shows a flat disc migrating across a plane containing randomly-spaced point obstacles: bigger discs will contact more obstacles, and hence go through the gel more slowly. Similar concerns govern the motion of DNA through a gel in three dimensions. As an overall rule, one expects that flat discs in two dimensions with the largest areas, or DNA cylinders in three dimensions with the largest volumes, will contact the most gel fibers and hence go most slowly.

One can estimate the volume of the DNA cylinder for a straight piece of DNA, by modeling it as a cylinder of radius $10\,\text{Å}$ and length $3.3\,\text{Å}$ per base-pair. But for a curved, superhelical piece of DNA, one has to consider the volume of a 'circumscribing' cylinder into which the superhelix can just fit.

Find the apparent volume for one superhelical turn of curved DNA with the repeating sequence A_6N_4, using the following parameters for its shape: superhelical radius $r = 18.1 \times 3.3\,\text{Å}$, contour length $N^* = 195.4 \times 3.3\,\text{Å}$, and pitch angle $\alpha = -54.5°$ (see Table 5.1). First, find the pitch height p from $N^*\sin\alpha$ as shown in Figure 5.4, and confirm that $2\pi r$ equals $N^*\cos\alpha$. Then add $10\,\text{Å}$ to the superhelical radius r, in order to account for half the thickness of the DNA itself. Finally, get the volume of the circumscribing cylinder from $\pi(r + 10)^2 p$.

Give your result as the ratio of two volumes, for curved DNA *versus* straight DNA of the same length N^*. Now repeat the same calculation, but using other curved sequences from Table 5.1, i.e. A_6N_2, A_6N_3, A_6N_5, A_6N_6, and A_6N_7. Which of these DNA sequences will go the most slowly through a gel, and which the most rapidly?

9.6   As explained in the text, closed circular DNA molecules run at a variety of speeds in a gel. That is because different linking numbers Lk can produce different interwound shapes, as shown, for example, in Figs 6.4(e) and 6.5(e) for Lk $= \pm 3$ or in Figs 6.4(a) and 6.5(a) for Lk $= 0$. Now it turns out that specimens with an open circular form (Wr $= 0$) run most slowly through a gel, while specimens with higher (whether positive or negative) values of Wr run more rapidly, since they are more compact and so present a smaller effective volume to the interfering gel fibers which retard forward motion.

This pattern of behavior enables one to pick out specimens having Lk = 0 from a gel; but one cannot easily pick out a specimen having, say, Lk = +3 from one with Lk = −3, because they both run at very similar speeds. Also, one cannot easily distinguish large values of Lk, say +19 and +20, from each other, because those molecules are not sufficiently different in shape to run at significantly different speeds.

Scientists get around these problems by a technique which involves the addition of ethidium bromide or some other dye (such as chloroquine phosphate) to the gel, so that in any given specimen the DNA untwists, thereby acquiring a more positive value of Wr, since Lk does not change.

Given that the addition of a certain amount of ethidium bromide imparts exactly Tw = −12 to all molecules in a sample, predict the values of Lk and Wr for the slowest-running specimen in the range Lk = 0 to −20. Which will be the fastest-running specimen?

(Use the equation on p. 122, and assume that the slowest molecule always has Wr = 0, while the fastest molecule has maximal Wr, whether positive or negative.)

CHAPTER 10

DNA in Disease, Diagnostics, and Medicine

In the previous chapters, we have explained as simply as possible how DNA works in biology, according to various aspects of its three-dimensional structure. For example, the local pairing of bases in a Watson–Crick fashion across the double helix, as A with T or G with C, allows a DNA molecule to be copied from generation to generation, providing the general mechanism of inheritance for all life on Earth. Also, we have seen repeatedly how the structural properties of DNA as conferred by the sequence itself may be used in biology. For instance, the preferred unwinding of a double helix at sequences of the kind TATA defines many (but not all) of the start-sites for making more DNA or RNA, during replication or transcription. Again, the preferred bending of a double helix into the minor groove at certain sequences (e.g. AAA/TTT), or into the major groove at others (e.g. GGC/GCC), encourages the precise packaging of DNA as it curves around histone proteins in chromosomes, or as it curves about repressor and activator proteins that control transcription.

There is still much more to learn about DNA, than what we have explained so far. Unfortunately, however, not everything about the workings of DNA in biology is currently understood, especially in complex organisms such as humans, animals or plants. So while we cannot provide a comprehensive, authoritative account of how DNA works in biology, we can offer a snapshot of current understanding in a field that is growing rapidly. If you could sleep for a hundred years, like Rip van Winkle, you might awake to a world in which DNA and biology were understood completely and deeply, and you might find that this knowledge was being applied in ways that are unimaginable at the moment.

Here we shall survey in a concise way what is known about the workings of DNA at a medically relevant level in complex organisms such as man or mouse, subject to the limitations of knowledge as emphasized above. Our survey will cover four areas of research:

(a) important methods which have been developed since 1985, such as transgenics, the polymerase chain reaction (PCR), di-deoxy four-colour sequencing, oligo-nucleotide microarrays, and polymerase extension to measure point-mutations (SNPs);

(b) errors in DNA structure associated with human disease, for example certain point-mutations, frameshift-mutations and tri-nucleotide expansions;

(c) a type of genetic inheritance called 'imprinting', whereby the activity of a gene depends upon whether you have inherited it from mother or from father; and

(d) attempts to fix errors in human DNA that cause disease, by adding new full-length genes of the correct kind, or else by adding new DNA which may repair pre-existing genes: these strategies are known as 'gene therapy' and 'gene correction', respectively. Yet another strategy to suppress gene expression uses short double-stranded RNA; such techniques will be described in Appendix 3.

However, before proceeding to a survey of such diverse subjects, we need to explain why scientists are devoting so much effort to the study of DNA and its associated genes at a detailed level in complex organisms.

Many scientists and physicians in the early 21st century have come to the conclusion that traditional forms of medicine are inherently inadequate for dealing with certain ailments. Those traditional forms include surgical operations, where the doctor cuts away infected tissue with a knife and then repairs the damage by sewing, until nature has healed the wound – a dangerous and lengthy procedure; and the prescription of some small-chemical drug which may be ingested, inhaled, or injected into the body. Usually, the doctor hopes that such a drug will spread quickly throughout the body, so as to find some specific biological target, often either an enzyme or a signalling protein. By binding tightly to its specific target, such a drug may then inhibit the aberrant biological function of the enzyme or the signalling protein, in order to cure a disease, or at least to slow its progress.

We now know that surgical approaches are unlikely to cure most forms of cancer, or to cure infection by viruses such as hepatitis or HIV (which causes AIDS), since the cancerous cells or deadly viruses may spread rapidly throughout the body so as to become essentially inoperable. Furthermore, a cancer-causing cell often looks much like an ordinary human cell on a molecular scale; and it seldom offers any unique biological target, against which a small-chemical drug may specifically act. This difficulty in distinguishing normal cells from cancerous and virus-infected cells also makes it hard to treat certain diseases through immunization – that is, the activation of the body's natural immune response. And most chemotherapeutic agents that target DNA have powerful but adverse side-effects on healthy cells.

By contrast, the recent development of many new and exciting techniques in molecular biology has allowed scientists and physicians to do new kinds of experiment, which may provide for entirely new forms of medical therapy in the future – as useful say as the discovery of immunization in 1800, or the discovery of penicillin in 1930. As the first part (a) of our survey, let us describe briefly some of the more important techniques which have been developed since 1985, and which have made new kinds of medicine possible, if not always practical, by 2004.

First, scientists can often determine the function of any normal human gene, or else test if a mutant human gene is defective, by injecting the complete DNA for such a gene into a mouse embryo, as shown in Fig. 10.1. The injected DNA will then be incorporated stably into a mouse chromosome by a process called 'illegitimate recombination', which is essentially a semi-random joining of injected human DNA onto pre-existing mouse DNA at rare chromosomal breaks. Thus, wherever the chromosomal DNA of a mouse breaks, infrequently and at random places in an egg cell, it will be repaired by several different enzymes, which usually include a 'ligase' or DNA-joining enzyme to re-connect the broken ends. Any injected human DNA may then take part in such mending, if each end of the human DNA is joined by a repair-ligase enzyme onto both broken ends of pre-existing mouse DNA.

When a mouse with this extra human DNA grows to an adult, it may show differences in physiology or behavior relative to a normal

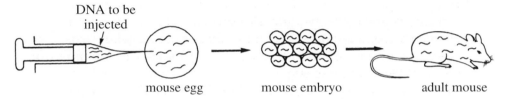

DNA to be injected

mouse egg   mouse embryo   adult mouse

Figure 10.1 New DNA can be added in a permanent fashion to mice or other animals or plants, in order to provide extra genes for these organisms, or else to impair the function of pre-existing genes.

mouse. For example, some genetically-altered mice become especially fat, or grow to a larger size than normal, because their genes have been altered slightly. The additional human DNA may introduce new genes, or else it may impair the function of existing genes.

These altered animals are known in general as 'transgenic' or 'knockout' mice, where certain genes have been added to or deleted from, respectively, the normal mouse set of chromosomes. Often very small changes to the total DNA of an organism will produce large changes in its physical appearance, behavior, or intelligence. Transgenic mice have proved useful for creating animal models of human disease, for instance: prostate cancer, thyroid deficiency, obesity, lateral sclerosis, or Alzheimer's disease. Such genetically-altered mice may also be used to test new drugs which might potentially cure disease in humans.

Similar kinds of transgenic experiment are now being conducted extensively worldwide to make 'improved' animals or plants. For example, various genes have been added to pigs; not only to make them grow fatter or faster for agricultural purposes, but also (and somewhat incredibly) to modify cell surfaces along their bodily organs, so that pig organs will not be rejected by the human immune system, if they are transplanted by surgery into human patients! Genetically-modified pig hearts have already been tested successfully in monkeys, as a step before giving them to humans who might need an organ transplant.

Progress has also been made at cloning adult pigs, sheep and cows from their mature cellular tissues, rather than by normal reproduction using sex cells. Using a microscope, the scientist simply transfers by micro-dissection (or by disruption of the membranes by an electric field, a process known as 'electroporation') a nucleus from a mature cell into an empty egg-cell, while trying not to disrupt any of the mature chromosomes along the way.

Additions of extra DNA to plants have provided for fruit or vegetables of improved quality, for example: tomatoes or strawberries which ripen more slowly than normal; cotton which is more resistant to insects; rice with extra iron as a nutritional supplement; strawberries with extra vitamin C as an antioxidant supplement; and even carrots which carry a measles protein, as a possible and inexpensive way of 'vaccinating' people against the measles virus.

By a second new class of techniques, scientists can now amplify and determine the base sequence of practically any DNA or RNA molecule which might be found within a cell, starting from just tiny amounts of material (say 1 to 100 individual molecules). This means that they can easily isolate new, full-length genes of therapeutic or industrial interest. Also, they can quickly detect aberrant changes in

DNA or RNA sequence as a diagnostic for cancer, or for genetically-inherited diseases. Finally, they can detect low-level infection by micro-organisms such as viruses, bacteria or mycoplasmae; because such micro-organisms will add their own distinctive DNA to the cellular mix, which would not be present in a healthy individual. The genetic composition of the micro-organisms can also be ana-lyzed for drug-resistance profiles, which permits the tailored use of theraupeutic compounds such as antibiotics.

These new amplification and sequencing methods have proven useful as well for the identification of individuals in police forensic work. Some of the DNA within our chromosomes is quite specific to certain individuals, just as are fingerprints. Hence those highly-variable parts of human DNA (known as 'micro-satellites') may be used to identify suspects in a criminal case, from trace amounts of blood, hair, semen or skin left at the scene of a crime.

How exactly do these new amplification methods work? Why are they so sensitive to small amounts of starting material, yet so easy to carry out? In the past, in order to determine the base sequence of any small fragment of DNA, scientists first had to 'clone' that fragment into a circular carrier-molecule or 'plasmid'. Next, the plasmid and its extra DNA would be copied many times over, by the process of DNA replication within a fast-growing culture of bacterial cells. Finally, a large culture of cells would yield large amounts of the desired plasmid, from which the extra DNA insert could be obtained in sufficient amounts, to sequence by chemical or enzymatic means: see Fig. 10.2(a).

That carrier-plasmid method is still used today, when tens of micrograms or even milligram quantities of DNA are desired. Yet it was never reliable for cloning trace amounts of DNA, on the order of 1 to 100 molecules. Futhermore, it was always quite tedious to grow such carrier-plasmids overnight in bacterial cells, and then to extract and purify their small DNA insert, for sequence analysis by chemical or enzymatic means.

Hence the major breakthrough was to do all of that copying and amplification of DNA outside of the cell and in a test-tube. A schematic overview of the new, improved method is shown in Fig. 10.2(b). To any single DNA molecule in solution, one simply adds a heat-stable polymerase enzyme (e.g. *Taq* as taken from the bac-terium *Thermus aquaticus*, which grows in hot springs); along with two short pieces of chemically-synthesized DNA which act as 'primers' for the polymerase; plus an abundance of four nucleotide *tri*-phosphates A, T, C, and G. So far in this book, we have described only the nucleotide *mono*-phosphates, which are the units that make up a completed DNA chain. But the preliminary units that

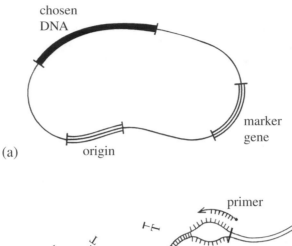

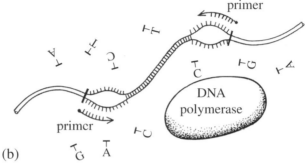

Figure 10.2 (a) Any small piece of DNA may be ligated into a carrier plasmid, that may be prepared on a large scale if bacterial cells which contain the plasmid divide and copy all of their DNA repeatedly. Most plasmids contain an 'origin of replication' to assist in DNA copying by polymerase enzymes; and also a 'marker gene' for resistance to some antibiotic such as ampicillin, chloramphenicol, kanamycin or tetracyclin. This marker gene ensures that all cells will contain the plasmid, when grown in a medium that includes antibiotic. (b) Any small piece of DNA may also be prepared on a large scale, outside of the cell and in a test-tube, by copying it repeatedly with heat-stable DNA polymerase enzymes – a process known as the polymerase chain reaction or 'PCR'. Two short pieces of chemically-synthesized DNA of selected base sequence may be chosen as 'primers', to tell the polymerase where to start copying along each strand.

the polymerase adds successively to any growing chain actually contain *three* phosphate groups, two of which are cut off during polymerization.

When the starting DNA molecule is heated strongly at 95°C, so as to separate its two strands, and then cooled to 55°C, the two short pieces of 'primer' DNA will bind or 'anneal' specifically at each end of the long, starting DNA molecule. Those primers will then form two specific binding-sites for the polymerase enzyme, with one adhering to each of the separated strands. Such double-stranded complexes between short primer DNA and long template DNA will next inititate the synthesis of more double-stranded DNA, when the reaction is warmed to 72°C in the presence of *Taq* DNA polymerase and the four deoxy-nucleotides A, T, C, and G. Those deoxy-nucleotides will rapidly 'extend' or add sequentially to the 3'-end of each primer, until the 5'-end of each strand of the template DNA is reached.

Hence, our heat-stable polymerase enzyme will make a new, full-length double helix from each of the two single strands which were present in the original sample after heating. Now when those two double-stranded products of the polymerase reaction are heated again to 95°C, then cooled to 55°C and warmed to 72°C, they will yield four single strands in total, which may in turn act as templates for further DNA synthesis. Next, when those four products of the second polymerase reaction are heated and cooled and warmed, they will yield eight single strands which may serve as templates for making more DNA, and so on. After 20 cycles of heating-cooling-warming and polymerase action, our starting DNA molecule will have been amplified from just one to nearly 10 million copies (i.e. by a factor of 2^{20}), or from picogram to microgram amounts.

Because such cyclic amplification of DNA in the test-tube resembles a chain reaction in physics, this new method is called the 'polymerase chain reaction' or 'PCR'. It reminds one of the old story about a king who loses his kingdom, by promising a loyal servant one kernel of corn for the first space on a chessboard, then two kernels of corn for the second space, four kernels for the third space, and so on for all 64 spaces on the board. By the twentieth space, the king has to give his servant more than a million kernels of corn; or by the fortieth space more than a million million, or enough to fill several large ships!

The two primers for each PCR reaction must be chosen carefully, so that they will each contain a short sequence of bases (typically 15 to 30) which binds specifically to two unique parts of the total chromosomal DNA for any organism, using Watson–Crick base-pairs. The template fragments of DNA which may be amplified by this method range in length from several hundred to several thousand base-pairs; or even to tens of thousands of base-pairs using a special method known as 'long PCR'.

This PCR method was the first to be developed, around 1985, and is still the most versatile; yet other useful refinements to methods of test-tube DNA amplification have been developed recently. Two of these are called SDA ('strand displacement amplification') and MDA ('multiple displacement amplification'). Neither of these requires any heating or cooling cycles; they can be performed isothermally at 55°C or 30°C respectively, which may offer some advantages under certain circumstances.

Next, following amplication of some biologically relevant DNA to microgram amounts by the PCR method (or the carrier-plasmid method, as described above), one would traditionally analyze its base sequence by chemical or enzymatic means; and then display or 'read' that base sequence using an electrophoretic polyacrylamide-urea gel, as discussed in Chapter 9. However, most scientists today use a third general class of methods: they send their amplified DNA to a commercial sequencing facility, where special machines are able to analyze hundreds or even thousands of different DNA fragments overnight.

The basic idea behind this procedure is to create, for each sample supplied by the user, a large number of DNA molecules of different length, and then to separate them out by electrophoresis, as explained in Chapter 9. Suppose, for the sake of definiteness, that a sample of 500 base-pairs is to be sequenced. Then the plan is to create a range of samples starting at the same end, and of length 1, 2, 3, ... up to 500 base-pairs. Each of these samples must end or 'terminate' in one of A, T, C or G; and if the end-letter can be identified on the gel 'ladder', then the sequence can, in principle, be read off.

But how can the various samples be made to stop at all of these different lengths? And how can the end-letter be identified? The key to the situation is to look at the DNA's sugar–phosphate 'backbone', as shown in Fig. 2.8(b). There we saw the chemical structure of a piece of backbone containing bases and sugar rings, joined by phosphate groups. Now the central picture of Fig. 10.3(a) shows one base in the same sort of backbone, but with two carbon atoms of the sugar ring labelled 2′ and 3′, respectively. Also, on the left is shown the corresponding portion of an RNA backbone; the only difference being that the RNA sugar ring has an OH group attached at position 2′. DNA is called *Deoxy*-ribo-Nucleic Acid precisely because it lacks this oxygen, which is a feature of Ribo-Nucleic Acid's sugar ring.

Finally, look at the right-hand picture of Fig. 10.3(a). This shows the special *di-deoxy* form of the sugar ring: it has no oxygen atom either at position 2′ or at position 3′. In this case the lower phosphate group is not shown, precisely because it cannot attach itself to the ring if an oxygen is absent from position 3′. In other words, the sugar–phosphate chain cannot continue to extend further, once a di-deoxy base has been added by the polymerase enzyme.

In this method, therefore, the surrounding medium in which the polymerase operates contains not only the usual C, A, T and G deoxy nucleotides, just as in the PCR method, but also some of these special di-deoxy nucleotides of all four kinds. And like PCR, the sequencing process requires a short 'primer' DNA for the polymerase to start the copying. Thus, when the polymerase is proceeding along a

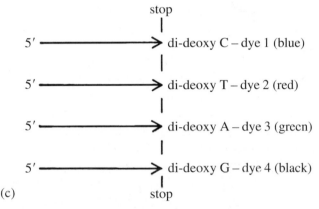

Figure 10.3 Key concepts from the di-deoxy, four-colour method of DNA sequencing: (a) comparison of three forms of nucleotide on an atomic scale; (b) comparison of chain extension by 2′-deoxy cytosine *versus* 2′,3′-di-deoxy cytosine under the action of DNA polymerase; the chain extends normally if 2′deoxy-C is added (and pairs with the guanine on the template), but stops or terminates if 2′,3′-di-deoxy-C is added; (c) fluorescent dyes of four different colours can be attached to the four different di-deoxy nucleotides C, T, A and G, so that any particular product of chain termination will show just one color by capillary electrophoresis.

particular DNA molecule it will occasionally, and at random, add a di-deoxy nucleotide; and the chain-building will abruptly termi-nate right there. This is shown schematically in Fig. 10.3(b): the poly-merase will stop if a di-deoxy C happens to be added, but not if an ordinary, deoxy C is put there. Now in a large sample, we can rely on this random event – the addition of a di-deoxy nucleotide – occurring at every possible stopping-point between 1 and 500, and hence a full set of truncated DNA molecules being created.

The only problem remaining is to identify the four kinds of di-deoxy nucleotide. This is done by chemically modifying these nucleotides with fluorescent dyes of different colour – typically rhodoamine or fluorescein derivatives – as shown in Fig. 10.3(c). These dyes are attached to the 7-position of A and G or to the 5-position of C and T, through long flexible linkers, so that these large and bulky compounds do not interfere with the ability of the polymerase to add each unit to the end of the terminated chain (see Fig. 11.1 for the numbering convention of the 7- and 5-positions). After a mixture of new chains has been made, the DNA-sequencing machine will separate them by size, using capillary electrophoresis. Separation of DNA chains by size in a capillary filled with some gel-like polymer, is precisely analogous to separating DNA fragments by size using a large polyacrylamide or agarose gel; but for a tiny capillary much higher voltages may be applied (without excess heating), which speeds the process greatly.

Finally, the DNA-sequencing machine uses a fluorescent light-source and detector, to look for the distinctive 'colour' which is associated with the last base of each chain , as it flows through the capillary under the influence of an electric field.

Some typical results of di-deoxy, four-colour DNA sequencing are shown in Fig. 10.4. Part (a) shows the partial sequence of a PCR-derived fragment. It reads TGGAA-GTCGT-TTGGC-TTGTG-GTGTA-ATTGG-G going left to right across the page. C (cytosine) bases are shown in blue; T (thymine) bases are shown in red; A (adenine) bases are shown in green; and G (guanine) bases are shown in black. Below each base is a curve which tells how strongly each di-deoxy-fluorescent signal was detected, when any particular DNA fragment proceeded through the capillary under the influence of an electric field.

Figure 10.4(b) shows a similar result for a PCR-derived fragment which is slightly different from that of (a). Here a single location 119 on the right-hand side has been read by the computer as 'N' to signify 'uncertain base'. Such a result could be due conceivably to experimental error. However, in this case we know that the PCR-derived fragment shown in (b) contains a mixture of two different bases at position 119, namely T (red) and G (black), owing to a genetic mutation on one of a patient's two homologous chromosomes. What a boon for medical science, to be able to detect single-copy (i.e. heterozygous) mutations in DNA which can cause disease, within a hospital laboratory in just a short period of time, from inconsequential amounts of a patient's blood!

TG G A A G T C G T T T G G C T T G T G G T G T A AT T G G G
  100   110   120

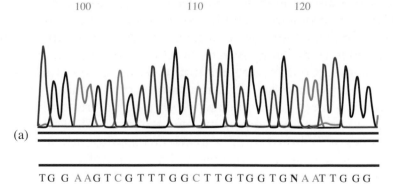

(a)

TG G A A G T C G T T T G G C T T G T G G T G N A AT T G G G
  100   110   120

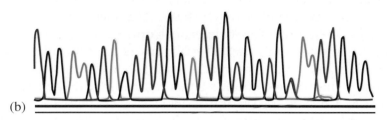

(b)

Figure 10.4 Two examples of di-deoxy, four-colour sequencing using PCR fragments as amplified from the DNA of two human patients. The sequencing plot shown in (a) comes from a healthy individual, whereas the sequencing plot shown in (b) comes from an individual affected with the neurological disease Amyotrophic Lateral Sclerosis or 'ALS'. A point-mutation, shown by 'N' on the right-hand side of (b) at position 119, changes 'T' on both homologous chromosomes to 'T' on one chromosome but 'G' on the other; and thereby changes amino-acid 148 in the protein 'superoxide dismutase' from valine to glycine, so as to produce this ailment. Courtesy of G. Nicholson.

In fact, the four-colour method for determining the base sequence of DNA is so powerful and fast, that it has enabled the full sequence determination of many different bacterial, plant and animal genomes, including human. Such data provide an excellent 'map' of the genome to facilitate genetic or medical research. Most of these fast-accumulating results (20 million bases per day at large facilities) can be accessed *via* the Internet at **www.ncbi.nlm.nih.gov**. Who would have thought, in 1960, that medical and biological scientists would spend much of their time accessing DNA sequences from a worldwide computer network? In the future, those extensive sequence data may provide for new cloned proteins as therapeutic compounds; already they have proven useful for the diagnosis of many kinds of disease.

The di-deoxy, four-colour method represents the most common means by which scientists today analyze their DNA sequences, as obtained from PCR or carrier-plasmid amplification. Yet it should be stressed that a fourth general class of methods is currently under development, which may even replace PCR and four-colour sequencing to some extent in the near future. Thus, many workers have made various kinds of DNA microarray or 'biochip', which can detect any short DNA sequence which may be present in a mixed chromosomal DNA sample, by means of specific Watson–Crick base-pairs or 'hybridization'; and which can also measure the relative abundance of some particular DNA or RNA molecule, in a sample which may contain all of the different DNA or RNA molecules found in a living cell.

As shown schematically in Fig. 10.5(a), those workers first prepare a glass slide or nylon filter, to which they attach chemically many different single-stranded DNA oligo-nucleotides of varied sequence in a regular microscopic array. The oligomers are typically about 25 nucleotides long. Next, they add a sample of cellular DNA or RNA, melted so that its two strands separate, and allow the two single strands to form specific Watson–Crick base-pairs with any matching oligo-nucleotides on the chip. Usually one adds the cellular sample under stringent conditions such as high temperature, and in the presence of moderate amounts of urea or formamide, to select against non-Watson–Crick base-pairs during the binding step. Finally, if each DNA or RNA sample is labelled beforehand with a fluorescent dye (or radioactive atom), by either enzymatic or chemical means, then each cell-derived sample will give a reproducible pattern of fluorescent (or radioactive) light signals on the biochip, which may be recorded under a light microscope (or by a phosphor screen).

Oligonucleotide microarrays have been useful to some extent for determining variations of base sequence between DNA from different individual samples (i.e. 'genotyping'); but they have been more useful for measuring the relative amounts of messenger-RNA present in any living cell (i.e. 'expression'). Thus, researchers can now compare the relative transcription of DNA into RNA for thousands of different genes in any complex set of biological samples: say before and after a patient has been given a therapeutic drug, or before and after certain cells have turned cancerous.

By a fifth new class of techniques, scientists have been able to speed up the process of medical diagnostics, by looking just for certain naturally occuring genetic variations in people. These are small mutations and are sometimes known as 'SNPs' (single nucleotide polymorphisms) which may sometimes be related to human disease. Once a point-mutation or SNP is determined from sequencing

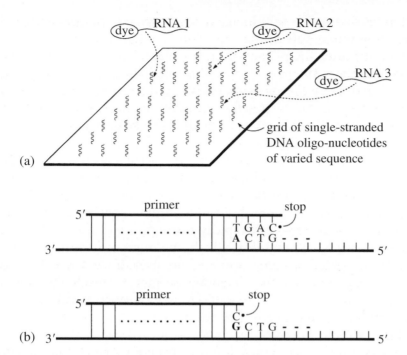

(a)

(b)

Figure 10.5 (a) A schematic illustration of how many different single-stranded DNA oligonucleotides may be attached in a regular microscopic array to a surface, so as to create a microarray or 'biochip' to which long cellular DNA or RNA molecules may bind in a specific fashion, and be detected through attached dyes. Here, only a small lattice is shown, but in practice the arrays may contain thousands of elements. (b) A hypothetical example of SNP determination: DNA polymerase will extend a short primer from its 3'-end, along a template DNA which may read as either **ACTG** or **GCTG** for the next four bases. If three deoxynucleotides T, G and A are added to the reaction, along with one di-deoxynucleotide C, that short primer will increase in length by four bases if the sequence reads **ACTG**, *versus* just one base if the sequence is instead **GCTG**. Hence the SNP A/G may be determined, by measuring an overall length for the primer once the polymerase reaction is completed.

several variants of a human genome, how do physicians find out whether their sick patient might contain such a SNP, as a possible cause of the ailment?

First one amplifies by PCR certain discrete sections of human DNA, which are thought to be of possible medical interest (typically 100 to 1000 base-pairs). Next, one can use each of those PCR products as a template for a special *polymerase extension reaction*, which includes only three of the normal deoxy-nucleotides A, T, C or G; with one of the special di-deoxy nucleotides which terminate a chain. For example as shown in Fig. 10.5(b), a specific primer of 20 bases (which has been chosen to stop just short of the suspected SNP) has been added to our PCR-derived DNA, and then a polymerase enzyme will increase the length of that primer by different amounts, depending on whether it detects a template A or else G just after the primer 3′-end. Hence if the first few template bases to follow the primer 3′-end are **A**CTG, the polymerase will add di-deoxy-C after the primer grows longer by four nucleotides to 24 bases total. But if the first few template bases are **G**CTG, the polymerase will add di-deoxy-C after the primer grows longer by just one nucleotide, to 21 bases total.

By choosing which primer to synthesize, and which di-deoxynucleotide to add (along with the appropriate three normal deoxy-nucleotides), practically any SNP may be identified with ease. Indeed, hundreds or even thousands of polymerase extension reactions can now be analyzed rapidly through use of *time-of-flight mass spectrometry* ('MALDI-TOF'), which can measure the mass of the DNA to very high precision and with great accuracy. There the length of time required for any extended primer to reach the detector depends sensitively on its overall size, say 24 *versus* 21 nucleotides in the example given above. For scientists without access to such a costly mass spectrometry machine, an inexpensive alternative is to use two primers, each with a different terminal 3′ base, and each with a different-color dye attached to its 5′-end; then each SNP (say **A**CTG *versus* **G**CTG) can be amplified by normal PCR to yield a product of distinctive color in a capillary or gel.

So far scientists have mapped an astonishing 2 to 3 million SNPs within the human genome, and are looking for ways to use that abundant, but often meaningless, information. Most SNPs are simple changes such as T to C, or A to G (i.e. pyrimidine to pyrimidine, or purine to purine): those are called 'transitions'. Some are changes from T to G or A, or from C to G or A: those are called 'transversions'. Many or most of these changes are probably harmless, and their distribution often reflects genetic variation among different races, or among individuals within a race. Scientists are also examining polymorphisms in other organisms, such as fruit flies, to compare the naturally occuring variations within populations of breeding animals, when certain subgroups of any species cannot interbreed due to separation by geography. These have provided some insight into how genes 'flow' and change in the wild.

While most of the SNPs in humans are probably harmless, a few may be dangerous in themselves, or may be linked with defective genes. By statistical means, scientists have associated certain of those SNPs with human disease: for example, asthma, lung cancer, high blood pressure, diabetes, lupus, or migraine. Yet SNP-disease associations remain very controversial at present, since they are based only on weak statistical evidence. Furthermore, they do not address any environmental factors which might influence whether or not someone becomes ill.

By a sixth new class of methods, scientists can now determine whether any C base in human chromosomal DNA might represent 'normal' cytosine, or else a chemically modified form known as 'methyl' cytosine. There a $-CH_3$ methyl group is added to the 5-position of the cytosine base, just as for thymine *versus* uracil (see Fig. 2.13), by special 'methylase' enzymes which are present in any cell. A full discussion of cytosine methylation will be deferred until Chapter 11.

With all of these novel technologies in hand – transgenics, PCR, four-colour sequencing, microarrays, SNPs, detection of methylation – scientists have been able to make rapid progress at finding the underlying genetic origins of many diseases. These usually involve tiny but specific defects in human chromosomal DNA. So, as the second general part (b) of our survey, let us see how the techniques described in part (a) are used in modern medicine.

Medical researchers will typically identify some small group of people who possess a particular genetic disease, and then screen their DNA within probable protein-coding regions to search for a deleterious point-mutation or slight alteration of length. Researchers can also identify a broad region of probable genetic defect under the light microscope, by studying visual abnormalities in human metaphase chromosomes. But the new molecular methods described above provide greater resolution still, of the disorders which we either can or cannot see by use of light microscopy.

Once a probable mutation is found by PCR, cloning and sequencing, the next step will often be to use rapid methods of DNA diagnostics, to determine which other members of the human population carry the same mutation. Once we examine a large number of people from diverse ethnic backgrounds, who happen to carry that same mutation (or SNP), will they appear healthy or diseased? Finally, some researchers may choose to make transgenic mice, to which a defective human gene has been added; or else make a similar deleterious mutation within the normal mouse. Will such slight genetic alterations of the mouse cause symptoms of disease identical to those seen in humans? Transgenic mice can also be used to test which drugs might cure any genetic ailment, before embarking on a human trial.

Most genetic defects as found in this way have turned out to be errors within the coding region of some essential protein, which impair its function. Sometimes a single base of the patient's DNA which has been altered by a 'point mutation', say from A to G or from C to T, makes the mutant gene code for an altered amino-acid (or else 'stop') at one location along a protein chain. In other cases, one or several bases may be added or deleted, so that the mutant DNA codes for a dramatically altered protein by means of a 'frameshift mutation' (see Chapter 1).

By trial-and-error, scientists have slowly begun to associate various mutations in human DNA with many kinds of disease, such as cystic fibrosis, breast cancer, or neurological ailments. For example, it is now possible to predict whether a young woman will have a high chance of getting breast cancer, or whether a young man will have a high chance of getting prostate cancer, from a study of their DNA. These discoveries have prompted much debate about the ethics of medical diagnosis among physicians, as well as a keen interest in DNA from life insurance companies!

The rate of progress has been remarkable: as of 2003, over 200 different genetic or infectious diseases may be diagnosed using PCR and sequencing, by the few medical researchers working in Sydney alone. One finds for example tests for: antitrypsin deficiency, auto-immune syndrome, five different forms of cancer, Charcot-Marie-Tooth disease, chicken pox, cystic fibrosis, viral encephalitis, fragile-X syndrome, Friedrich's ataxia, haemophilia, hepatitis, and herpes simplex virus. The tissues from which DNA may be extracted to perform such tests include: blood, amniotic fluid, spinal fluid, bone marrow, mouth swabs, nasal swabs and mucus (see **www.cs.nsw.gov.au/csls**, Handbook of DNA Testing).

What a cornucopia of DNA technology, to spring up in just twenty years since the PCR method was developed! Furthermore, this abundance of medical information does not even include the numerous PCR tests used for veterinary work (e.g. dogs, sheep or horses), forensic work (e.g. rape or murder), or archeological studies (e.g. ancient humans).

Two examples of human genetic defect are shown in Figs 10.4 and 10.6. The four-color, automated sequencing plots shown previously in Fig. 10.4 were derived from the DNA samples of real patients with neurological disease. We looked previously at Fig. 10.4 in order to explain the general nature of DNA sequencing results; but now let us try to understand those results in their full medical context.

The upper sequencing plot in Fig. 10.4(a) shows DNA from a healthy patient, whereas the lower sequencing plot in Fig. 10.4(b) shows DNA from a neurologically diseased patient, who suffers from a disease called Amyotrophic Lateral Sclerosis or 'ALS1'. Thus, the single-base change shown at position 119 of Fig. 10.4(b) corresponds to a single-copy, heterozygous point-mutation (i.e. the two copies of the same gene in any cell are different): from T (thymine) normally found on both chromosomes of an individual, to T on one chromosome of the patient but G (guanine) on the other.

Because that point-mutation lies within an important protein-coding region of chromosomal DNA, namely part of the gene for superoxide dismutase 1 or 'SOD1', it changes a key amino acid at position 148 of that protein from GTA (valine) to GGA (glycine). While the patient will make both the mutated and the normal version of the enzyme – one from each chromosome – the mutant protein will disrupt the biochemistry of human nervous function, although details of the mechanism are not yet clear.

Other point-mutations within SOD-1 also cause neurological disease, typically associated with wasting of the muscles, leading to paralysis and sometimes death. In order to learn more about ALS1 or related diseases, the reader may wish to consult an excellent general reference on the Internet at **www.ncbi.nlm.nih.gov**, subsection 'OMIM' for 'Online Mendelian Inheritance in Man'. If you proceed to that site, and search for reference number 105400, it will tell you much concerning ALS and its genetic causes; plus a history of the disease and its prevalence in ethnic populations. The OMIM website is a valuable reference for doctors who treat patients with genetic disease, and use PCR and/or sequencing to diagnose inherited ailments.

Next, the electrophoretic gel shown in Fig. 10.6 reveals detailed information about another kind of neurological ailment known as Machado-Joseph disease or Spino-cerebellar ataxia 3 ('SCA-3'), which is also described on the OMIM database under reference number 109150. This particular genetic disease comes about because one small part of human DNA gets bigger or 'expands' during early human development, at a repeated sequence CAG-CAG-CAG-CAG within the gene for a protein called 'ataxin-3'. Within normal individuals the $(CAG)_n$ triplet varies in length from $n = 10$ to 40, but in affected individuals it varies in length from $n = 60$ to 80. Such dramatic expansion of coding DNA at a chromosomal level causes affected individuals to make a larger-than-normal protein, which contains 60 to 80 successive glutamine amino acids. That mutant protein then causes, through indirect means – for example, impairment of mitochondrial function – loss of balance, and wasting of muscles in the hands, feet or face.

Looking closely at Fig. 10.6, one can see a series of PCR-derived products which measure the overall length of that CAG repeat, using DNA samples as supplied by different individuals. All PCR products were labelled with radioactive phosphorous to permit easy detection in a polyacrylamide gel. Four lanes on the left show DNA sequencing reactions for G, A, T or C as produced by the di-deoxy method, but using a gel rather than a capillary: these are there to provide markers of DNA length within the gel. Nine lanes on the right show PCR products of the ataxin gene as derived from real

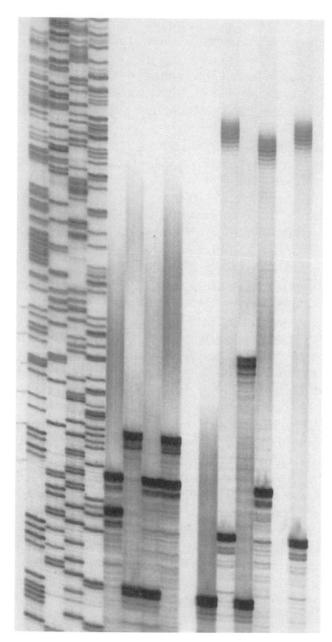

Figure 10.6 An electrophoretic gel shows large size-expansions of a triplet $(CAG)_n$ within genomic DNA, as derived from human patients, all of whom have a neurological disorder, but three of whom have the specific disease Spinocerebellar ataxia or 'SCA3'. Part of the coding region for a protein 'ataxin-3' was measured in each case, using PCR primers which anneal to base sequences on either side. Three of the patients have long $(CAG)_n$ repeats, which run slowly through the gel at top right. The remaining lanes show a normal range of lengths for $(CAG)_n$ repeats. All patients show PCR products from two homologous genes of different length, or one from each chromosome.

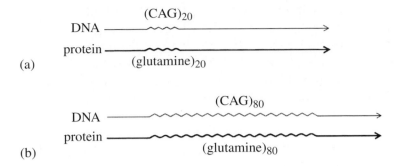

Figure 10.7 (a) A short $(CAG)_n$ repeat of $n = 20$ in healthy individuals codes for a protein containing 20 successive glutamine amino acids, whereas (b) an expanded $(CAG)_n$ repeat of $n = 80$ in diseased individuals codes for a larger protein containing 80 successive glutamine amino acids. It is that long poly-glutamine tract which seems to cause neural defects, by mechanisms which are still not clear.

patients. Out of those nine samples, six show short CAG repeats $n = 10$ to 40 typical of healthy individuals, whereas three show expanded CAG repeats $n = 60$ to 80 typical of diseased individuals.

The biochemical effects of triplet-expansion are illustrated schematically in Fig. 10.7. There we can see how a normal CAG triplet of $n = 20$ gives rise to an ataxin protein with 20 successive glutamine amino acids; whereas an expanded CAG triplet of $n = 80$ gives rise to a larger protein with 80 successive glutamine amino acids.

How many other kinds of genetic disease might be due to triplet expansion, as opposed to a simple point or frameshift mutation? So far more than a dozen different human diseases, mostly neurodegenerative, have been associated with the expansion of triplet repeats; as well as three easily breakable or 'fragile' sites within human chromosomes. For example, Huntington's disease is associated with expansion of a triplet $(CAG)_n$ in a different protein, while myotonic dystrophy is caused by expansion of a triplet $(CTG)_n$. Fragile-X syndrome is caused by expansion of a triplet $(CGG)_n$; while Friedreich's ataxia is caused by expansion of a triplet $(GAA)_n$ (to a size as large as $n = 1000$ base-pairs).

In summary, the vast majority of genetic diseases as detected within the past 20 years correspond to point or frameshift mutations, or else triplet expansions, within some part of human chromosomal DNA that codes for protein. There do exist, however, exceptions to this rule. For example, as we have seen, Spinocerebellar ataxia and Huntington's disease are both due to expansion of a triplet CAG, within the coding region for some protein. Yet myotonic dystrophy and fragile-X syndrome are due to expansion of triplets CTG or CGG respectively in *non-coding* regions of DNA, which make messenger-RNA but are not translated into protein.

How might triplet expansion at a transcribed but *non-coding* sequence influence human health? The triplet expansion of $(CGG)_n$ which causes fragile-X, a disease of mental retardation, is located on the X chromosome, and causes that chromosome to become easily broken or 'fragile'. Hence its genetic locus has been called 'FMR1' for 'Fragile-X Mental Retardation 1'. The FMR1 gene codes for a protein which is necessary for brain function (specifically, regulated translation of messenger-RNA in nerve cells); yet the protein made from FMR1 remains identical for both normal and affected individuals. Instead, it is the overall production of this protein which is lost in fragile-X syndrome. How might that be accomplished?

Near the 5'-end of FMR1, where the DNA has already started to make messenger-RNA, but is not yet coding for protein, a healthy individual contains a base sequence which repeats as $(CGG)_n$ for *n* from 6 to 50, as shown schematically in Fig. 10.8(a). But in mentally retarded individuals, that $(CGG)_n$ triplet expands to roughly 600 base-pairs as shown in Fig. 10.8(b). So for mentally retarded individuals, the long repeat of 600 base-pairs somehow impairs production of the FMR1 protein.

Do we have any clues as to how this impairment might work? It is usually observed that the expanded CGG triplet of FMR1 becomes modified by cellular methylase enzymes, which convert certain bases within that particular small region of DNA from normal cytosine to 5-methyl-cytosine (see Chapter 11). Since 5-methyl-cytosine can prevent the binding of proteins known as 'transcription factors' to promoter DNA, which enable the transcription of DNA into

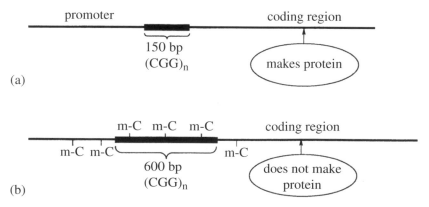

(a)

(b)

Figure 10.8 (a) When the FMR1 gene contains a repeat of $(CGG)_n$ less than 150 base-pairs long, that DNA still works normally to make RNA and protein. (b) But when the FMR1 gene contains a repeat of $(CGG)_n$ more than 600 base-pairs long, the expanded repeat somehow triggers methylation of nearby promoter DNA (i.e. conversion of cytosine to methyl-cytosine) by cellular methylase enzymes; and so the gene no longer works to make RNA or protein in the normal amounts.

RNA, this may explain how an extended CGG repeat impairs pro-duction of FMR1 protein: by blocking messenger-RNA synthesis.

Certain kinds of cancer have been associated, but only in an indi-rect fashion, with the expansion of *di*nucleotide repeats such as TATATA…, CACACA… or CGCGCG…. Those and other highly-repetitive sequences (such as triplets (as above) or tetramers) may be found at many different places within a human genome. In the majority of cases, however, they do not code for protein; and they seldom cause disease, apart from the few triplets mentioned above. Still, patterns of length variation in non-essential repeats or 'microsatellites' have proven useful for the genotyping (i.e. 'finger-printing') of individuals within any species. They are now used for example in the identification of individuals in police forensic work (as mentioned earlier); the establishment of animal pedigrees; the diagnosis of colorectal cancer from stool, or liver cancer from blood; and the avoidance of genetic defects when a couple chooses *in vitro* fertilization.

The wide variety of lengths seen for repetitive microsatellites in humans, as well as for triplet-repeats which cause disease, raises a question concerning the molecular mechanism of such expansions. Why should certain trinucleotides, when repeated over and over again, expand in number during early human development so as to cause disease? Why should microsatellites vary in length throughout the chromosomes, typically in non-coding regions?

Most workers believe that certain DNA molecules of highly-repeated sequence can 'slip out' so as to form hairpin or stem-loop structures, during DNA copying or replication; which in turn may induce expansion on a molecular scale. As shown in Fig. 10.9(a), DNA polymerase will usually copy both strands of any pre-existing DNA molecule, so as to form two new double helices. Almost all polymerase enzymes work in a specific 5′ to 3′ direction while mak-ing new strands. Thus the lower polymerase shown in part (a) can copy the lower strand continuously, as it works from right to left; but the upper polymerase can only copy the upper strand discon-tinuously, in small parts of roughly 300 base-pairs. Those small parts – known as 'Okazaki fragments' – will be joined together later by a 'ligase' enzyme. Once sealed and ligated, each new double helix will go to one new cell. It may be a normal somatic cell, or else a sex cell (egg or sperm).

Now if the DNA that is being copied includes a long repeat of $(CTG)_n$, within either the new strand or the old, that particular sequence may slip out to form a hairpin as shown in Fig. 10.9(b) and (c). Suppose a CTG hairpin forms within the newly-made strand: then some CAG triplets may be copied twice as shown in

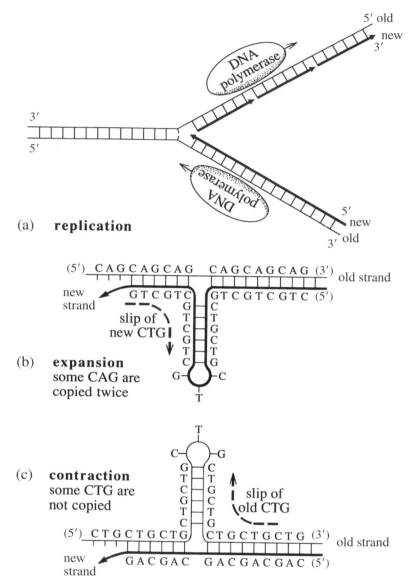

Figure 10.9 (a) During normal replication, a DNA polymerase enzyme can make new strands in a 5′ to 3′direction; hence it can copy one strand continuously, but has to copy the other strand in small pieces of size around 300 base-pairs, which will be joined together subsequently by a ligase enzyme. (b) DNA of repeated sequence $(CAG)_n$ may expand in size during replication, if the new CTG strand slips backward, so that some CAG are copied twice. (c) DNA of repeated sequence $(CTG)_n$ may contract in size during replication, if some CTG on the old strand slip out into a hairpin, so that not all CTG are copied.

Fig. 10.9(b); and the DNA will get longer. But suppose a CTG hairpin forms within the old strand: then some CTG triplets may not be copied if the polymerase skips over them as shown in Fig. 10.9(c); and the DNA will get shorter. Such potential expansion-contraction events seem especially favored for a cellular DNA polymerase of the kind 'beta' (as opposed to 'alpha'), which 'stalls' at stable hairpins that are long enough for slippage to occur.

This general mechanism for DNA expansion or contraction, which invokes slippage of a double helix to form hairpins during replication, has been supported by many kinds of evidence. For example, sequences of the kind $(CTG)_n$ expand when they replicate in bacteria; whereas other DNA sequences which do not show expansion in humans, do not show expansion in bacteria either. Hence, DNA hairpins as formed from the six triplets which cause human disease – CGG, CCG, CTG, CAG, GAA and TTC – could perhaps be more stable than hairpins made from the other 58 triplet DNA sequences. All of those favored-hairpins would rely on non-Watson–Crick base-pairs for their stabilities. For example, the $(CTG)_5$ hairpins shown in Fig. 10.9(b) and (c) might contain T–T base-pairs of a 'wobble' variety (as for a G–T pair), plus C–G pairs of a Hoogsteen type: see Figs 2.12 and 2.13. A stable hairpin made from $(TTC)_3$ would contain T–T, C–T or C–C pairs; and similarly for $(GAA)_3$ with G–G, G–A or A–A pairs. These would appear to be 'crazy' forms of DNA; yet the available evidence suggests that they exist, at least transiently during replication!

In summary, much has been learned from point mutations and triplet expansions, about the structure and dynamics of DNA in chromosomes. Now let us proceed to the third general part (c) of our survey, which concerns the inheritance of genes in a non-Mendelian fashion, by a poorly-understood mechanism known as 'imprinting'. Gregor Mendel found from his long-term study of plants in 1860, that his specimens could inherit two slightly different genes of each particular kind from its two parents, in an apparently random fashion. We know now that this is because we inherit two slightly different chromosomes for any homologous pair, with slight variations of DNA sequence between them. Some genes when inherited in that way appear to 'dominate' the others, while other genes remain passive or 'recessive'; and only show their corresponding physical attributes ('phenotype') when no dominant gene is present.

Still, it should make no difference by Mendel's scheme whether you inherit a gene from your mother or from your father. A dominant gene from your mother will act in the same way as a dominant gene from your father, according to early 20th-century rules of genetics.

Yet it has been known for a long time that those 'rules' can be at most partly true. For some genes it matters greatly whether you inherit them from your mother or from your father. For example, when you mate a horse with a donkey, the kind of hybrid mule that you get depends greatly on whether it has a horse father and a donkey mother, or vice-versa. And in humans, the CGG repeat of FMR1 only expands to a large size, causing mental retardation, when it is inherited from your mother in an egg, and not from your father in a sperm. Such non-Mendelian inheritance is called 'imprinting', because one parent or the other seems to place an imaginary mark or imprint on the activities of certain genes, when they pass those genes onto their children in sex cells. For example, an imprinted gene may always be active if received in the sperm from your father, and inactive if received in the egg from your mother, or vice-versa.

About 60 imprinted genes have been found so far in mouse and man, while 200 more are expected to be found in the future. Some of these imprinted genes cause genetic disease – often a dysfunction of a long-distance signalling process (known as 'endocrine') – if the natural imprinting mechanism fails: for example Prader-Willi/ Angelman syndromes, Silver-Russell syndrome or Beckwith-Wiedeman syndrome. When studied closely by genetic or biochemical means, all of those diseases seem to be inherited from only one parent, either the mother or the father, but not equally from both.

The molecular mechanism by which genes may be imprinted remains poorly understood. It seems to be related to all of: (a) differences in cytosine methylation between active and inactive genes (see Chapter 11); (b) differences in histone protein methylation or acetylation between active and inactive genes (see Chapter 7); and (c) the synthesis of certain non-coding RNA molecules which adhere to specific chromosomal regions within the nucleus. These long, non-coding transcripts of RNA then repress or 'shut down' genes wherever they adhere, and also for some distance on either side. In other words, imprinted genes are usually found in large clusters along any chromosome: if a single gene becomes imprinted, then its neighbors are likely to become imprinted as well.

Figure 10.10 shows a general 'working hypothesis' that may perhaps be useful, when contemplating the data from imprinted gene systems. In part (a), we see that certain regions of human DNA may contain base sequences which form 'direct repeats'. In other words, those regions show approximately the same sequence of 50 to 200 base-pairs over and over again, repeated for 5 to 20 copies in a tandem array. In part (b), we see that directly-repetitive regions of DNA may 'slip out' to form large staggered loops during DNA replication, just as certain triplet-repeat sequences may slip in steps of three, as shown in Fig. 10.9. Then in part (c), we see that if RNA happens to be transcribed from those directly-repetitive regions of DNA, such RNA molecules can potentially bind to the large staggered loops using Watson–Crick base-pairs.

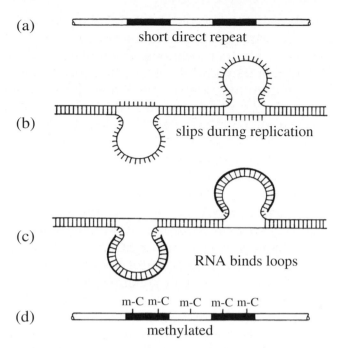

Figure 10.10 A 'working hypothesis' for the combined role of DNA direct repeats, non-coding RNA and cytosine methylation in gene imprinting. As shown in (a), certain regions of chromosomal DNA may be highly repetitive with respect to base sequence in a direct sense. Next (b) some of those regions may potentially 'slip out' to form large staggered loops during DNA replication (cf. the triplets in Fig. 10.9). Then (c) if such directly-repetitive DNA is transcribed into RNA, the RNA molecules so formed may potentially bind to those large staggered loops of DNA using Watson–Crick base-pairs. Finally, as shown in (d), the binding of RNA to DNA direct-repeats seems somehow to trigger cytosine methylation, which is known to silence genes during imprinting.

Finally, in Fig. 10.10(d), we see that the binding of RNA to DNA within such direct repeats may potentially induce a DNA-modifying activity of cellular methylase enzymes, which can convert cytosine 'C' bases to 5-methyl-cytosine or 'm-C'. It is still not known how step (d) is accomplished, but such are the data as typically observed. Perhaps a methyl-dependent, gene-repression activity evolved long ago as a defense against RNA viruses, some of which try to infect mammalian or plant cells by copying their genomes into a host chromosome? If any such insertion were successful, would the host cell want those foreign genes to remain active? It seems worthwhile to note here, that a surprisingly large 45% of human DNA consists of ancient (but now inactive) infectious elements, whether based on RNA or DNA. So we really do need a defense system of that sort!

Many different RNA molecules of the kind shown in Fig. 10.10 (c) are transcribed from DNA inside any cell nucleus, but never leave the nucleus to make protein – in contrast to the ribosomal and transfer-RNA, which are exported from the nucleus. Instead, they travel within the nucleus to other locations, where they bind to DNA or RNA molecules that they find there, and so effect regulatory functions such as imprinting. We see then that the phenomenon of imprinting could depend critically on which kinds of RNA a cell includes inside itself. Studies to date have only begun to reveal the enormous complexity of genetic inheritance by this novel mechanism, which Mendel could never have foreseen.

Not much is known about RNA-based imprinting, yet in one case a mechanism seems clear. The best-characterized example of an RNA molecule which causes the shutdown of nearby genes is called *Xist* for 'X inactivation specific transcript'. That long RNA molecule is made from the female X chromosome in humans and other mammals as a single strand of 18 000 bases, but it does not code for protein. Instead, *Xist* RNA binds within the general vicinity of its own DNA template; and somehow it causes many genes on the X chromosome to become inactive due to induced cytosine methylation. As its repressive effect spreads widely, through means as yet unknown, eventually the entire X chromosome loses gene activity.

Within *Xist*, the DNA or RNA sequence contains many directly-repeated regions of length 50 to 200 base-pairs. Those regions are called 'A, B, C and D' respectively going in a 5' to 3' direction: the 'C' region in particular contains 14 direct repeats of 110 to 120 bases. Some experiments suggest that the 'C' region of *Xist* mediates its binding to DNA on the X-chromosome. Might the directly-repetitive DNA slip out there into staggered loops as shown in Fig. 10(b) and (c), and thereby facilitate binding of *Xist* RNA? Unfortunately, no one has yet determined a structural mechanism for the specific binding of *Xist* RNA to DNA within living cells. Region 'A' near the 5'-end of *Xist* seems important to trigger gene silencing and perhaps cytosine methylation, once *Xist* binds using 'C' or other sequences.

Now for some simple genetics: it hardly matters whether females lose all gene activity from one of their two X chromosomes, since the other X will continue to function normally. Fortunately the single X chromosome in males remains completely active; and since the single active X in females produces proteins in the same relative amount as for males, the biochemistry of cells may remain the same for both males and females. This levelling-effect on protein synthesis is called 'dosage compensation', and it explains why 'trisomies' – i.e. getting three copies of any non-sex chromosome rather than two – are often harmful.

By a strange twist of Nature, it seems that each individual cell chooses by chance which of its two X chromosomes will be shut down by *Xist* RNA. Hence, any mature female may contain a 'mosaic' of two kinds of cell, depending on which X chromosome remains active. For example, the variegated colors seen in the coat of a tortoiseshell cat are due to that effect, where hairs of the cat may have several different colors, depending on their choice of X chromosome locally.

In another example of RNA-induced imprinting, a long repetitive RNA molecule known as *'Air'* has been shown to control silencing of genes on a non-sex chromosome or 'autosome'. Also, a protein called *'Eed'* – for 'embryonic ectoderm development' – has been implicated as helping to maintain the repressed state. Changes of DNA methylation are thought to proceed by the joint action of two methylases called 'DNMT3L' and 'Dnmt3a' (see Chapter 11). Finally, the addition of a transcriptional activator called "VP16" to a repressed gene can reverse the effects of imprinting, and cause that gene to work normally again. What a mess! How could any field be more descriptive than the study of gene imprinting at present?

Finally let us proceed to the fourth general part (d) of our survey, which concerns attempts by scientists to correct errors in human DNA – either by adding extra functional DNA into human cells, or else by repairing an undesired mutation within pre-existing DNA. These efforts are known in general as 'gene therapy' or 'gene correction' respectively.

As shown in Fig. 10.11, one may add extra double-stranded DNA to human cells by wrapping it in either: (i) special viruses or else (ii) fat-DNA complexes known as 'liposomes'. One can also repair an undesired mutation in pre-existing DNA, by adding a short single-strand (typically 30 to 500 bases) of the correct DNA sequence, that may be copied into a defective chromosomal region by enzymes such as 'recA' or 'RAD51', which mix or 'recombine' one DNA strand with another. Those recombination enzymes are used normally by the cell to repair defects on one chromosome during cell division, using sequence information found on the other homologous chromosome; but scientists can 'trick them' into correcting general mutations as well.

The double-stranded DNA as added by viruses or liposomes will often carry a functional gene, to supply a necessary protein which the person does not have on account of genetic disease. In other cases, that extra DNA may carry the information for making 'anti-sense' or 'silencing' RNA, which can bind to the 'sense' or messenger-RNA that codes for some defective protein, so as to block its template action at ribosomes. Alternatively, short segments of single-stranded DNA may fix a point or frameshift mutation, which will simply return a person's DNA to the human consensus.

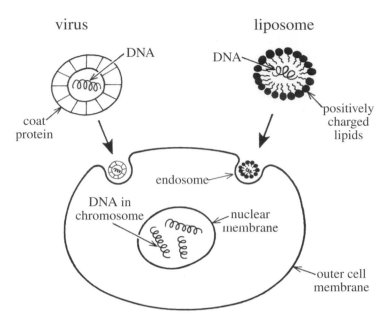

Figure 10.11 Extra DNA may be added to cells for 'gene therapy', after packaging into either viruses or liposomes.

Unfortunately, none of these promising methods works well at present, for several reasons. First, both gene addition and gene correction act much more efficiently on DNA within cultured cells, than on DNA within the whole body of a living animal. Hence one may have to extract bone-marrow or embryonic stem-cells from a donor, in order to carry out such therapies efficiently. Second, the extra double-stranded DNA may cause cancer or have other undesirable effects, if it inserts at many random places within a chromosome. Thirdly, short segments of single-stranded DNA which repair an existing defect, would seem to be much safer than long pieces of double-stranded DNA which add an entirely new gene; however, the efficiency of fixing a point or frameshift mutation by short single-stranded DNA is currently so low, that little therapeutic effect is obtained.

Let us summarize nevertheless some of the standard approaches for DNA delivery to cells or patients, which have been used so far. The liposome method is the most common: there one adds to any long piece of double-stranded DNA (or even short pieces of single-stranded DNA), certain chemicals – 'lipids' – which contain water-insoluble fat on one end, and positively-charged nitrogen N–H groups on the other. Such chemicals, of which there are many varieties, will rapidly enclose a therapeutic DNA molecule inside a protective, fatty coat: see Fig. 10.11.

Other methods for adding DNA within cells, or to humans, animals and plants include: injection into solid tissue; electroporation (i.e. the disruption of membranes by electric fields); the 'gene gun' (i.e. high-velocity particles which may penetrate tissues); ultrasound disruption; positively-charged polymers; and positively-charged peptides.

Those fat-DNA complexes or 'liposomes', once formed, can pass readily through the outer membrane of most cells. Thus our long DNA molecule, once wrapped into a liposome, will typically enter a cell by means of an 'endosome', where a small part of the outer membrane folds into a bubble, as shown in Fig. 10.11. But then trouble begins. Because a typical cell does not wish to take foreign DNA inside of itself, it has developed several defense mechanisms to degrade any DNA which does enter. Almost immediately, the cell begins to pump strong acid and various enzymes which digest protein and DNA into this endosome bubble, so as to depolymerize the molecules inside! The DNA as carried by our liposome has little chance to escape, and so most of it becomes degraded before it can enter the main part or 'cytoplasm' of a cell. By contrast, certain viruses have devised clever mechanisms for escaping from the endosome, for example by making a long 'protein tunnel' through which the DNA can pass.

Next, if some small amount of DNA does escape from the endosome, and proceeds into the general cell fluid or cytoplasm, that extra DNA still has to negotiate many small, tight passages which lead through the nuclear membrane into the cell nucleus, before it can begin to make the messenger-RNA which enables it to synthesize protein. Once again, certain viruses have evolved clever strategies to get their DNA into a cell nucleus: say by attaching DNA to specific viral proteins, which possess the specific recognition codes needed to go in and out. Currently however, only one out of every million DNA molecules that enter the cell by means of a liposome, eventually enters the cell nucleus and makes a desired protein.

Even if we succeed in getting our extra DNA into the cell nucleus by use of a liposome, or else a human virus such as adenovirus or adeno-associated virus, still we have no satisfactory way of adding such extra DNA permanently, and at specific locations, within the DNA of normal chromosomes. The best method so far uses the replication or *rep* protein of adeno-associated virus (AAV), to insert extra DNA by means of a particular base sequence to a specific region on human chromosome 19.

The general mechanism of insertion by AAV seems quite interesting, and is shown schematically in Fig. 10.12. There we can see that the *rep* protein of AAV makes a specific 'nick' within one

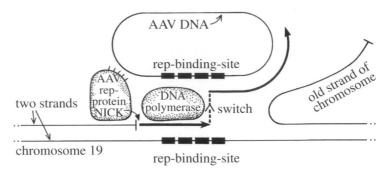

Figure 10.12 A general mechanism for insertion of new genes *via* adeno-associated virus (AAV), specifically at one region along human chromsome 19. First the DNA of chromosome 19 is nicked specifically by the AAV rep protein, which also pulls AAV and chromosome 19 together in space. Next, a cellular DNA polymerase copies from that nick in a rightward sense, until it reaches a repeated sequence known as the rep-binding-site (RBS), typically $(GCTC)_4$. Once stalled at that RBS, the polymerase can easily switch templates; so that it begins to copy from another RBS which lies on AAV nearby. Later, the polymerase returns to the human chromosome.

strand of the human DNA, just at one precise location on chromosome 19, as shown on the left. Next, starting from that nick, a new copy of the DNA is made by DNA polymerase (thick line with arrow), while the old strand is displaced, as indicated on the right in the figure. The DNA polymerase then slows down or stalls once it reaches a repeated sequence known as the *rep-binding-site* ('RBS'). The *rep* protein meanwhile, in addition to making the nick, also pulls the single-stranded viral DNA into close proximity of the chromosomal nick site. While stalled therefore, there is an increased chance that the polymerase will hop onto the nearby viral DNA, and continue to extend the new DNA chain (second thick line with arrow) onwards from the viral RBS, until it returns to the chromosomal DNA somewhat later.

Hence, strand-switching by DNA polymerase, as initiated by a specific *rep* nick, and enhanced by a repetitive sequence RBS, all combine to add new AAV genes specifically at one location in the entire human genome. Other DNA-integrating viruses (e.g. lentoviruses and retroviruses) have thus far performed less well for medical purposes. For example, they have caused cancer in stem-cell trials, by inserting their DNA at many non-specific places within a patient's chromosomes. Further work in this active field will eventually determine which DNA delivery system might be best.

Recall how scientists can add foreign DNA to mouse or fly eggs, by use of a fine needle through which the DNA is injected, directly into the cell nucleus without having to pass through any cell membrane.

That simple technique will not work for gene therapy, however, because the doctor has to add extra DNA to millions of different cells in culture or within a living patient, in order to make enough therapeutic protein to be useful. Then such extra DNA has to be maintained as an active gene which produces protein for many years, without being lost through mutation, or being repressed through imprinting or cytosine methylation.

So scientists today remain highly divided over whether gene addition or gene correction will really work. Still, that has not stopped them from trying! Over the next ten years, we may expect to see many new discoveries which should advance the field substantially.

Further Reading

Transgenic mice, animals and plants

Jimenez-Bermudez, S., Redondo-Nevado, J., Munoz-Blanco, J., Caballer, J.L. *et al.* (2002) Manipulation of strawberry fruit softening by antisense expression of a pectate lyase gene. *Plant Physiology* **128**, 751–9. The genetic design of strawberries which ripen more slowly than normal.

Marquet-Blouin, E., Bouche, F.B., Steinmetz, A., and Muller, C.P. (2003) Neutralizing immunogenicity of transgenic carrot-derived measles virus hemagglutinin. *Plant Molecular Biology* **51**, 459–69. Genetic modification of carrots to carry a measles protein for vaccine purposes.

Murray-Kolb, L.E., Takaiwa, F., Goto, F., Yoshihara, T. *et al.* (2002) Transgenic rice is a source of iron for iron-depleted rats. *Journal of Nutrition* **132**, 957–60. Genetic modification of rice to carry more iron for nutritional purposes.

Nicoll, J.A., Wilkinson, D., Holmes, C., Steart, P., Markham, H., and Weller, R.O. (2003) Neuropathology of human Alzheimer disease after immunization with amyloid-beta peptide. *Nature Medicine* **9**, 448–52. Testing of a therapy for Alzheimer disease in transgenic mice *versus* humans.

Qaim, M. and Zilberman, D. (2003) Yield effects of genetically modified crops in developing countries. *Science* **299**, 900–2. Cotton which has been genetically modified to be pest-resistant assists in its propagation.

Walker, S.C., Shin, T., Zaunbrecher, G.M., Romano, J.E. *et al.* (2002) *Cloning Stem Cells* **4**, 105–12. Technical procedures for reproductive cloning of adult pigs from mature-tissue cells.

DNA amplification technologies

Christian, A.T., Garcia, H.E., and Tucker, J.D. (1999) PCR *in situ* followed by microdissection allows whole chromosome painting probes to be made from single microdissected chromosomes. *Mammalian Genome* **10**, 628–31. The use of semi-random PCR to amplify trace DNA from single copies of a chromosome.

Corsaro, D., Valassina, M., Venditti, D., Venard, V., Le Faou, A., and Valensin, P.E. (1999) Multiplex PCR for rapid and differential diagnosis of *Mycoplasma pneumoniae* and *Chlamydia pneumoniae* in respiratory infections. *Diagnosis in Microbiology and Infectious Disease* **35**, 105–8. The use of PCR to detect infectious organisms as a cause of asthma.

Dean, F.B., Hosono, S., Fang, L., Wu, X. *et al.* (2002) Comprehensive human genome amplification using multiple displacement amplification. *Proceedings of the National Academy of Sciences, USA* **99**, 5261–6. The isothermal method MDA for amplifying DNA which can be done overnight at 30°C.

Gold, B., Bergeron, J., Lachtermacher-Triunfol, M., and Dean, M. (2001) Human duplex sex determination PCR. *Biotechniques* **31**, 28–35. The use of PCR to identify male *versus* female DNA samples using X or Y chromosomes.

Koetz, K., Bryl, E., Spickschen, K., O'Fallon, M., Goronzy, J.J., and Weyand, C.M. (2000) T-cell homeostasis in patients with rheumatoid arthritis. *Proceedings of the National Academy of Sciences, USA* **97**, 9203–8. The use of PCR to monitor dysfunction of the immune system in rheumatoid arthritis.

Nuovo, G.J. (2000) *In situ* strand displacement amplification: an improved technique for the detection of low-copy nucleic acids. *Diagnostic Molecular Pathology* **9**, 195–202. Tiny amounts of DNA may be amplified at 55°C by an isothermal method known as SDA.

DNA sequencing and array technologies

Cantor, C.R. and Smith, C.L. (1999) *Genomics: the science and technology behind the human genome project*. John Wiley and Sons, New York. A summary of methods used for the sequence analysis of entire genomes.

Elkin, C.J., Richardson, P.M., Fourcade, H.M., Hammon, N.M. *et al.* (2001) High-throughput plasmid purification for capillary sequencing. *Genome Research* **11**, 1269–74. Methods for sequencing 18 million bases of DNA per day at one facility in the USA.

Gupta, V., Cherkassky, A., Chatis, P., Joseph, R. *et al.* (2003) Directly labeled mRNA produces highly precise and unbiased differential gene expression data. *Nucleic Acids Research* **31**, e13. Messenger-RNA for microarray applications may be labeled for detection either by enzymes or chemicals.

Petricoin, E.F., Hackett, J.L., Lesko, L.J., Puri, R.K. *et al.* (2002) Medical applications of microarray technologies: a regulatory science perspective. *Nature Genetics Supplement* **32**, 474–9. A summary of DNA micro-array technology.

Rosenblum, B.B., Lee, L.G., Spurgeon, S.L., Khan, S.H. *et al.* (1997) New dye-labeled terminators for improved DNA sequencing patterns. *Nucleic Acids Research* **25**, 4500–4. Synthesis and testing of di-deoxynucleotides to which fluorescent dyes have been attached for sequencing purposes.

Single-nucleotide-polymorphisms

Jordan, B., Charest, A., Dowd, J.F., Blumenstiel, J.P. *et al.* (2002) Genome complexity reduction for SNP genotyping analysis. *Proceedings of the National Academy of Sciences, USA* **99**, 2942–7. The use of semi-random PCR to analyse single-nucleotide-polymorphisms (SNPs).

McClay, J.L., Sugden, K., Koch, H.G., Higuchi, S., and Craig, I.W. (2002) High-throughout SNP genotyping by fluorescent competitive allele-specific polymerase chain reaction. *Analytical Biochemistry* **301**, 200–6. Inexpensive analysis of SNPs using two PCR primers of different sequence and colour.

Pusch, W., Wurmbach, J.H., Thiele, H., and Kostrzewa, M. (2002) MALDI-TOF mass spectrometry-based SNP genotyping. *Pharmacogenomics* **3**, 537–48. A summary of time-of-flight mass spectroscopy as applied to detection of DNA single-nucleotide-polymorphisms.

Shastry, B.S. (2002) SNP alleles in human disease and evolution. *Journal of Human Genetics* **47**, 561–6. A review of the latest progress on SNPs, and their implications for medical diagnosis.

Shi, M.M. (2002) Technologies for individual genotyping: detection of genetic polymorphisms in drug targets and disease genes. *American Journal of Pharmacogenomics* **2**, 197–205. The latest techniques used to measure slight genetic variation in human patients.

Triplet expansions and microsatellites

Chang, Y.C., Ho, C.I., Chen, H.H., Chang, T.T. *et al.* (2002) Molecular diagnosis of primary liver cancer by microsatellite DNA analysis in the serum. *British Journal of Cancer* **87**, 1449–53. Over 100 different microsatellites were screened to diagnose liver cancer from tumor DNA in blood.

Girardet, A., Hamamah, S., Anahory, T., Dechaud, H. *et al.* (2003) First preimplantation genetic diagnosis of hereditary retinoblastoma using informative microsatellite markers. *Molecular and Human Reproduction* **9**, 111–6. Microsatellites may be used to avoid disease when couples choose to have a child by *in vitro* fertilization.

Heidenfelder, B.L., Makhov, A.M., and Topal, M.D. (2003) Hairpin formation in Friedreich's ataxia triplet repeat expansion. *Journal of Biological Chemistry* **278**, 2425–31. Even GAA and TTC repeats form stable hairpins, presumably by non-classical G–A or T–C base-pairs.

Jin, P. and Warren, S.T. (2003) New insights into fragile-X syndrome: from molecules to neurobehaviors. *Trends in Biochemical Science* **28**, 152–8. A general review of fragile-X syndrome and its biochemistry.

Sinden, R.R., Potaman, V.N., Oussatcheva, E.A., Pearson, C.E. *et al.* (2002) Triplet repeat DNA structures and human genetic disease: dynamic mutations from dynamic DNA. *Journal of Bioscience* **27** (supplement), 53–65. An overview of triplet expansion in terms of DNA structural dynamics.

Sutherland, G.R. and Baker, E. (2000) The clinical significance of fragile sites on human chromosomes. *Clinical Genetics* **58**, 157–161. A clinical review of fragile-X syndrome and other fragile chromosomal sites.

DNA imprinting and regulatory RNA

Beletskii, A., Hong, Y.-K., Pehrson, J. Egholm, M., and Strauss, W.M. (2001) PNA interference mapping demonstrates functional domains in the noncoding RNA *Xist*. *Proceedings of the National Academy of Sciences, USA* **98**, 9215–20. The repetitive "C" domain of *Xist* RNA mediates its binding to the X-chromosome.

Jouvenot, Y., Ginjala, V., Zhang, L., Liu, P.Q. *et al.* (2003) Targeted regulation of imprinted genes by synthetic zinc-finger transcription factors. *Gene Therapy* **10**, 513–22. A zinc-finger protein can carry a transcription activator to silence genes in cultured cells, and thereby reverse repression by imprinting.

Lee, J.T. (2003) Molecular links between X-inactivation and autosomal imprinting. *Current Biology* **13**, R242–54. The latest molecular and genetic data on imprinting of sex and non-sex chromosomes.

Mattick, J.S. (2001) Non-coding RNAs: the architects of eukaryotic complexity. *EMBO Reports* **2**, 986–991. A survey of non-coding RNA molecules and their possible biological functions.

Polychronakos, C. and Kukuvitis, A. (2002) Parental genomic imprinting in endocrinopathies. *European Journal of Endocrinology* **147**, 561–9. A summary of human genetic diseases caused by faulty imprinting.

Rougeulle, C. and Heard, E. (2002) Antisense RNA in imprinting: spreading silence through *Air*. *Trends in Genetics* **18**, 434–7. A long non-coding RNA called *Air* controls imprinting of a gene cluster on a non-sex chromosome.

Szittya, G., Silhavy, D., Molnar, A., Havelda, Z. *et al.* (2003) Low temperature inhibits RNA silencing-mediated defence by the control of *si*-RNA generation. *EMBO Journal* **22**, 633–40. Small, interfering RNA molecules of size 21 to 26 bases can protect plants against viral infection.

Viral and non-viral gene therapy

Hirata, R.K. and Russell, D.W. (2000) Design and packaging of adeno-associated virus gene-targeting vectors. *Journal of Virology* **74**, 4612–20. Gene correction at frequencies of 1% in cultured cells, using AAV to deliver single-stranded DNA into the cell nucleus.

Hirata, R., Chamberlain, J., Dong, R., and Russell, D.W. (2002) Targeted transgene insertion into human chromosomes by adeno-associated viral vectors. *Nature Biotechnology* **20**, 735–8. Gene addition at frequencies of 1% in cultured cells, using AAV to deliver single-stranded DNA into the cell nucleus.

Liu, L., Rice, M.C., and Kmiec, E.B. (2001) *In vivo* gene repair of point and frameshift mutations directed by chimeric RNA/DNA oligonucleotides and modified single-stranded nucleotides. *Nucleic Acids Research* **29**, 4238–50. The use of electroporated oligonucleotides to repair a variety of genetic mutations in yeast.

Niidome, T. and Huang, L. (2002) Gene therapy progress and prospects: nonviral vectors. *Gene Therapy* **9**, 1647–52. Physical or chemical methods which may be used to add new DNA to human cells or patients.

Recchia, A., Parks, R.J., Lamartina, S., Toniatti, C. *et al.* (1999) Site-specific integration by a hybrid adenovirus/adeno-associated virus vector. *Proceedings of the National Academy of Sciences, USA* **96**, 2615–20. Two adenoviruses, one carrying AAV and the other expressing *Rep78* protein, can add a new gene to a specific site on human chromosome 19.

Shayakhmetov, D.M., Carlson, C.A., Stecher, H., Li, Q., Stamatoyannopoulos, G., and Lieber, A. (2002) A high-capacity hybrid adenovirus/adeno-associated virus for stable transduction of human hematopoietic cells. *Journal of Virology* **76**, 1135–43. A mixed adenovirus/AAV vector can add a large globin gene permanently within cells.

Thorpe, P., Stevenson, B.J., and Porteous, D.J. (2002) Optimising gene repair strategies in cell culture. *Gene Therapy* **9**, 700–2. The use of lipofected single-stranded DNA to repair a point mutation in cultured cells.

Web-based Resources

Introduction to polymorphisms, and data-bases for SNPs
http://www-hto.usc.edu/~cbmp/2001/SNP

CHAPTER 11

Cytosine Methylation and DNA Epigenetics

In Chapter 10 we discussed many aspects of 'normal genetics' and its relation to modern medicine. Normal genetics comes about mainly from the pairing of two bases across any double helix, as A with T or G with C in a Watson–Crick fashion. We also pointed out that those same DNA bases may become modified reversibly to different chemical forms by cellular enzymes known as 'methylases'. Once we add DNA methylation to our list of unexpected phenomena which have been observed in biology, we enter another realm of science known as 'epi-genetics', where 'epi' means here 'outside of' or 'in addition to'.

Thus, epigenetics describes an unexpected extension of classical 20th-century genetics, where information may be stored not only within the primary sequence of DNA, i.e. A, T, C or G; but also within small chemical changes to that primary sequence, for example by adding a carbon atom and its associated hydrogens to any base ring. For all animals and plants on Earth, it turns out that methylation of cytosine C bases will provide the key to understanding how most epigenetic effects work: for example the sex-cell-specific 'imprinting' of genetic inheritance between different generations; or the low incidence of CG dinucleotides in any genome from animals or plants; or a subtle means of defence whereby animals or plants can deal with the unwanted invasion of foreign DNA. But before we introduce such advanced topics, we first need to describe the 'nuts and bolts' of DNA methylation, and something of the brief history of the subject.

The most common forms of methylated DNA are shown in Fig. 11.1(a) and (b). Cytosine may be modified reversibly by adding a methyl group CH_3 to its carbon 5-position; while adenine may be

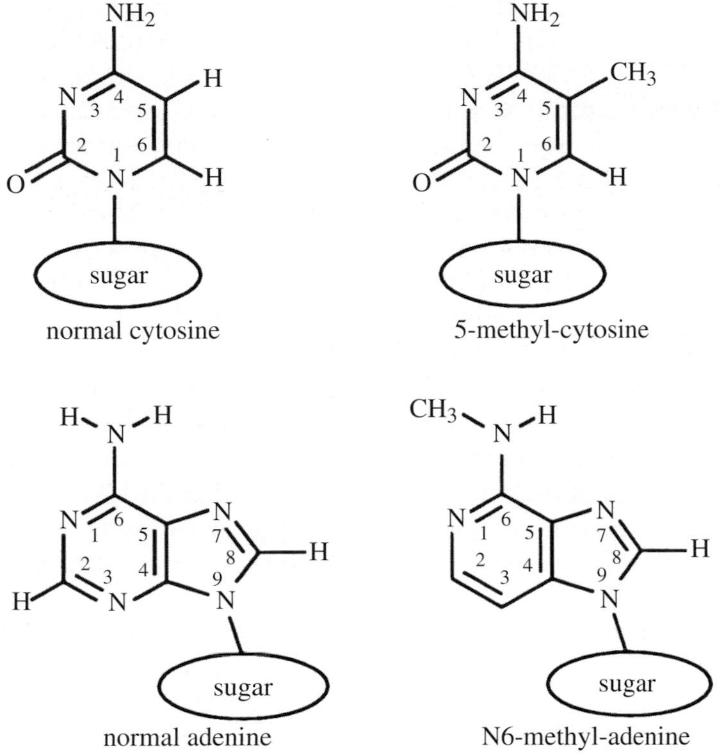

(a)   normal cytosine   5-methyl-cytosine

(b)   normal adenine   N6-methyl-adenine

	most animals	worm, fly	plants	bacteria
5-methyl cytosine	CG	–	CG, CNG	$CC_{T}^{A}GG$ and others
N6-methyl adenine	–	–	–	GATC and others

(c)   major sites of DNA methylation

Figure 11.1 Two common forms of methylated DNA bases. In (a) we show the structures of normal cytosine and 5-methyl-cytosine. In (b) we show the structures of normal adenine and N6-methyl-adenine, the latter of which is found in *E. coli* or other bacteria at sequences *dam* Gm-ATC; *M. Eco* RI GAm-ATTC or *M. Taq* I TCGm-A. The table in (c) gives a simplified summary of the major sites of DNA methylation in different, representative organisms.

modified by adding a methyl group CH_3 to its nitrogen 6-position. Other modifications, such as 5-hydroxy-methyl-cytosine (add CH_2-OH) or 8-hydroxy-guanine (add OH), are found only in bacterial viruses, or in oxidatively damaged DNA.

These methylated bases are usually found at specific sequences in DNA, and not just at any general location. For example, as shown in Fig. 11.1(c), 5-methyl-cytosine is found mainly at the dinucleotide CG for animal DNA, or at the two short sequences CG or CNG for plant DNA, where N can be any base. (The CG dinucleotide is often referred to in the literature as 'CpG', where p stands for the phosphate group; see Chapter 1). Here a special enzyme methylates nearby cytosine bases on both strands. We do not see cytosine methylation at other sequences (that is, CC, CA or CT) except in special cases, where slight amounts of methyl-CA or methyl-CT may be found in fly embryos, or in mammalian embryonic stem cells. The adenine base may also be methylated, as shown in Fig. 11.1(b). N6-methyl-adenine is found usually at the tetranucleotide GATC in many bacteria including *E. coli*, as well as in other minor locations: it is important for bacterial biochemistry, but we shall not discuss adenine methylation further here. Finally, certain bacteria and viruses often have special enzymes that add 5-methyl-cytosine at very precise locations, for example at CCAGG or CCTGG for the *dcm* methylase of *E. coli*; but those enzymes are not found in animals or plants.

Now it is not only the bases in DNA which may be chemically modified, since some of the amino acids in proteins may be modified reversibly as well, by enzymes known as 'protein acetylases' or 'protein methylases'. Such modifications typically involve a change of normal lysine to N-acetyl-lysine (add an acetyl $CO-CH_3$), or else to di-methyl or tri-methyl-lysine (add two or three methyl CH_3), within the abundant histone proteins of eukaryotic chromatin. Those histone proteins wrap long pieces of DNA around themselves into compact ball-like structures known as nucleosomes, as described in Chapter 7. Hence a change of lysine chemical structure within histones might perhaps influence the tightness of DNA wrapping around these proteins; and it might perhaps affect the entire structure of the chromatin, in particular chromosomal locations, and thereby influence the activities of specific genes. We shall return to that subject below, in a discussion of DNA methylases and their interactions with enzymes that modify histones.

Biologists had been aware of the existence of 5-methyl-cytosine within DNA for nearly fifty years; but such slight variations of DNA content were not regarded as significant. One reason for dismissing the potential significance of DNA methylation in biology was that simple organisms such as worm, fly or yeast contain little or no 5-methyl-cytosine. Nevertheless those organisms eat, move, reproduce and thrive. What is good enough for the worm must be good enough for humans!

The inadequacy of such a simple-minded philosophy did not become clear until 1975, when two workers, Art Riggs and Robin

Holliday, suggested independently that 5-methyl-cytosine could play an important role in the normal biology of humans or other mammals. Riggs suggested that 5-methyl-cytosine would prove important for X-chromosome inactivation in the early embryo – where only one female X-chromosome remains active, while the other becomes repressed, as mentioned in Chapter 10. Holliday suggested that 5-methyl-cytosine would prove important for regulating gene expression in the development of a complex organism.

Back then there was hardly any interest in the field, especially when compared with the explosion of interest that we have seen recently. For example, Riggs would visit the laboratory of one of the authors of this book (H.R.D.) in the late 1970's, every Wednesday for lunch; and one day the following conversation ensued. 'Should I work on a crystal structure of DNA with 5-methyl-cytosine?' asked the young student. 'Not many people would be interested, so it might hurt your career. But it would interest me!' – Art Riggs, 1978.

In fact, research before 1990 in the field of cytosine methylation was often not particularly incisive, because it was limited to only a few simple techniques. Most commonly, total genomic DNA would be digested with either of two bacterial enzymes, *Hpa* II or *Msp* I. The enzyme *Hpa* II cuts only unmethylated CCGG, whereas the enzyme *Msp* I will cut both unmethylated and methylated CCGG equally; see Fig. 11.2(a) and its legend. Then the products of such digestion would be applied to an agarose gel, where they could be separated by size. Next, all DNA fragments from the gel would be transferred to a nylon filter, where they could be visualized by hybridizing to some 'radioactive probe' which would be made from some small part of the genome. (Such DNA filter-hybridization is known as a 'Southern blot', after Ed Southern who invented it.) When the pattern of radioactive bands was examined afterwards, scientists could determine which bases near some particular gene might contain normal cytosine or else 5-methyl-cytosine.

As a variant of that same technique, genomic DNA samples from various organisms could be digested with *Hpa* II or *Msp* I, and then applied, as above, to an agarose gel. Next, all of the many different DNA fragments could be studied in total by staining with ethidium bromide dye, rather than by using a sequence-specific probe. The unexpected finding from such work was that a small (1%) fraction of vertebrate or animal DNA seemed to contain many unmethylated CG sequences clustered together, over short regions of 500 to 1000 base-pairs: see Fig. 11.2(b). Those short regions were therefore called 'CG islands', and they were detected first by their sensitivities to the enzyme *Hpa* II, which cuts only unmethylated CCGG.

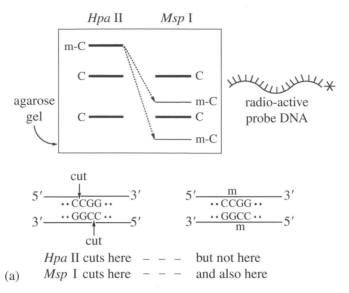

Hpa II   Msp I

m-C ━━━━

C ━━━━ ━━━━   ━━━━ C

agarose   ━━━━ m-C   radio-active
gel   C ━━━━ ━━━━   ━━━━ C   probe DNA

C ━━━━ ━━━━   ━━━━ m-C

                                    cut
                        5'━━━┃━━━━3'   5'━━━m━━━3'
                           ‥CCGG‥   ‥CCGG‥
                        3'━━━━━━━━5'   3'━━━━━━━━5'
                           ‥GGCC‥   ‥GGCC‥
                              ┃   m
                            cut

                Hpa II cuts here   – – –   but not here
(a)   *Msp* I cuts here   – – –   and also here

Figure 11.2 Three common kinds of experiment to study methylated DNA that were performed in the 1980s. In (a) we show an idealized agarose gel, which has been used to separate by size two different samples of genomic DNA; one of which has been digested with *Hpa* II, and the other with *Msp* I. After the gel has been run, it is treated with radioactive DNA of specific sequence which attaches itself to the few DNA fragments in the gel with complementary sequence. Here, three fragments have been picked out by the radioactive probe, and one can see that a large band 'm-C' on the left becomes split into two small bands 'm-C' on the right, because *Hpa* II cannot cut a methylated sequence Cm-CGG whereas *Msp* I can – as shown in the lower part of the diagram. In (b) we show the same kind of experiment using an agarose gel which has been stained with ethidium bromide, rather than hybridized to a specific radioactive probe. The genomic DNA contains many sequences CCGG, most of which are methylated, but some of which are not. The *Msp* I enzyme cuts all of them indiscriminately, breaking the DNA into a continuum of fragment sizes ranging from long (at the top of the gel) to short (at the bottom). In contrast, the *Hpa* II enzyme cuts only the un-methylated CCGG sequences . The sample on the left is therefore not cut into nearly so many pieces as the sample on the right; but the fuzzy series of bands on the lower left shows that some of the DNA fragments are short, indicating a tight clustering of the un-methylated cutting sites of HpaII into 'CG islands'. The lower diagram shows approximately the overall distribution of CG steps in the two parts of the genome. In (c) we show the treatment of human cells with a chemical 5-aza-cytosine, which inhibits enzymes that add methyl groups to cytosine, and thereby leads to loss of 5-methyl-cytosine across the entire genome; by studying the DNA extracted from such cells with *Hpa* II or *Msp* I, and the RNA extracted from such cells to look for messenger-RNA, scientists can determine whether loss of methylation might activate any particular gene.

The remaining (99%) of the genomic DNA had its CG sequences methylated; and so one could see a striking contrast in the gel for DNA digested with *Hpa* II: between the short pieces resulting from cutting *within* CG islands, *versus* the long uncut pieces of DNA

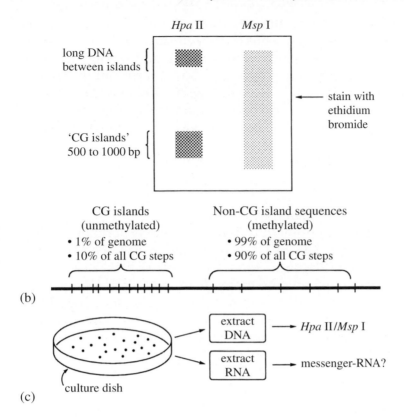

(b)

(c)

found *between* the islands. A striking contrast could also be seen between the high CG-dinucleotide content of the 'island' DNA – 10 times more than expected randomly – and a low CG-dinucleotide content of DNA found elsewhere in the same genome. That contrast will be explained below.

We know now that the DNA within any CG island often contains a promoter sequence for some nearby gene, and is usually protected from methylation by proteins that facilitate transcription; however, that was not known at the time. All of this shows that methylation of cytosine can indeed affect the activity of a gene – which may be related to cancer and other diseases, as we shall see below.

A third early technique was to treat cells in culture with a chemical known as '5-aza-cytosine', in order to *remove* methyl groups from many different cytosine bases across the whole genome: see Fig. 11.2(c). Then scientists could extract the total DNA from those cells, and digest it with *Hpa* II/*Msp* I to see whether any particular gene had lost its methyl groups. They could also extract total RNA, to see whether any particular gene had begun to make more messenger-RNA. The chemical 5-aza-cytosine is an analogue of normal cytosine, in which the 5-carbon of normal cytosine has been replaced by nitrogen. Thus 5-aza-cytosine acts as a strong inhibitor of most of the cellular enzymes, which add methyl groups CH_3 to cytosine in that same 5-position. Now it turns out that when mammalian or plant cells are treated with 5-aza-cytosine, certain genes do indeed become de-methylated and make more messenger-RNA. Hence, loss of DNA methylation was shown in some cases to correlate with increased gene activity.

Research involving DNA methylation in the years 1980–1990 seemingly became bogged down on account of lack of technical innovation. Thus the *Cold Spring Harbor Symposium* volume 47 for 1982 shows, for example, endless patterns of *Hpa* II/*Msp* I digestion along with numerous 5-aza-cytosine experiments! Since 1990 however, that field of DNA methylation has been reborn due to many new techniques, some of which will be described below. Furthermore, when new methods in methylation have been combined with innovations from normal genetics such as PCR, four-colour sequencing or microarrays (see Chapter 10), a great variety of new approaches to the understanding of cytosine methylation and its effects in biology have become available.

The most important breakthrough came in 1992, when Marianne Frommer, Geoff Grigg, Douglas Millar (and later, Susan Clark) developed a simple chemical method for determining whether any particular C base in total genomic DNA might be normal cytosine or else 5-methyl-cytosine. First, they treated their genomic sample with a reactive chemical called sodium bisulfite or $NaHSO_3$ (used also to clean swimming pools). When applied to single-stranded DNA, sodium bisulfite adds a sulfite group (SO_3) to the 6-position of cytosine: see Fig. 11.3(a). But bisulfite does not react with methyl-cytosine on account of steric hindrance from the nearby 5-methyl group; see Fig. 11.3(b). Next, they subjected their modified sample to high pH, so that all sulfonated cytosines would convert to uracil U, which is almost like thymine T: see Figs 11.3(a) and 2.13. Finally, they amplified one small part of their converted sample by PCR using two specific primers, and analyzed its base sequence by the di-deoxy method: see Chapter 10. They found that every methylated cytosine continued to read as C, whereas every unmethylated cytosine had been converted by bisulfite from C to U, which is read as T in the PCR and sequencing reactions.

The bisulfite method has since proven to be very valuable for analyzing 5-methyl-cytosine content in clinical DNA samples, in order to determine whether or not someone has a particular disease. For example, certain genes will lose or gain a particular series of methyl groups when any cell turns cancerous, with sufficient

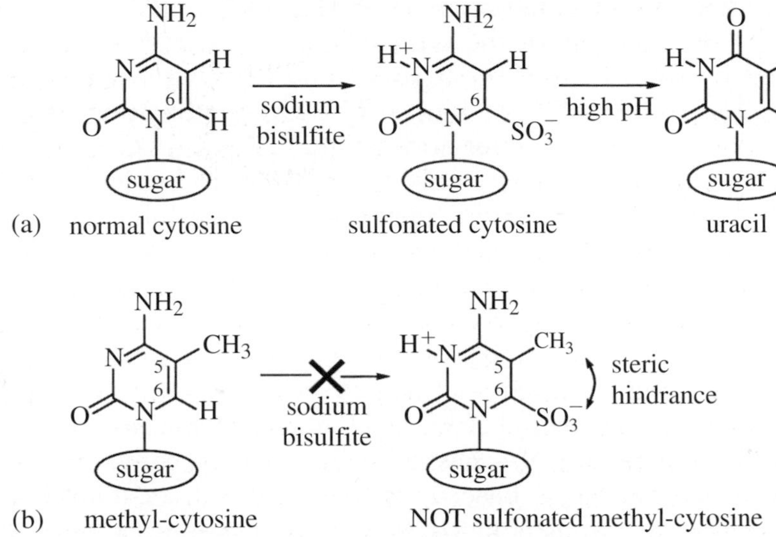

Figure 11.3 A summary of the bisulfite method that has proven very useful to distinguish normal cytosine from 5-methyl-cytosine. In (a) normal cytosine reacts well with sodium bisulfite ($NaHSO_3$) so as to leave sulfonate (SO_3) at the 6-ring position; then it converts to uracil once treated with high pH. In (b) 5-methyl-cytosine will not react with sodium bisulfite, due to steric hindrance between the 6-ring position and a large 5-methyl group.

reliability to enable this small molecular marker to be used as a diagnostic method. Soon, indeed, we should see new, non-invasive tests for many common forms of cancer (e.g. prostate, colon or breast), based on bisulfite-PCR analysis of low-level tumour DNA methylation in the blood or stool. Such DNA-based tests could eventually replace surgical or needle biopsies which, in addition to being painful, can also spread cancer cells undesirably through the lymph system when the needle disturbs the cancerous growth.

Scientists are even contemplating today an 'Epigenome Project', where cytosine methylation would be studied in many different human tissues or clinical samples, by comprehensive four-colour sequencing of both normal and bisulfite-treated DNA. Some people are actually discussing a 'methyl-SNP' project, where millions of different changes in cytosine-to-methyl-cytosine ('single-nucleotide-polymorphisms') would be used for the prediction of long-term human disease.

In addition to the chemical bisulfite method for detection and location of methyl-cytosine within a complex genome, we have also seen the discovery of more bacterial enzymes which are sensitive to 5-methyl-cytosine content in their cutting of DNA. Recall that the DNA-cutting action of *Hpa* II but not of *Msp* I is blocked by CG methylation at CCGG tetramers (Fig. 11.2(a)). Similarly, we know now that the activity of the enzyme *Sma* I but not of *Xma* I is blocked by CG methylation at CCCGGG hexamers; while all of *Hha* I, *Xho* I and *Not* I are blocked by CG methylation at their respective tetramer, hexamer and octamer sites GCGC, CTCGAG and GCGGCCGC. These are only a handful of examples from the hundreds of different bacterial 'restriction enzymes' that have now been discovered. Many of them show a sensitivity to cytosine methylation at a particular sequence; and they may be used to probe 5-methyl-cytosine within DNA having a wide range of sequences.

As well as the large number of 'cutting' enzymes, a broad variety of bacterial enzymes called 'methylases' have also been discovered, which will add new methyl groups CH_3 onto cytosine bases at many specific sequences of DNA. Here, we shall use the convention that m-C is a methylated C base. From analyzing diverse strains of bacteria, scientists were able to discover for example: *M. Sss* I, that methylates all CG to m-CG; *M. Hpa* II, that methylates CCGG to Cm-CGG; *M. Msp* I, that methylates CCGG to m-CCGG; and *M. Hha* I, that methylates GCGC to Gm-CGC. In all cases, the methylation occurs on both strands.

Lastly, we have learned much over the past ten years, about the various enzymes which add methyl groups to cytosine bases in human, mammalian or plant cells. The most abundant of these is called Dnmt1 in animals or MET1 in plants, as shown in Fig. 11.4(a). These convert DNA which has been methylated on only one strand into DNA that has been fully methylated on both strands. But how does it come about that DNA is sometimes found to be methylated only on one strand?

After replication, all bases newly added by the polymerase enzyme to the template strand are ordinary C, A, T and G, without modification (see Fig. 10.2(b)). Consequently, although the template strand may be methylated, the new strand will not be. Such DNA is said to be 'half-methylated' or 'hemi-methylated'; and so the replication process will create two half-methylated 'daughter' DNAs from the fully methylated 'parental' DNA. The purpose of Dnmt1 (or MET1) is to restore, or 'maintain', the original pattern of methyl groups that was present before replication.

Another abundant enzyme, which is found only in plants, is called CMT3 in *Arabadopsis*, or ZMET2 in maize. It converts half-methylated CNG to fully-methylated CNG where N may be any base. All of those enzymes are called 'maintenance methylases', because they serve to maintain some pre-existing pattern of methyl groups within the DNA, but cannot start a new pattern *de novo*. Maintenance methylases are found often in the vicinity of replication

```
        m                         m
5' ——————————— 3'         5' ——————————— 3'  ⎰ Dnmt1, animals
    ·· CG ··                   ·· CG ··        ⎱ MET1, plants
    ·· GC ··                   ·· GC ··
3' ——————————— 5'         3' ——————————— 5'
                                       m


        m                         m
5' ——————————— 3'         5' ——————————— 3'  ⎰ CMT3, plants
    ·· CNG ··                  ·· CNG ··       ⎱ ZMET2, maize
    ·· GNC ··                  ·· GNC ··
3' ——————————— 5'         3' ——————————— 5'
                                       m
```

(a) maintenance methylases

```
                                  m    m    m
5' ——————————— 3'         5' ——————————— 3'  ⎧ Dnmt3a, 3b
   · CG · CA · CT ·           · CG · CA · CT ·  ⎨ mammalian
   · GC · GT · GA ·           · GC · GT · GA ·  ⎩  embryo
3' ——————————— 5'         3' ——————————— 5'
                                  m


                                  m      m
5' ——————————— 3'         5' ——————————— 3'  ⎰ Dnmt2
   ··· CA ·· CT ·            ··· CA ·· CT ·    ⎱ fly embryo
   ··· GT ·· GA ·            ··· GT ·· GA ·
3' ——————————— 5'         3' ——————————— 5'
```

(b) *de novo* methylases

```
                                  m    m    m
5' ——————————— 3'         5' ——————————— 3'  ⎧ Dnmt3a
   · CG · CG · CG ·           · CG · CG · CG ·  ⎨  and
   · GC · GC · GC ·           · GC · GC · GC ·  ⎩ Dnmt3L
3' ——————————— 5'         3' ——————————— 5'
   ⟨ 3a ⟩═⟨ 3L ⟩                  m    m    m
```

(c) imprinting when 3a, 3L bind together

```
    (RNA)                     (RNA)
3' ············· 5'        3' ············· 5'  ⎫
   · ACGAGGUAGA ·             · ACGAGGUAGA ·     ⎪
5' · TGCTCCATCT · 3'       5' · TGCTCCATCT · 3'  ⎬ plants
    (DNA)                     ↑ m mm    m        ⎪
                              (DNA)              ⎪
                                                 ⎪
3' · ACGAGGTAGA · 5'       3' · ACGAGGTAGA · 5'  ⎭
    (DNA)                     (DNA)
```

(d) RNA-induced *de novo* methylation

Figure 11.4 A summary of the enzymes that add methyl groups to cytosine in animals and plants. In (a) we show the 'maintenance methylases' which will add a second methyl group to the strand opposite where one already exists, but will not add two methyl groups *de novo*; these are called Dnmt1 in animals or MET1 in plants for the sequence CG; or else CMT3 in plants or ZMET2 in maize for the sequence CNG where N, N can be any base and its complement. In (b) we show the '*de novo* methylases' which will add two new methyl groups to any unmethylated site; these are called Dnmt2 in fly embryos for the sequences CA or CT, or Dnmt 3a or 3b in mammalian embryos for the sequences CG, CA or CT in order of prevalence (high on left, low on right). In (c) we show an 'imprinting protein' Dnmt3L which has no methylase activity of its own, yet helps *de novo* methylase Dnmt3a to methylate genes in mammalian embryos in particular circumstances. In (d) a short, single-stranded region of RNA binds to a complementary strand of double-helical DNA, thereby displacing the other DNA strand – here shown below the RNA-DNA hybrid; which is a signal for the methylase to methylate *every* cytosine of the bound DNA strand.

forks in dividing cells. There some old DNA strand carries an established pattern of methyl groups, whereas the newly-made strand does not, at least until a maintenance methylase goes to work on it!

Another general class of enzymes are known as *de novo* methylases, because they can create a new pattern of methyl groups within the DNA where none previously existed – for example in early embryos. As shown in Fig. 11.4(b), fly embryos contain a *de novo* methylase called Dnmt2, which methylates CA or CT rather than CG, yet has little activity in humans. Mammalian embryos and embryonic stem-cells contain two important *de novo* methlyases called Dnmt3a and Dnmt3b. Both will convert unmethylated CG to fully-methylated CG, and also will methylate CA or CT weakly. A related protein Dnmt3L shows no activity of its own, yet helps Dnmt3a to make a methylation 'imprint' in female pro-nuclear DNA (i.e. the DNA contribution from an egg cell prior to nuclear formation), as shown in Fig. 11.4(c).

Lastly, there seems to exist in plants a special *de novo* methylase which adds methyl groups to *all* cytosine bases, within a small region of RNA-DNA binding as shown in Fig. 11.4(d). That methylase is somehow directed by a short, 25-to-50 nucleotide segment of RNA, which binds to specific sequences of DNA within a plant chromosome; yet so far its amino-acid identity remains unknown. The existence of such an RNA-directed, *de novo* activity has nevertheless enabled scientists to manipulate the DNA methylation patterns of plants in a directed fashion, using short pieces of added non-cellular RNA; but not yet in animals, where the enzymes seem to be different.

In principle, those methylation enzymes might methylate *all* CG steps in the genome; but in practice, only a subset of CG steps are actually methylated. How might these enzymes, whether maintenance or *de novo*, know where to add methyl groups within a total genome? It is now known that such methylases will be targeted or 'recruited' to particular genomic regions by other proteins which already happen to be binding there. For example, as shown in Fig. 11.5(a), the mammalian protein 'methyl-CG-2' or MeCP2 binds preferentially to CG methylated DNA of any kind; and also binds to the maintenance methylase Dnmt1 (or to the imprinting protein Dnmt3L), as a way of localizing methyl-adding activity where it is most needed.

Mammalian or plant cells also contain a wide range of enzymes which can modify the histone proteins of chromatin. It is known from recent work that such histone-modifying enzymes can localize DNA methylases to particular genes. For example, the most abundant

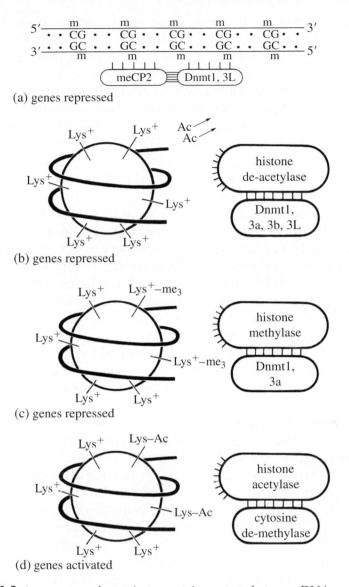

(a) genes repressed

(b) genes repressed

(c) genes repressed

(d) genes activated

Figure 11.5 A summary of protein-to-protein contacts between DNA methylases and other cellular enzymes. In (a) we see that the major methyl-cytosine-binding-protein 'methyl-CG-2' (or MeCP2) will bind to both methylases Dnmt1 or Dnmt3L, thereby linking methylation with maintenance or imprinting activities. In (b) we see that the major histone de-acetylase that removes acetyl groups from lysine on histone proteins, will bind to all of Dnmt1, 3a, 3b and 3L, thereby linking histone deacetylation with all forms of cytosine methylation. In (c) we see that histone methylase (which adds methyl groups to lysine 9 of histone H3) will bind to either Dnmt1 or Dnmt3a, thereby linking histone methylation with maintenance or imprinting. Finally in (d) we wonder whether histone acetylase (which adds acetyl groups to lysines) might bind to an unpurified cytosine de-methylase. All of phenomena (a, b, c) are associated with repressed genes, whereas (d) is associated with active genes.

cellular enzyme which *removes* an acetyl group from lysine is known as 'histone de-acetylase 1'. It binds well to all of the cytosine methylases Dnmt1, Dnmt3a, 3b and 3L, as shown in Fig. 11.5(b). Such protein-to-protein associations serve to link histone de-acetylation with cytosine DNA methylation, as two redundant factors which can both repress genes. There also exist 'histone methylases' which can add methyl groups to lysine (e.g. lysines 4 or 9 of histone H3), especially within repetitive human DNA sequences (some of which have been given names, such as 'Alu', 'SINE' or 'LINE'). Those histone methylases also bind to the Dnmt1 or Dnmt3a methylases, and repress genes similarly, as shown in Fig. 11.5(c).

When histone de-acetylase is inhibited by a small molecule such as sodium butyrate, valproate or trichostatin A, the activity of other enzymes known as 'histone acetylases' (which *add* acetyl groups to lysine) become predominant. Then we see highly-acetylated histone proteins in living cells, which readily activate genes as shown in Fig. 11.5(d). Such histone acetylases may bind in principle to (hypothetical) cytosine de-methlyase enzymes, which would remove methyl groups from nearby CG sequences.

So much for the enzymes which control DNA methylation, or histone acetylation/methylation: they seem very complicated! But what about the *effects* of such epigenetic changes on biological activity? When studied in a broad perspective, DNA methylation is now known to regulate five general kinds of biological activity in higher animals or plants: (i) formation of specific tissues in the embryo (e.g. *de novo* methylation by Dnmt3a or Dnmt3b); (ii) imprinting of genes in a newly fertilized egg cell (see Chapter 10); (iii) transcriptional repression of an entire chromosome (e.g. one of the two female X chromosomes in mammals); (iv) transcriptional repression of highly selected genes (e.g. by blocking the binding of transcription factors to a promoter, or by recruiting MeCP2); and (v) suppression of invading foreign DNA.

As shown in Fig. 11.6, the male contribution to the genome of the embryo (or 'male pronucleus') becomes actively demethylated by unknown enzymes (which we may call 'sperm de-methylases') just after fertilization, and before the male and female pronuclei join together as homologous chromosome pairs. All DNA molecules of both male and female origin then become passively demethylated by dilution of the parental strand through successive cell-divisions, for a brief period of time when the methylase enzymes remain absent. Finally, once *de novo* methylases Dnmt3a and Dnmt3b as well as maintenance methylase Dnmt1o (a variant of Dnmt1) begin to be expressed around the four-to-eight-cell stage, the total DNA of both genomes slowly begins to gain methyl groups again,

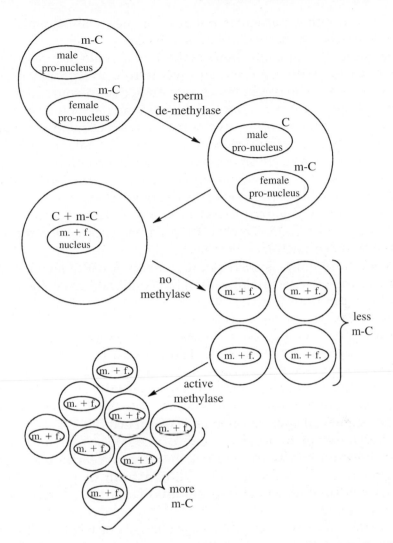

Figure 11.6 A summary of changes in DNA cytosine methylation that have been observed within early mammalian embryos. The fertilized egg (upper left) contains initially both male-origin and female-origin chromosomes from sperm and egg respectively, both of which are highly methylated. Next a sperm-demethylase activity removes most of the cytosine methylation from male-origin chromosomes but not the female, within 4 to 6 hours as shown at upper right. Then both half-sets of chromosomes merge to form homologous pairs (middle left), after which the newly-created cell undergoes limited growth and division in the absence of any methylase enzyme activity. Hence even female-origin chromosomes will become passively demethylated through dilution, when methyl-cytosine is replaced by normal cytosine during DNA replication (middle right). Finally, all of the cytosine methylases Dnmt1o (a variant of Dnmt1 found in embryos), Dnmt3a and Dnmt3b are expressed at a late stage of development: thus various genomic sites on both male and female-origin chromosomes will undergo *de novo* DNA methylation, which seems essential for the formation of correct body tissues.

thereby adopting particular patterns of methylation which may help to determine particular kinds of tissue.

'Imprinted' regions of both male- and female-origin genomes (recall Chapter 10) represent an exception to this general picture, because only male-origin genes remain active in some cases, and female-origin in others. Another exception is seen just in female embryos, where one of the two homologous X chromosomes shuts down and becomes transcriptionally inactive or 'silent'. The choice of which of the two chromosomes will be turned off appears to be random; and it is seems that once a 'choice' has been made, the closing down must spread cooperatively throughout one and only one of the two chromosomes. By a poorly understood process, the *Xist* RNA triggers DNA methylation on one copy of X but not the other (see Chapter 10).

When farm animals such as pigs, sheep or cattle are cloned by 'nuclear transfer' technology, then the combined male and female DNA contributions from a mature cell-nucleus are added all at once to an empty pig, sheep or cow egg-cell. Often the well-organized process of de-methylation and subsequent re-methylation (see Fig. 11.6) then goes badly wrong, and so the vast majority of nuclear-transfer embryos will be aborted. Obviously the male pronucleus in such a case will not be de-methylated specifically *versus* the female; while the carefully-timed expression of cytosine methylases in the four-to-eight cell embryo often becomes aberrant.

Similarly, when transgenic mice are prepared so that most of their Dnmt1 methylase activity is absent, the effects on early development are profound. Such mice appear as runts at birth; they carry chromosomal abnormalities; and by six months of age they are replete with aggressive cancers.

In adult animals or plants, many changes of DNA methylation patterns can be detected. Those changes reflect both a normal, time-dependent aging of the organism, as well as an abnormal conversion of normal cells into cancer cells. It is not certain yet whether changes of cytosine methylation are a primary event in cancer, or are secondary to other phenomena which silence genes, for example histone de-acetylation. In any case, such changes of DNA methylation seem reliable enough to diagnose cancer for medical purposes, without the need for surgical biopsies (as evidenced by ongoing clinical trials).

How precisely might changes of DNA methylation be related to aging or cancer in mammals, at a deep biochemical level? A large part of the aging and cancerous processes has to do with imperfect enzymatic repair of mutations in which a C–G base pair has become a T–G base-pair; these mutations arise naturally and frequently in methylated DNA. As mentioned already, it is well known that animal or plant genomes, which contain mostly methyl-cytosine at their CG steps, also show a reduced incidence of those CG steps, of only 10% to 20% relative to the incidences of GC, CC or GG. Why should such a profound deficiency exist? It is simply because cytosine bases C are continually 'de-aminating' their base rings at a constant, slow rate, to yield the RNA base uracil U as shown in Fig. 11.7(a). Such ongoing de-amination changes CG steps to UG, or GC steps to GU, or CC steps to CU, once every minute or so, at some location in a large genome! But in fact, most cells contain a repair enzyme known as 'uracil glycosylase', which goes around cutting out U bases wherever it can find them, and then a repair polymerase pairs the opposite G base with C in order to fix the damage (see Fig. 11.7(c)).

Yet if the original base happens to be methyl-cytosine rather than cytosine, such deamination will leave the normal base thymine T, rather than the abnormal base uracil U as shown in Fig. 11.7(b). (This process has the same end result as that shown in Fig. 11.3(a); but without the mediation of sodium bisulfite.) De-amination will thus change any m-CG step to normal TG, which uracil glycosylase cannot repair. There does exist another repair enzyme called 'thymine glycosylase', that recognizes the base-pair mismatch of TG on one strand with CG on the other (i.e. where G pairs with T), then cuts out the T and replaces it by C with the help of a repair polymerase (see Fig. 11.7(d)). But thymine glycosylase is far less efficient than uracil glycosylase at removing defects. Hence any individual m-CG step will decay to TG in mammals with a half-life of perhaps 1000 to 10 000 years, unless that particular m-CG happens to be essential for the working of the cell. All of these processes taken together result in a long-term deficiency of CG steps within a mammalian genome, while the frequency of the decay products TG and CA will often be enhanced.

Now when human DNA from adult tissues is compared with human DNA from embryos, scientists often find numerous CG to TG step mutations that correlate with age, auto-immunity or cancer. For example, most mutations in colon or brain cancer are natural de-aminations of that kind. Other mutations may result from damage by environmental sources, for example sunlight or carcinogens. In any case, it is likely that there will soon be diagnostic assays for colon cancer, aging, or infertility, which will be based on the analysis of specific CG to TG step mutations which accumulate naturally in our tissues as we age, but at different rates in different people. If the normally unmethylated CG steps within a 'CG island' become methylated aberrantly, for example, that is usually very serious; because it may shut down some essential gene for a tumor-suppressor protein, and thereby cause cancer directly.

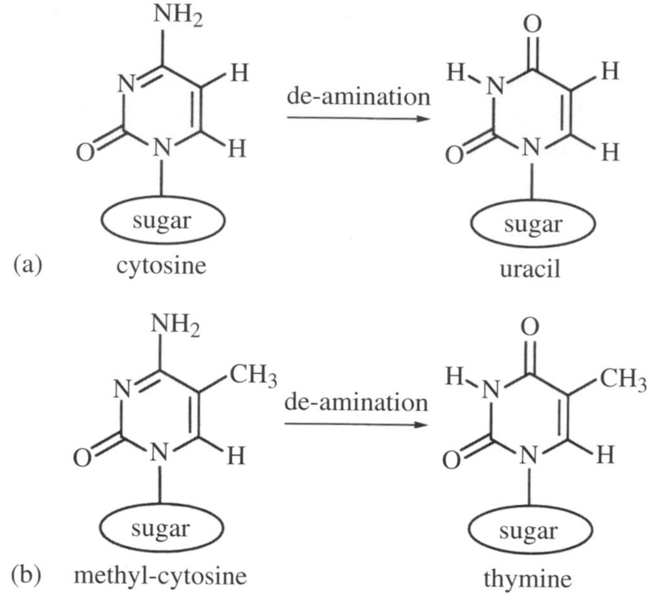

Figure 11.7 Two processes of chemical decay, followed by enzymatic repair, influence the di-nucleotide step composition of animal and plant genomes. In (a) we see that cytosine de-aminates continually to uracil at a tiny but finite rate. In (b) we see similarly that 5-methyl-cytosine de-aminates continually but slowly to thymine. In (c) we see that a change of step CG to UG may be fixed by a repair enzyme called 'uracil glycosylase', that seeks any aberrant uracil U bases (which are not normally present in DNA) and cuts them out entirely. A repair polymerase then restores the missing base. But (d) changes of the kind m-CG step to TG cannot be fixed so efficiently, since thymine T is a normal DNA base. Another repair enzyme 'thymine glycosylase' removes certain m-CG to TG step changes, based on recognition of a G–T base-pair mismatch between the mutated TG sequence and its partner CG on the other strand; yet this latter process is not entirely efficient, and so some m-CG to TG step changes will accumulate, despite repair. The overall result will be to deplete CG sequences from animal and plant genomes, wherever cytosine bases might typically be methylated.

All of a sudden this discussion has taken an unpleasant turn, because cytosine methylation seems to make us old and sick, without any possible cure! Why should such a hazardous system have evolved? If the fly or worm can do without cytosine methylation, why cannot we? When we study this question more closely, we can see that the biochemical system of DNA methylation, when combined with a decay of methyl-cytosine to thymine, actually serves as an excellent defence against foreign invading DNA; which was probably essential during our long evolutionary history.

For example, the human race (or ancestors thereof) have been invaded roughly 10 000 times during the past fifty million years,

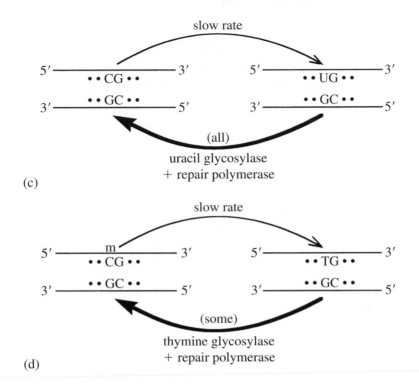

slow rate

5'————————3'   5'————————3'
    •• CG ••   •• UG ••
3'————————5'   3'————————5'
    •• GC ••   •• GC ••

(all)
uracil glycosylase
+ repair polymerase

(c)

slow rate

5'————————3'   5'————————3'
      m
    •• CG ••   •• TG ••
3'————————5'   3'————————5'
    •• GC ••   •• GC ••

(some)
thymine glycosylase
+ repair polymerase

(d)

with foreign DNA elements known as 'mariner transposons', that inhabit specialized, single-cell micro-organisms known as 'trypanosomes', which in turn live inside insects such as the tsetse fly. Now when an infected tsetse fly bites a human, it will transfer some of its trypanosomes into the human bloodstream, where they can multiply and cause illnesses such as 'sleeping sickness' and 'Chagas disease'. Once in the blood, those trypanosomes can also infect sex cells (usually sperm) with their own mobile DNA (or fly DNA), and thereby transfer mariner transposons into the human germline, and consequently into subsequent generations!

How could our ancestors have defended themselves against such ancient DNA invasions, which now make up an astonishing 45% of the human genome? They defended themselves largely by means of a methylation defence system, which works as follows. Since most invading DNA elements remain high in CG dinucleotide content, because they come from micro-organisms that lack cytosine methylation, the cells of our ancestors could add methyl groups almost immediately to any unprotected CG within the foreign DNA, as shown in Fig. 11.8(a) and (b). Those cells could thereby silence any foreign genes, that might be responsible for inserting or 'jumping' foreign DNA within a chromosome, or else for making an RNA copy to help the foreign DNA to reproduce.

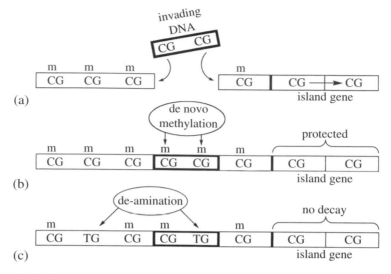

Figure 11.8 Cytosine methylation actually serves as a system of defence against invading foreign DNA, say from transposons or retrotransposons. In (a) we see that the invading DNA contains typically unmethylated CG sequences, especially if it comes from some simple animal such as a worm or fly. On the right, we see also a 'CG island' and its associated gene that remain typically unmethylated. Next in (b) we see that *de novo* methylases such as Dnmt3a or 3b will quickly add new methyl groups to the foreign DNA, whereas any nearby CG island will remain protected from methylation by bound proteins. Then in (c) we see that all methylated CG steps will undergo slow but continual deamination to TG, both in the animal genome and also in the foreign DNA; but not at any unmethylated CG island nearby. Hence foreign genes will be silenced immediately but reversibly by cellular methylases; and then will be silenced slowly but permanently by methyl-CG to TG mutations in their coding regions.

By this system, most CG sequences in the human chromosome would also become methylated and inactive. However, the human genome is designed so that it contains a large number of 'CG islands', where CG sequences remain protected from methylation by tightly-bound proteins which facilitate transcription. Almost all of those CG islands remain protected from methylation, so that they can specify active genes where needed: see Fig. 11.8(b).

Next, once any invading DNA has been silenced by CG methylation, all of its coding regions will begin slowly to decay, due to continual deamination of methyl-CG steps to TG without repair as shown in Fig. 11.8(c). Hence most foreign genes will become mutated and inactive within just a few generations. One can study in fact various foreign genes within the human genome, and estimate their original date of invasion by the ratio of CG-to-TG steps, just as for a radioactive half-life in physics.

Somewhat amazingly, DNA oligonucleotides which have been tested in humans for medical purposes (see Appendix 3) can sometimes induce an immune response in human blood, if they happen to contain one or more unmethylated CG step or steps. The human body seems to be designed so that it possesses several layers of defense against invading DNA from micro-organisms: whether by an immunological response at the whole-organism level; or by methylation response within individual cells.

In summary, we have learned here how cytosine DNA methylation works biochemically in humans and plants, through a variety of different enzymes, and through specific protein-to-protein interactions at a chromosomal level. Also, we have discussed its great importance in biological or medical phenomena such as embryonic development, aging or cancer. How then might specific patterns of DNA methylation be established in early embryos? And how might specific patterns of methylation be changed in a controlled fashion, so as to alleviate aging or cancer in adults?

No one knows today how new patterns of DNA methylation might be laid down in early embryos; we can only speculate over the possibilites. We know that the male pro-nucleus loses most of its methyl-cytosine (except at imprinted regions) just after fertilization; and that the female pro-nucleus loses some of its methyl-cytosine (except at imprinted regions) due to a limited number of cell divisions, before any methylase becomes active. But once expressed, the *de novo* enzymes Dnmt3a and Dnmt3b are somehow directed to add new methyl groups within a growing embryo, with great specificity across the entire genome, so as to create different tissues. Not all of the old methyl groups would be lost from the original sperm or egg DNA, especially within imprinted regions and on female-origin chromosomes. Perhaps those prior patterns form the initial loci of further *de novo* methylation?

Other epigenetic factors, say histone acetylation or methylation, might direct such *de novo* DNA methylation to certain regions of the genome but not others. Another contributing factor might be small RNA molecules that serve as 'guides' for *de novo* cytosine methylation in animals as well as plants. We know that small fragments of RNA can guide 2'-ribose methylation of ribosomal RNA, and can guide the degradation of messenger-RNA by nucleases (see Appendix 3). Also, long, repetitive RNA such as *Xist* or *Air* might induce *de novo* cytosine methylation all across the genome, as well as at the X-chromosome and imprinting sites.

Finally, how might one alter specific patterns of methylation for medical purposes, for example to alleviate aging or cancer? The de-methylating agent 5-aza-cytosine is too toxic to give to humans, although a safer chemical known as 'valproate' shows some activity of that kind. Overall methylase activity may be reduced by using 'anti-sense' oligomers which degrade the messenger-RNA for Dnmt1; while methyl groups may perhaps be added to specific genes by using oligomers that contain 5-methyl-cytosine. Some workers have used anti-sense oligomers to degrade RNA molecules such as *Xist*, which control methylation imprinting, and that show their own anti-sense in living cells called *'Tsix'*. Most work here, however promising, still seems to be a long way from clinical applications.

To conclude: if as a student you have reached the end of this Chapter, and have also read the other ten Chapters in this book, you should have gained much knowledge concerning the structure of DNA and its role in biology. Scientists of the past were no less clever than the ones today, yet they had less technology by which to do good experiments. Also they perceived the world differently in each era: many of their well-established ideas turned out to be faulty! When contemplating your own future in science, therefore, be sure to keep an open mind. Also try to remember what Aldous Huxley wrote about scientific research in 1960: 'What we perceive depends on the conceptual lattice through which it has been filtered'.

Further Reading and Bibliography

Early hypotheses, recent reviews

Holliday, R. and Pugh, J.E. (1975) DNA modification mechanisms and gene activity during development. *Science* **187**, 226–32. An early article suggesting the importance of cytosine methylation for gene activity.

Jaenisch, R. and Bird, A. (2003) Epigenetic regulation of gene expression: how the genome integrates intrinsic and environmental signals. *Nature Genetics Supplement* **33**, 245–54. A summary of the many factors and/or proteins which influence patterns of cytosine methylation.

Millar, D.S., Holliday, R., and Grigg, G.W. (2003) Five not four: history and significance of the fifth base. In *The Epigenome: Molecular Hide and Seek*, eds. S. Beck and A. Olek, Wiley Academic Press, New York, pp. 1–18. A historical review of 5-methyl-cytosine and its role in biology.

Riggs, A.D. (1975) X inactivation, differentiation and DNA methylation. *Cytogenetics and Cellular Genetics* **14**, 9–25. An early article suggesting the importance of cytosine methylation for X-chromosome inactivation.

Turner, B.M. (2001) *Chromatin and Gene Regulation*: *Molecular Mechanisms in Epigenetics* Blackwell Science, Oxford. A survey of chromatin as it relates to epigenetic phenomena such as DNA or histone methylation.

Chemical or enzymatic methods of methylation analysis

Fraga, M.F. and Esteller, M. (2002) DNA methylation: a profile of methods and applications. *Biotechniques* **33**, 632–49. A survey of methods used to analyze cytosine methylation at specific sites in chromosomal DNA.

Gonzalgo, M.L. and Jones, P.A. (2002) Quantitative methylation analysis using methylation-sensitive single-nucleotide primer extension. *Methods* **27**, 128–33. Extension of a short primer by DNA polymerase on bisulfite-treated DNA can determine whether any position contains 5-methyl-cytosine or else normal cytosine.

Grigg, G. and Clark, S. (1994) Sequencing 5-methyl-cytosine residues in genomic DNA. *Bioessays* **16**, 431–6. Invention of the bisulfite method for determining whether any base might be cytosine or 5-methyl-cytosine.

Li, J., Protopopov, A., Wang, F., Senchenko, V. *et al.* (2002) *Not* I subtraction and *Not* I-specific microarrays to detect copy number and methylation changes in whole genomes. *Proceedings of the National Academy of Sciences, USA* **99**, 10724–9. An enzymatic method for profiling normal *versus* cancerous cells, based on the sensitivity of *Not* I (GCGGCCGC) to methylation at CG steps.

Millar, D.S., Warnecke, P.M., Melki, J.R., and Clark, S.J. (2002) Methylation sequencing from limiting DNA: embryonic, fixed and microdissected cells. *Methods* **27**, 108–13. Procedures for bisulfite analysis of cytosine methylation from small quantities of DNA.

Cytosine methylation in whole genomes

Amoreira, C., Hindermann, W., and Grunau, C. (2003) An improved version of the DNA methylation database. *Nucleic Acids Research* **31**, 75–7. An Internet-based compilation of all data regarding cytosine methylation at **http://www.methdb.net.**

Bird, A.P. (1986) CpG-rich islands and the function of DNA methylation. *Nature* **321**, 209–13. A small fraction of human DNA consists of small CG-rich regions known as 'islands', which usually remain unmethylated and may contain the regulatory sequence for a nearby gene.

Ramsahoye, B.H., Biniszkiewicz, D., Lyko, F., Clark, V. *et al.* (2000) Non-CpG methylation is prevalent in embryonic stem cells and may be mediated by DNA methyltransferase Dnmt3a. *Proceedings of the National Academy of Sciences, USA* **97**, 5237–42. DNA sequences other than CG may contain 5-methyl-cytosine in embryonic stem cells, where Dnmt3a adds methyl groups occasionally to CA and CT.

Tomso, D.J. and Bell, D.A. (2003) Sequence content at human SNPs: over-representation of the CG dinucleotide at polymorphic sites, but suppression of variation in CpG islands. *Journal of Molecular Biology* **327**, 303–8. CG dinucleotides are more polymorphic than others at general methylated sites in human DNA, due to chemical instability without repair caused by deamination; but are less polymorphic in 'CG islands', where cytosine bases remain unmethylated.

Cytosine methylation in cancer, embryonic development and cloning

Cezar, G.G., Bartolomei, M.S., Forsberg, E.J., First, N.L. *et al.* (2003) Genome-wide epigenetic alterations in cloned bovine fetuses. *Biology of Reproduction* **68**, 1009–14. Cattle embryos which are prepared by nuclear transfer show reduced levels of cytosine methylation, so as to cause abortion.

Clark, S.J. and Melki, J.M. (2002) DNA methylation and gene silencing in cancer: which is the guilty party? *Oncogene* **21**, 5380–7. Gene silencing prior to cytosine methylation may be the cause of human cancers.

Gaudet, F., Hodgson, J.G., Eden, A., Jackson-Grusby, L. *et al.* (2003) Induction of tumors in mice by genomic hypomethylation. *Science* **300**, 489–92. Transgenic mice with low but non-lethal amounts of Dnmt1 acquire chromosomal abnormalities and cancer at an early age.

Hardeland, U., Bentele, M., Lettieri, T., Steinancher, R. *et al.* (2001) Thymine DNA glycosylase. *Progress in Nucleic Acid Research and Molecular Biology* **68**, 235–53. Thymine glycosylase repairs mutations (methyl-CG to TG), along with uracil glycosylase (CG to UG).

Pfeifer, G.P. and Denissenko, M.F. (1998) Formation and repair of DNA lesions in the p53 gene: relation to cancer mutations? *Environmental and Molecular Mutagenesis* **31**, 197–205. Some of the mutations that cause cancer are biochemical (methyl-CG to TG), whereas others are environmental (sunlight, carcinogens).

Santos, F., Hendrich, B., Reik, W., and Dean, W. (2002) Dynamic reprogramming of DNA methylation in the early mouse embryo. *Developmental Biology* **241**, 172–82. Loss of methylation in the male pro-nuclear DNA proceeds within six hours after fertilization, perhaps because the de-methylase is sperm-specific; whereas the female pro-nuclear DNA remains methylated initially, followed by passive loss of methylation due to cell division; finally the methylases Dnmt1, Dnmt3a and Dnmt3b become active at a later stage.

Pharmaceutical strategies involving cytosine methylation

Goffin, J. and Eisenhauer, E. (2002) DNA methyltransferase inhibitors – state of the art. *Annals of Oncology* **13**, 1699–716. 5-azacytidine and 5-aza-2′deoxycytidine which inhibit methylases, as well as an antisense oligonucleotide which binds to the messenger-RNA for Dnmt1, are being tested as anti-cancer drugs.

Holliday, R. and Ho, T. (2002) DNA methylation and epigenetic inheritance. *Methods* **27**, 179–83. Certain genes may be activated using 5-aza-cytidine (a methylase inhibitor), or else may be silenced using 5-methyl-cytosine triphosphate (when incorporated during replication).

Krieg, A.M. (1999) Mechanisms and applications of immune stimulatory CpG oligonucleotides. *Biochimica et Biophysica Acta* **1489**, 107–16.

Certain DNA oligonucleotides will produce an immune response in humans, if they contain unmethylated CG sequences that resemble DNA of microbial origin.

Yao, X., Hu, J.F., Daniels, M., Shiran, H. *et al.* (2003) A methylated oligonucleotide inhibits IGF2 expression and enhances survival in a model of hepatocellular carcinoma. *Journal of Clinical Investigation* **111**, 265–73. A DNA oligonucleotide containing 5-methyl-cytosine may induce altered DNA methylation at specific sites in cells.

Enzymes or proteins which regulate cytosine methylation

Cervoni, N. and Szyf, M. (2001) Demethylase activity is directed by histone acetylation. *Journal of Biological Chemistry* **276**, 40778–87. Increased acetylation of histones leads to active de-methylation of nearby DNA, although the enzyme responsible has not been characterized.

Chedin, F., Lieber, M.R., and Hsieh, C.-L. (2002) The DNA methyltransferase-like protein Dnmt3L stimulates *de novo* methylation by Dnmt3a. *Proceedings of the National Academy of Sciences, USA* **99**, 16916–21. The protein Dnmt3L, which is responsible for methylation 'imprints' in the female genome, stimulates *de novo* cytosine methylation by Dnmt3a but not by Dnmt3b.

Fuks, F., Hurd, P.J., Wolf, D., Nan, X. *et al.* (2003) The methyl-CpG-binding protein MeCP2 links DNA methylation to histone methylation. *Journal of Biological Chemistry* **278**, 4035–40. The mammalian protein MeCP2, which binds specifically to CG-methylated DNA, also binds specifically (perhaps through Dnmt1) to histone de-acetylases and histone methylases.

Huang, N., Banavali, N.K., and MacKerell, A.D. Jr. (2003) Protein-facilitated base-flipping in DNA by cytosine-5-methyl-transferase. *Proceedings of the National Academy of Sciences, USA* **100**, 68–73. A cytosine base may flip out from a double helix to become methylated by M.*Hha* I from bacteria, which resembles Dnmt2 from humans.

Kimura, H. and Shiota, K. (2003) Methyl-CG-binding protein MeCP2 is a target for the maintenance DNA methyltransferase Dnmt1. *Journal of Biological Chemistry* **278**, 4806–12. The mammalian protein MeCP2, which binds specifically to CG-methylated DNA, also binds specifically to Dmnt1.

DNA methylation in plants, fungi, flies and bacteria

Aufsatz, W., Mette, M.F., van der Winden, J., Matzke, A.J., and Matzke, M. (2002) RNA-directed DNA methylation in *Arabidopsis*. *Proceedings of the National Academy of Sciences, USA* **99**, 16499–506. Short double-stranded RNA of 21 to 24 nucleotides in plants can degrade messenger-RNA in the cytoplasm, or induce *de novo* methylation of DNA in the nucleus: every cytosine in the RNA-DNA is methylated by an unknown enzyme.

Hamilton, A., Voinnet, O., Chappell, L., and Baulcombe, D. (2002) Two classes of short interfering RNA in RNA silencing. *EMBO Journal* **21**, 4671–9. Short RNA of 21 nucleotides guides degradation of messenger-RNA in the cytoplasm, while short RNA of 25 nucleotides guides cytosine methylation of DNA in the nucleus.

Kato, M., Miura, A., Bender, J., Jacobsen, S.E. *et al.* (2003) Role of CG and non-CG methylation in immobilization of transposons in *Arabidopsis*. *Current Biology* **13**, 421–6. Two maintenance methylases MET1 for CG or CMT3 for CNG act to suppress 'jumping' of DNA transposons in the plant *Arabidopsis thaliana*.

Tamaru, H. and Selker, E.U. (2003) Synthesis of signals for *de novo* DNA methylation in *Neurospora crassa*. *Molecular and Cellular Biology* **23**, 2379–94. Repeat sequences composed of the easily unwound $(TAAA)_n$ or $(TTAA)_n$ induce *de novo* methylation at nearby CG sequences in a bread mould.

Tamaru, H., Zhang, X., McMillen, D., Singh, P.B. *et al.* (2003) Trimethylated lysine 9 of histone H3 is a mark for DNA methylation in *Neurospora crassa*. *Nature Genetics* **34**, 75–9. Adding methyl groups to lysine 9 of histone H3 is associated with chromatin condensation (in yeast or flies) or cytosine methylation (in mould, mammals or plants).

Urig, S., Gowher, H., Hermann, A., Beck, C. *et al.* (2002) The *Escherichia coli dam* DNA methyltransferase modifies DNA in a highly processive reaction. *Journal of Molecular Biology* **319**, 1085–96. N6-adenine methylase from bacteria controls replication, gene expression and mismatch-repair by modifying a sequence GATC.

Postscript

Our book is a small one, but the investigation of DNA is clearly a large subject, and potentially very large indeed. We have included eleven chapters on what we consider to be the most fundamental and well-understood aspects of DNA research; but we have omitted several potential chapters that might have dealt with other, less well-understood subjects such as: (a) the chemical theory of base-stacking in the context of DNA bound to antibiotics or proteins; (b) 'recombination' of the DNA molecule by various enzymes which cross-over and then re-connect its two strands; (c) the possible use of DNA for constructing tiny molecular machines or novel computational devices, generally known as 'nanotechnology'; and (d) the impact of genomic sequencing on our understanding of molecular evolution, and the apparent historical relationships among many different forms of life on Earth.

The chemical theory of base-stacking (a) in free DNA is discussed briefly in Appendix 2, with special reference to the recent work of Chris Hunter. We anticipate that other aspects of DNA chemical theory will soon be better understood, particularly in the context of its specific interactions with proteins, since new protein-DNA X-ray structures are now appearing almost weekly.

With regard to (b) recombination, it should be emphasized that the long-range ordering of base-pairs along any chromosomal DNA molecule need not always be rigidly fixed, but can change somewhat due to various enzymatic processes, as the cell grows and divides. Such *recombinatorial* changes may occur even before the fertilization of an egg cell by a sperm, when homologous pairs of chromosomes join together to make the precursor to a sex cell. Recombination of a type called 'V-D-J' is also very important for generating immunological diversity in adult human cells, so as to achieve T-cell or antibody recognition of foreign particles. (Changes of base sequence such as CG to TG as noted in Chapter 11 may also contribute to such diversity). Finally, recombination occurs frequently in mature cells as part of the general mechanism for DNA repair: where a severely damaged stretch of bases on one chromosome may be cut away, and replaced by intact bases from a nearby homologous chromosome.

In relation to (c) nanotechnology, some scientists and engineers (for example the visionary Ned Seeman) are now trying to use the specific base-pairing properties of DNA in order to assemble electronic, mechanical or computing devices on a very small, near-atomic scale. Some progress has already been made at fabricating tiny rotary devices, by engineering DNA in the form of specific and controllable 'swivels'. But there are still fundamental difficulties to be overcome, when trying to build real, useful devices from such highly flexible building-blocks.

With regard to (d) evolutionary biology, it seems that recent developments in sequencing whole genomes of DNA have brought about a huge accumulation of information from hundreds of diverse species. This information can be used in principle to deduce historical relationships among many different species, on the assumption that genes move only from parent(s) to children; in other words, if we assume that hardly any genes have moved 'horizontally' *between* species. DNA sequencing can also provide details of the slight genetic variation found within any species, such as the human polymorphisms mentioned in Chapter 11. These data can be used to study recent historical events: for example the human settlement of different continents, or the strange 'bottleneck' of population size which occurred when *Homo sapiens* emerged about 100 000 to 200 000 years ago.

In summary, whether we consider the chemical basis of DNA structure, recombination, nanotechnology or molecular evolution, there is one overriding feature that distinguishes a well-understood field of science from a poorly understood one. In a well-understood field, there is a well-developed sense of the relations between things, and how they are structured with respect to one another. Often a single theory may explain hundreds of experimental observations. By contrast, in a poorly understood field there are many 'effects' and 'factors' and long, hard-to-remember words, but few relations among them in a theoretical sense. As helpful advice to a young student: the practitioners of any poorly-understood field may often be distinguished by their refusal to say the words 'I don't know!'

Our goal, therefore, should be to distinguish what is known from what is not known, and not ever to pretend that word-knowledge is a substitute for structural-knowledge. This important idea was expressed clearly long ago by John Locke (1690), as a comment on his *Essay Concerning Human Understanding*:

> We cannot but think that angels of all kinds much exceed us in knowledge; and possibly we are apt sometimes to envy them that advantage, or at least to repine[1] that we do not partake with them in a greater share of it. Whoever thinks of the elevation of their knowledge above ours, cannot imagine it lies in a *playing with words*, but in the contemplation of things, and having true notions about them; a perception of their habitudes and relations one to another. If this be so, methinks we should be ambitious to come in this part, which is a great deal in our power, as near them as we can.

Note

1. Repine: to fret, be discontented.

Further Reading

Protein-DNA theory

Oobatake, M., Kono, H., Wang, Y., and Sarai, A. (2003) Anatomy of specific interactions between lambda repressor and operator DNA. *Proteins* **53**, 33–43. A computer analysis may identify key features of the lambda repressor-operator DNA complex which are responsible for specific recognition.

Parry, D., Moon, S.A., Liu, H.H., Heslop, P., and Connolly, B.A. (2003) DNA recognition by the *Eco RV* restriction endonuclease using base analogues. *Journal of Molecular Biology* **331**, 1005–16. Chemical changes to base-pairs within the *Eco RV* recognition site GATATC are used to probe its specificity.

Tateno, M., Yamasaki, K., Amano, N., Kakinuma, J. *et al.* (1997) DNA recognition by beta-sheets. *Biopolymers* **44**, 335–59. Studies of general inter-atomic contact when protein beta-sheets are used to recognize DNA.

Recombination

Chen, L., Huang, S., Lee, L., Davalos, A. *et al.* (2003) WRN, the protein deficient in Werner syndrome, plays a critical structural role in optimizing DNA repair. *Aging and the Cell* **2**, 191–9. A human defect of recombination-repair enzymes can cause a disease known as Werner syndrome, which involves cancer and premature aging.

Mucha, M., Krol, J., Goc, A., and Filipski, J. (2003) Mapping candidate hotspots of meiotic recombination in segments of human DNA cloned into yeast. *Molecular and Genetic Genomics* **270**, 165–72. The most recombi-natorial human sequences may be identified by cloning into yeast as CG islands (Chapter 11) and repetitive 'minisatellites' (Chapter 10).

Seidman, M.M. and Glazer, P.M. (2003) The potential for gene repair *via* triple helix formation. *Journal of Clinical Investigation* **112**, 487–94. Third-strand-forming DNA oligonucleotides can induce novel recombination and repair as a possible medical therapy.

Nanotechnology

Okamoto, A., Tanaka, K., and Saito, I. (2003) Rational design of a DNA wire possessing an extremely high transport ability. *Journal of the American Chemical Society* **125**, 5066–71. Incorporation of a modified base 'benzo-deaza-adenine' enables better 'hole transport' through a DNA helix.

Sussman, H.E. (2003) Nanodevices hold promise for gene therapy. *Drug Discovery Today* **8**, 564–5. New inventions from nanotechnology which could be used for DNA gene medicine.

Zhu, L., Lukeman, P.S., Canary, J.W., and Seeman, N.C. (2003) Nylon/DNA: single-stranded DNA with a covalently stitched nylon lining. *Journal of the American Chemical Society* **125**, 10178–9. A novel composite of DNA and nylon polymer that may be useful for the construction of molecular devices.

Molecular evolution

Anzai, T., Shiina, T., Kimura, N., Yanagida, K. *et al.* (2003) Comparative sequencing of human and chimpanzee MHC class regions unveils insertions/deletions as the major path to genomic divergence. *Proceedings of the National Academy of Sciences, USA* **100**, 7708–13. Detailed comparison of human and chimp DNA in an important genomic region.

Cooper, G.M., Brudno, M., Green, E.D., Batzoglou, S. *et al.* (2003) Quantitative estimates of sequence divergence for comparative analyses of mammalian genomes. *Genome Research* **13**, 813–20. Use of DNA sequencing data to study the possible historical evolution of mammals.

Thomas, J.W., Touchman, J.W., Blakesley, R.W., Bouffard, G.G. *et al.* (2003) Comparative analyses of multi-species sequences from target genomic regions. *Nature* **424**, 788–93. A study of molecular evolution by DNA sequencing among diverse vertebrates.

Appendix 1: Notes on the Derivation of Some Technical Terms

All of the terms in the following list have been introduced and explained in the text. The sole point of these notes is to explain the philological background of the words.

Archaea (or archae-bacteria). Micro-organisms, similar to ordinary bacteria, but different in their molecular organisation. Greek: *archaeo* = ancient, primitive.

Bacteria (singular: bacterium). From the Greek, a diminutive of *baktron*, a stick or staff.

Cell. A term first used by Robert Hooke in 1665, on seeing the structure of angular spaces in a thin section of cork in a microscope. The structure was similar to that of cells in a honeycomb, monastery or prison.

Circular. Like a circle in the sense of being endless: not necessarily even a plane figure (cf. linear).

Chromosome. Filamentous body in the cell nucleus which appears colored (Greek: *chromos* = color, *soma* = body) in the light microscope after treatment with chemical stains.

Enzyme. An organic catalyst, usually a protein. From the Greek *zumosis* = leaven, used to describe the process of fermentation.

Eukaryote. Organism in whose cells the DNA is contained within a nucleus. From the Greek: *eu* = truly, *karyon* = kernel, nut.

Gene. From the Greek *gen*, to produce. Cf. generator, generation, Genesis.

Homologous. Corresponding. From the Greek: *homo* = same, *logos* = word.

Interphase. From two Greek words: *phasis* = appearance and *inter* = between; used to describe an appearance in the middle of the cell cycle.

Linear. Like a line in the sense of having two ends and no branches: not necessarily straight, or even planar (cf. circular).

Metaphase. From two Greek words: *phasis* = appearance and *meta* = beyond or after; used to describe an appearance near the end of the cell cycle.

Nucleus. From the Latin *nux* = a nut; the central kernel of the cell. A term coined by the botanist Robert Brown in 1831.

Oligomer. A short example (Greek: *oligo* = few) of a long-chain molecule.

Polymorphic. Having many shapes. From two Greek words: *poly* = many, *morphe* = form.

Prokaryote. Organism (e.g. bacteria, archaea) in which the cell does not have a nucleus. Greek: *pro* = before, as if the nucleus came later.

Protein. From the Greek *proteios* = of the first rank. First applied to the large group of compounds by Muldner, acting on a suggestion made in 1838 by the famous chemist Berzelius.

Topoisomerase. From the Greek *topos* = a place and *isos* = equal. The ending –ase denotes an enzyme. An enzyme that changes the linking number Lk of DNA, to make different forms of DNA equal.

Toroid. A ring of the kind put through the nose of a bull: Latin *taurus* = bull.

Translation. From the Latin: *trans* = across, *latum* = past participle of 'bring'. Hence, movement of a body without rotation.

Appendix 2: The Chemical Theory of Base-stacking Interactions in DNA

In Chapter 2 we explained how the water-repelling hydrophobic quality of the bases could push a 'ladder' of DNA into a twisted, double-helical shape; and we called that a first-order effect on the structure of DNA. Then in Chapter 3 we invoked the same hydrophobic property in order to account for the 'propeller twist' which is usually found in the Watson–Crick base-pairs; and we described that as a weaker, 'second-order' effect. Propeller twist makes the stacking of base-pairs onto one another much less straightforward than for planar base-pairs, and it thereby provides a key to understanding some aspects of the conformational behavior of double-helical DNA. In Chapter 3 we also mentioned the contribution made to stacking by partial electric charges within the base-pairs, in order to explain more accurately some of the second-order structural effects. And we said that a fuller treatment of them would be given in Appendix 2.

Here, then, we shall explain more about the nature of partial electric charges in the organic bases of DNA, as they influence the overlap or stacking of adjacent base-pairs; and we shall describe some observations on stacking that can only be explained by their use. We shall also describe briefly some recent observations from the crystal structures of DNA, that show the important role of propeller twist in determining stacking arrangements.

A new detailed theory of chemistry, by which the base-pairs of DNA stack on one another specifically due to several kinds of chemical energy, including weak interactions between partial charges in the base-pairs, has been developed over the past decade by Chris Hunter. He had used the same theory previously with some success to describe the stacking behavior of large, flat molecules called 'porphyrins', which are used by the body for many different purposes, such as carrying electrons in the mitochondria and carrying oxygen in the blood. But the same theory is also very helpful in the study of DNA.

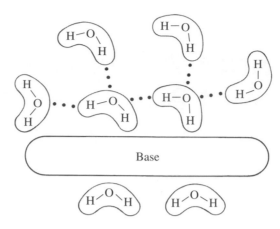

Figure A1 Schematic diagram showing how H···O bonds between water molecules are disrupted by a hydrophobic base. The loss of such hydrogen bonds can be minimized when two base-pairs stack directly on top of one another, as in Figs 2.4 and 2.5.

Hunter says that there are three general kinds of electrical force that control the interactions between molecules such as porphyrins, or among the base-pairs of DNA. The first force is the one that we have described already, namely a tendency for the base-pairs to stack on one another as fully as possible, in order to escape from contact with water. What is the basis of this effect, in molecular terms? As shown schematically in Fig. A1, water molecules in solution have a strong tendency to associate with other water molecules by means of hydrogen-to-oxygen or H···O bonds, which are electrical in nature. Normally any water molecule will have a total of four partners: two that place their hydrogen atoms in contact with the water oxygen, and two that place their oxygen atoms in contact with each of the water hydrogens. This intricate and ever-changing network of hydrogen-bonding between water molecules is responsible for the 'stickiness' of water, for example its tendency to form drops when falling slowly out of a tap. But when those water molecules contact an oily, water-insoluble DNA base, many hydrogen bonds of the kind H···O are lost, as shown in the diagram. Hence, the optimal arrangement of base-pairs in water is as shown previously in Figs 2.4 and 2.5, where the bases stack on one another as completely as possible in order to exclude water. In a solution of water mixed with ethanol (as in strong whisky) or, for example, at low water content in fibers of DNA, this kind of water-exclusion force becomes rather weak, so the base-pairs no longer need to stack so firmly onto one another.

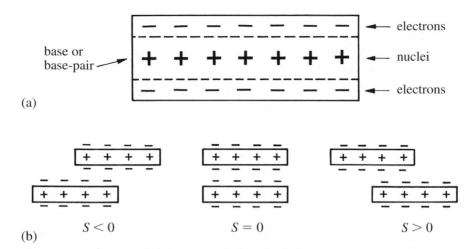

(a)

(b) $S < 0$ $S = 0$ $S > 0$

Figure A2 (a) Distribution of electric charge in a base or base-pair. (b) Positive or negative slide at a base-pair step reduces the repulsion between the negative charges, and produces some attraction between positive and negative charges in different base-pairs.

The second kind of force between base-pairs comes about because the upper and lower surfaces of any base (or base-pair) have a slight negative charge. As shown in Fig. A2(a), the negatively charged electrons that make up the 'aromatic' system of bonding between atoms in any base lie mainly above and below the main part of the ring. Yet the positively charged nuclei of the various atoms that join to form a base (such as carbon, nitrogen and oxygen) lie near the center of the ring, when viewed edge-on, as in Fig. A2(a). So the overall distribution of electric charge in any base looks somewhat like a 'sandwich', where the electrons are the bread and the nuclei are the cheese.

Now when two bases (or base-pairs) come into contact along their upper and lower surfaces, they must repel each other to some extent, because of a negative-to-negative charge repulsion. An idealized, vertical stacking of two base-pairs is shown in the center part of Fig. A2(b) for the case of slide $S = 0$. We said previously that a fully stacked geometry is favorable because it excludes water; but now we see that it is highly unfavorable when viewed in the context of our second force, which says that there is a strong negative-to-negative repulsion between base-pairs in a fully stacked geometry. Were it not for the water, the two base-pairs would repel one another like the North poles of two magnets. Generally, the base-pairs in DNA will slide away from one another in a left-to-right sense to escape this repulsion, because the flexibility of sugar–phosphate chains is greatest in that direction. Thus, as shown in

Fig. A2(b), the base-pairs can lie offset from one another at either negative slide S, as shown on the left-hand side of the drawing, or at positive slide S, as shown on the right-hand side. In either case, the negative-to-negative charge repulsion is lessened by the increased distance between pairs, and some positive-to-negative attraction may come about between nuclei in one base-pair and electrons in the other.

Thus the 'bread-and-cheese' effect is the exact opposite of the 'hydrophobic' effect, so far as the stacking of adjacent base-pairs on one another is concerned. Which of the two is the stronger? This depends critically on the amount of water which surrounds the bases. When conditions are wet, the hydrophobic effect wins out; but when conditions are dry, the bread-and-cheese effect is stronger.

These notions explain rather simply the structural behavior of double-helical DNA in fibers. Originally, in the 1950s, it was not possible to grow crystals of DNA, because scientists had not yet learned how to synthesize DNA chemically; so most investigators had to isolate chromosomal DNA from natural sources such as calf thymus or salmon sperm, and then pull it out into long fibers, in order to study its structure by X-ray diffraction (see Chapter 9). When these fibers were wet, the X-ray photographs showed a 'B' form structure with 10 base-pairs in a complete helical turn, and with base-pairs stacked vertically over one another at slide $S = 0$. But when the fibers were dry, either an 'A' form with 11 base-pairs per turn, or a 'C' form with 9.0 to 9.3 base-pairs per turn was observed. In the 'A' form, the base-pairs stack offset at a negative slide $S = -1.5\,\text{Å}$, while in the 'C' form they stack offset at a positive slide $S = +1.0\,\text{Å}$. Recall that slide S and twist T change in tandem, as described in Fig. 3.13.

Not all possible forms of double-helical DNA can be seen in fibers, because the packing of long, thin DNA molecules into fibrous bundles often restricts their helical repeat to integral values such as 10 or 11 base-pairs per turn. Yet when such structures are observed, we can understand their behavior as follows. When the fiber is wet the hydrophobic forces are dominant, so the base-pair stacking is one of low slide near $S = 0$, and the result is the 'B' form; but when the fiber is dry the hydrophobic forces are weaker, so the bread-and-cheese effect becomes the stronger of the two. Then the stacking involves a slide which is either negative – in which case the 'A' form appears; or positive – in which case the 'C' form appears. A left-handed 'Z' form, mentioned briefly in Chapter 2, can also be seen in fibers when the conditions are relatively dry; and in this 'Z' helix the base-pairs lie offset from one another, as expected from this theory.

Finally, according to Hunter, there exists a third important kind of electrical force between base-pairs, which depends on the identities and locations of individual atoms in the four different kinds of base ring. Certain atoms, mainly hydrogens that are attached to nitrogens as NH or NH_2, or to carbon as CH, can generate a partial positive charge in their close vicinity; while other atoms such as oxygen O or simple nitrogen N can generate a partial negative charge. Thus, the stacking of different base-pairs on one another can be 'fine-tuned' by the electrical interactions of individual atoms in the rings; and this feature will depend strongly on the sequence or ordering of bases at any step.

It follows from this that bases are not so free to slide over each other as we might suppose, on the basis of a model in which the surfaces of the bases are featureless, as we have assumed so far with our models in which base-pairs are represented by simple blocks. Instead, certain positions will be preferred – positions in which a positively charged atom from one base lies directly above or below a negatively charged atom from the other, and in which atoms with 'like' charges do not lie directly above or below each other. In other words, the preferred positions will maximize attractive juxtapositions of atoms and minimize repulsive ones. This so-called 'partial-charge' effect is well-known in chemistry; and it can have a striking influence in the positioning of large, flat molecules relative to one another.

Some partial charges for atoms in the DNA base-pairs, as calculated by Hunter, are shown schematically in Fig. A3. Roughly, they correspond to what you would expect from having oxygens and nitrogens negative, and hydrogens positive; but the computer calculations are much more accurate than any assignment of partial charges that could be made directly from chemical intuition. For clarity, no individual atom types are identified; but one may compare with Fig. 2.11(a) and (b) to identify the different kinds of atom at various places in the rings.

The most striking aspect of this figure is a dense accumulation of negative charge along the major-groove edge of the guanine ring, as shown at bottom left. Two negatively charged atoms, a nitrogen and an oxygen, lie there close together in space, and so generate a joint concentration of negative charge that is greater than that of any single atom elsewhere in the G–C or A–T base-pairs. There is also a concentration of positive charge on the major-groove edge of the cytosine ring. What are the implications of highly charged guanine and cytosine rings for the three-dimensional structure of DNA?

First, when two guanine rings stack over one another in a step GG/CC, they will not find it easy to stack vertically, one over the other at slide $S = 0$, owing to a strong repulsion between negatively

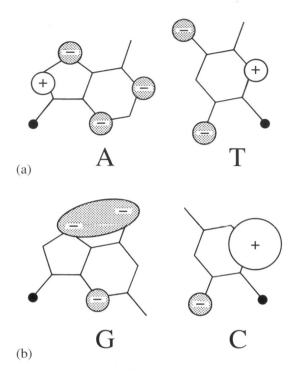

(a)   A   T

(b)   G   C

Figure A3 Regions of 'partial charge' for A–T and G–C base-pairs. The base-pairs have the same relative orientations as in Fig. 2.11(a) and (b), where atom types H, C, N, O can be identified.

charged regions; instead they will prefer to lie offset from one another, either at low slide, $S = -1$ to -2 Å or else at high slide, $S = +1$ Å, in order to minimize charge–charge repulsions. And indeed, we saw precisely that effect in Fig. 3.12(c), where the GG/CC step showed an unoccupied middle range of slide from $S = -1$ to 0 Å. Second, when a G–C base-pair stacks over another such pair in a step GC/GC or CG/CG it will favor certain stacking geometries over others, because of interactions between the strong negative and positive charges on the guanine and cytosine rings, as shown in Fig. A3.

In contrast to the G–C base-pairs, the A–T base-pairs have no strong, joint concentrations of partial electric charge. And so the strong stacking preferences that we have just described for steps made up from two G–C base-pairs should not act so strongly when there is an A–T base-pair in any step, say for the step AC/GT.

A careful and detailed examination of all kinds of dinucleotide step in many crystallized oligomers of DNA has been made by Mustafa El Hassan. Figure A4 shows his generalized plot of roll R against slide S, in the same style as the three examples shown in Fig. 3.12, but with preferred regions now marked for nine of the

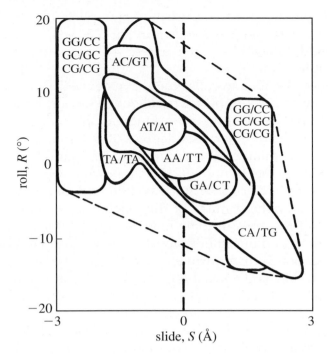

Figure A4 Schematic plot of roll *R versus* slide *S* for nine types of base-pair steps as seen in crystallized DNA oligomers. (Data for three particular step types are shown in more detail in Fig. 3.12.) Adapted from El Hassan and Calladine (1996) *Endeavour* **20**, 61–7.

ten kinds of sequence step, for which data are plentiful and clear. The partial electric charges, as described above, might be expected to play a clear part in the stacking of those steps where the constituent base-pairs are both G–C. Indeed, for all of the steps GG/CC, GC/GC and CG/CG, it is found that the conformation of the step prefers to lie in either of two separate regions of roll/slide, as may be seen from detailed inspection of Fig. A4.

Figure 3.12(a) also showed an example, AA/TT, of a step which is 'rigid' in the sense that each of roll *R* and slide *S* is fixed within narrow limits. Figure A4 shows that the steps AT/AT and GA/TC behave in the same way, but adopt somewhat different chosen conformations.

While the partial-charge effects decribed above thus enable us to understand some of the empirical features shown in Figs 3.12 and A4, they cannot provide predictions of the actual conformation of dinucleotide steps without the introduction, somehow, of constraints imposed by the backbones. Now the simple idea of a 'rigid' backbone link, as shown in Fig. 3.13, provides a first-order simulation of backbone constraints. But it is too crude to predict many of the details shown in Fig. A4, because the sugar–phosphate backbone, as explained in Chapter 2 and shown in Fig. 2.8, acts as a rather more complicated linkage, because some of the inter-atomic bonds can rotate a little.

However, Hunter and colleagues have devised a simple and successful way of accounting for backbone behavior. First, they take slide and 'shift' as their main working variables for a dinucleotide step. Shift is one of Euler's six degrees of freedom of a rigid block with respect to another: it is an in-plane translation, like *slide*, but measured along the 'front-to-back' axis shown in Fig. 3.7. In general, shift does not vary nearly so much as slide; and for many purposes – as indeed throughout much of this book – its variations may be ignored. But it turns out to be important in the calculation of the conformation of dinucleotide steps. Hunter's idea is to endow the simple backbone link with an empirically-derived energy function of slide and shift; and when this is coupled with the chemical-energy computations, the conformations of individual dinucleotide steps of given composition are predicted well.

These same workers have recently investigated a further important feature of dinucleotide step conformation, which we have not mentioned previously: it concerns a conformational 'communication' between neighboring steps. Now a detailed study of *tri*-nucleotide and *tetra*-nucleotide fragments from El Hassan's database shows that both slide and shift are strongly correlated from step to step, whereas all other conformational variables depend only on the composition of the step in question: for example, slide, which has an overall range of some 6 Å (see Fig. A4) rarely differs by more than 1 Å from one step to the next. Shift, on the other hand, tends to be anti-correlated, in the sense that its values generally go high-low-high-low along the molecule.

It is clear that any such communication of conformational features from one step to the next must be made *via* the backbones, and probably through the single sugar ring that is shared by two consecutive backbone segments. Again, by further endowing the backbone with simple, semi-empirical energy expressions that depend on the *difference* of slide from step to step (in order to penalize variation), and likewise the *sum* of the shift values, Hunter and colleagues have been able to predict, by economical computations, a range of observed context-dependent features in DNA oligomers. For example, G–C rich regions tend to be *bi*-stable between 'A' and 'B' or 'C' forms (low and high slide, respectively) – a feature that may indeed be seen qualitatively in Fig. A4 when it is recalled that slide varies but little from step to step. Also, the role of AA steps in inhibiting the 'A' form over the surrounding region is clarified by this work. Finally, these recent computations have given encouraging estimates of the sequence-dependent flexural/torsional stiffness of DNA.

Now we may recall that the more-or-less unique conformation of AA/TT steps was attributed in Chapter 3 to a 'locking' or stabilizing effect, on account of the high propeller twist in both of the A–T base-pairs (recall Fig. 3.6). More detailed data from DNA crystals show that there is in fact a very strong empirical correlation between the average propeller twist in the two base-pairs of any step, and the total range of slide which is allowed. Thus, all three steps in Fig. A4 which show the narrowest range of slide values prefer high values of propeller twist, while those steps that show the widest range of slide values, particularly CA/TG, tend to have the least propeller twist. One can make a simple physical model, as suggested in Exercise 3.5, to show that a small propeller-twist offers little hindrance to slide.

The data shown in Fig. A4 have all been derived from X-ray diffraction studies of 'free' DNA oligomers, which are typically packed with 40 to 60% water into a crystalline state. Different ranges of conformation in individual steps may therefore be induced in some cases, by the moderate but non-negligible intermolecular forces which arise when these oligomers are deformed so as to pack together into the regular arrangements known as 'crystal lattices'.

But when DNA is bent strongly around proteins, much wider ranges of roll-slide-twist are seen in X-ray studies. For example, AA/TT steps (which in Fig. A4 appear 'rigid') show high roll and low twist in the sequence TATATAAA, when it is bound to the protein TBP (Chapter 4); while certain CA/TG steps bound to the histone octamer (Fig. 7.2) again show a wider range of conformation than would be deduced from Fig. A4.

In summary, the stacking interactions of base-pairs in DNA are influenced by three kinds of electrical force: (1) maximizing base-to-base overlap to avoid contact of the bases with water; (2) reducing base-to-base overlap to avoid repulsion of negatively charged surfaces; and (3) maximizing attraction and minimizing repulsion between partial charges on individual atoms in the base-pair rings. And all of these effects are subject, of course, to various subtle constraints imposed by the sugar–phosphate chains of DNA, and by propeller twist in the base-pairs; and so the base-pairs will not always be able to attain their otherwise optimal local configurations.

There do exist, of course, a wide range of other constraints which may combine to influence local step conformations, and thereby produce severe helical distortions. These include imposed curvature: which may be mild as in the binding of 434 repressor, or severe as in the binding of TBP, and may involve interaction of the sugar–phosphate chains with the protein; and there is also the intimate intrusion of intercalating drugs, or hydrophobic amino acids from contacting proteins, into the space between neighboring base-pairs, which can produce curvature and striking local distortions of geometry.

We think that a deeper understanding of all of these effects will be of considerable use in understanding the actions of DNA in biology; but a complete discussion is beyond the current scope of our book.

Further Reading

El Hassan, M.A. and Calladine, C.R. (1995) The assessment of the geometry of dinucleotide steps in double-helical DNA: a new local calculation scheme. *Journal of Molecular Biology* **251**, 648–64. How to calculate roll, slide and twist accurately for any base-pair step, by a method which does not rely on an overall helix axis.

El Hassan, M.A. and Calladine, C.R. (1996) Propeller-twisting of base-pairs and the conformational mobility of dinucleotide steps in DNA. *Journal of Molecular Biology* **259**, 95–103. The propeller twist of base-pairs seems to influence their ability to slide along the long axis of any base-pair step.

El Hassan, M.A. and Calladine, C.R. (1998) Two distinct modes of protein-induced bending in DNA. *Journal of Molecular Biology* **282**, 331–43. When strongly bent by proteins, DNA dinucleotide steps move outside the range of conformations shown in Fig. A4.

Gardiner, E.J., Hunter, C.A., Packer, M.J., and Willett, P. (2003) Sequence-dependent DNA structure: a database of octamer structural parameters. *Journal of Molecular Biology* **332**, 1025–35. A systematic computation of seven-step oligomers, which provides a clear description of context-dependent behaviour, including bi-stability.

Hunter, C.A. (1993) Sequence-dependent DNA structure: the role of base stacking interactions. *Journal of Molecular Biology* **230**, 1025–54. A clear description of the chemical theory which underlies base-stacking preferences in DNA, and some calculations of these effects.

Matsumoto, A. and Olson, W.K. (2002) Sequence-dependent motions of DNA: a normal mode analysis at the base-pair level. *Biophysical Journal* **83**, 22–41. Accurate calculations of DNA sequence-dependent structure in solution, using preferred values of roll-slide-twist as seen in crystals.

Packer, M.J., Dauncey, M.P., and Hunter, C.A. (2000) Sequence-dependent DNA structure: dinucleotide conformational maps. *Journal of Molecular Biology* **295**, 71–83. Theoretical analysis of base-pair stacking energies in terms of roll, slide and twist at a dinucleotide level; and a prediction of rigidity for AA/TT, but flexibility for CG or CA/TG or TA, in accord with observed preferences of dinucleotide sequences.

Richmond, T.J. and Davey, C.A. (2003) The structure of DNA in the nucleosome core. *Nature* **423**, 145–50. Details of *roll, slide, twist*, etc. in DNA bent around proteins, showing even more extreme conformations of CA/TG steps than in Fig. A4.

Suzuki, M., Amano, N., Kakinuma, J., and Tateno, M. (1997) Use of a 3-D structure database for understanding sequence-dependent conformational aspects of DNA. *Journal of Molecular Biology* **274**, 421–35. A careful analysis of free-DNA crystal structures in terms of roll, slide and twist at individual base-pair steps.

Wang, L., Hingerty, B.E., Srinivasan, A.R., Olson, W.K., and Broyde, S. (2002) Accurate representation of B-DNA double helical structure with implicit solvent and counterions. *Biophysical Journal* **83**, 382–406. The development of refined energy parameters to calculate DNA sequence-dependent structure in solution, including water and cations.

Appendix 3: How to Modify Gene Expression using Anti-sense Oligonucleotides, Ribozymes, or si-RNA

In Chapter 10 we explained how specially designed DNA or RNA molecules may be used in 'gene therapy' to correct the aberrant expression of particular genes within diseased animals or plants. Here we shall elaborate on three current methods that use DNA or RNA to inhibit gene expression. First, there is the 'anti-sense' approach, where small RNA (or DNA) oligonucleotides are designed so that they will form 20 to 30 specific base-pairs with some small part of a long messenger-RNA, and thereby block its translation into protein by several possible mechanisms. Second, there is the 'ribozyme' approach, where small RNA oligonucleotides are again designed to form specific base-pairs with some long messenger-RNA; but in addition they now carry a catalytic domain for RNA cleavage, which cuts the message so that full-length protein cannot be made (similarly for 'DNAzymes', which are composed of DNA rather than RNA). And third, there is the 'small interfering RNA' approach, where small RNA oligonucleotides once again are designed to form specific base-pairs with some long messenger-RNA; yet in this case they guide to the message a cellular ribonuclease complex, which cuts the message so that it cannot be translated into protein.

Since many of these methods use RNA, we shall first describe a few of its relevant structural properties. As we showed earlier in Fig. 10.3, the only chemical difference between RNA and DNA backbones consists of an OH group on the sugar ring at carbon 2'. DNA usually forms a double-helical structure; and indeed RNA is also able to form a double-helical structure, which is similar to the

'A' form of DNA (see Chapter 2). But whereas DNA is nearly always found in Nature in continuous double-helical form through Watson–Crick base-pairing, cellular RNA is usually single-stranded. Nevertheless, it can still form base-pairs between different parts of the same single strand. Thus, RNA often folds into complicated structures like that of the transfer-RNA molecule shown schematically in Fig. 2.14. Another example is the elaborate fold of the RNA found within the large assembly of ribosomal RNA molecules (called 28S, 18S and 5.8S in animals). As mentioned in Chapter 1 (see Fig. 1.12), this molecular machine translates information encoded by base triplets in the messenger-RNA into a linear sequence of amino acids within a polypeptide chain.

Anti-sense therapeutics

Now let us explain what is meant by the word 'anti-sense'. At the level of double-stranded DNA, the coding sequence of any gene is contained within the 'sense' strand, whereas its base-paired complement lies within the 'anti-sense' strand. Thus when messenger-RNA is made from any DNA gene by RNA polymerase, the newly-made RNA represents a 'sense' strand, whereas the DNA template which was 'read' by the polymerase represents an 'anti-sense' strand. How then might some short piece of anti-sense RNA or DNA, which presumably binds to long messenger-RNA after it has left the nucleus and entered the cytoplasm, impair the expression of that message into protein?

Intuitively, one might suppose that steric hindrance by a short antisense molecule could prevent proper 'reading' of a long messenger-RNA at the ribosome, where proteins are made, as shown in Fig. A5(a) on the right. This expectation has proved true in some cases; yet an equally important influence seems to be the action of a cellular enzyme known as 'RNase H' which cuts RNA–DNA hybrids. The normal function of RNase H is to degrade small pieces of RNA in RNA–DNA hybrids, that serve as primers for replication of the DNA during cell growth. Yet when short anti-sense DNA binds to a long messenger-RNA, that process also creates an RNA–DNA hybrid, which may be degraded by the nuclease RNase H, as shown on the left of Fig. A5(a).

Any 'anti-sense' nucleic acid will be liable to quick destruction by cellular enzymes that cut nucleic acids: such enzymes are known as 'nucleases'. In order to protect the full-length anti-sense molecule from the action of nucleases, certain modifications must be made. One approach by chemists has been to replace an oxygen atom on the phosphate group with a nuclease-resistant sulfur (this

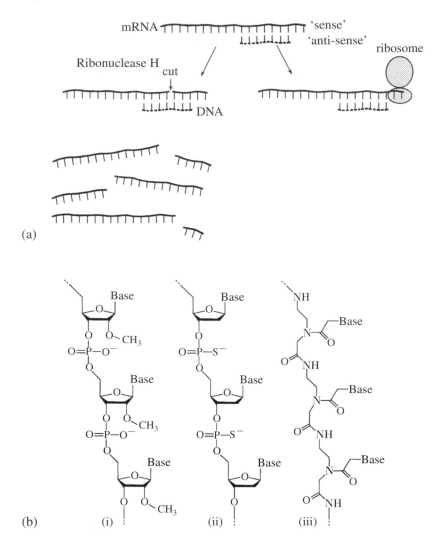

Figure A5 (a) Two common mechanisms by which anti-sense DNA can reduce gene expression: steric hindrance of translation at the ribosome (right), or destruction of the transcript by ribonuclease H (left). (b) Representative oligonucleotides used in the anti-sense approach. In the DNA molecule on the left, one of the oxygens on the phosphate group has been replaced by sulfur, to form a phosphoro-thioate that is resistant to nucleases. In the center, an anti-sense RNA molecule can be protected against destruction by placing a CH_3 group on the 2′ OH of the sugar. And on the right, the nucleic acid can be protected by replacing the sugar–phosphate backbone entirely by N-(2-amino-ethyl) glycine linkages, which are like peptides. There are many other stabilizing modifications, not shown here, that have also been useful for the anti-sense approach.

is called a phosphoro-thioate): see Fig. A5(b), center. Anti-sense RNA can also be protected against nucleases by adding a $-CH_3$ group to the 2′ OH on the sugar: see Fig. A5(b), left. Using these and other protective modifications, carefully designed anti-sense

oligomers can often exert a potent influence on specific gene activities in living cells. Indeed, some anti-sense molecules are now
undergoing clinical trials for the control of diseases such as viral
infection or cancer.

In order to achieve even greater nuclease resistance, Peter Neilsen
and colleagues have made a stable form of oligonucleotide by synthetic chemistry, which has *peptide* linkages between adjoining bases,
in place of the sugars and phosphates found in natural RNA and
DNA. That 'peptide nucleic acid' or 'PNA' is shown on the right in
Fig. A5(b). Because its backbone is electrostatically neutral, PNA
tends to form PNA-DNA or -RNA hybrids that are more stable under
low-salt conditions than natural DNA or RNA, whose phosphate
backbones contain many repulsive negative charges. PNA has turned
out to be very important for diagnostic work, where different DNA or
RNA molecules are detected on microarrays (see Fig. 10.5), and great
chemical stability is required. However, PNA has not yet become the
panacea which many people once expected for the control of gene
expression in cells, because the PNA-RNA hybrid is not cut by
RNase H; and also because simple steric hindrance is usually not
enough for any PNA to prevent the expression of some messenger-
RNA into protein at the ribosome. The ribosome is a very powerful
engine for making proteins, and it often manages to read through an
RNA message, even when some anti-sense DNA or PNA is present.

PNA and other anti-sense nucleic acids also show promise for
correcting the occasional mistakes that arise in the routine processing
of messenger-RNA. Normally, certain sections of the RNA that do
not code for protein (these are called 'introns') are 'spliced out'
before translation into the protein. But in certain diseases caused by
mutation, some of the splicing of a particular messenger-RNA may
not occur; or else an alternative set of splice sites within the RNA
might be used instead. This splicing of RNA at the wrong site has
been prevented in cells by use of anti-sense oligonucleotides that are
complementary to the incorrect splice junction. Also, proper splicing has been restored in cells where loss of a normal splice-site has
occurred, by linking to an anti-sense PNA that binds close to the
splice site a small peptide, which activates splicing.

RNA ribozymes and DNAzymes

Ribozymes were discovered in the 1980s, with the surprising finding that certain RNA molecules could cleave either themselves or
other RNA molecules. Subsequently it was found that ribozymes
can also catalyze other chemical reactions, in addition to cutting.
We can learn something about the catalytic potential of RNA from

the X-ray crystal structure of the ribosome, because it appears that the active site, where peptide units are linked, is constructed mostly or entirely from RNA; and that the RNA itself can catalyze the polymerization of peptide units. Hence, the ribosome itself may perhaps be regarded as a ribozyme – and a very ancient one too. The dual role of RNA both as a catalyst and also as an information carrier has supported much speculation that RNA was the first organic catalyst in the origin of life on Earth.

Certain plant and animal viruses contain RNA molecules that can cleave themselves. One particular example of such a naturally occurring, self-cleaving molecule is the *hammerhead ribozyme*, which may be isolated from small viruses that infect plants (known as 'viroids'); see Fig. A6(a). Several X-ray structures of the hammerhead ribozyme show that the molecule has a complex three-dimensional fold. The chemical mechanism of RNA-directed cleavage is thought to proceed by attack of a phosphate on the sugar 2'-OH group, as shown in Fig. A6(b).

Using the structural information from naturally-occuring ribozymes, scientists have designed artificial ribozymes that will recognize and cut specific messenger-RNA in cells. For instance, one such molecule uses a fragment of the hammerhead ribozyme which, like anti-sense RNA, recognizes a small section of the targeted messenger-RNA through complementary base-pairing, just as in Fig. A6(a). The messenger-RNA and ribozyme fragment co-fold to form an active ribozyme, which then cuts the messenger-RNA. For example, a hammerhead ribozyme has been designed to cleave the messenger-RNA that codes for a regulatory protein that is expressed aberrantly in breast cancer. When introduced into breast cancer cells in culture, that ribozyme restores a normal, non-cancerous lifetime to those cells. Other ribozymes have been used to target and destroy the messenger-RNA from HIV and hepatitis C virus; which stops these viruses from growing within the cell.

One advantage in using RNA to target the destruction of particular messenger-RNA is that it combines the catalytic potential of the molecule with the specificity provided by complementary base-pairing. Although the RNA ribozyme may have a rather complex, three-dimensional structure, procedures have been developed to isolate useful ribozymes by random selection and without the requirement to know the structure.

A complication of the ribozyme approach is that short RNA molecules are very sensitive to degradation by nucleases which are present in every cell. In order to get around that problem, some scientists have designed DNA-based ribozymes or 'DNAzymes', which can catalyze similar kinds of RNA-cleavage reactions. We mentioned above that RNA ribozymes use the 2′ OH group to attack and cleave a sugar-phosphate backbone; yet that 2′ OH group is missing in DNA (see Fig. 10.3(a)). The reactive OH group is therefore provided by the target RNA itself, while the DNA helps to orient it

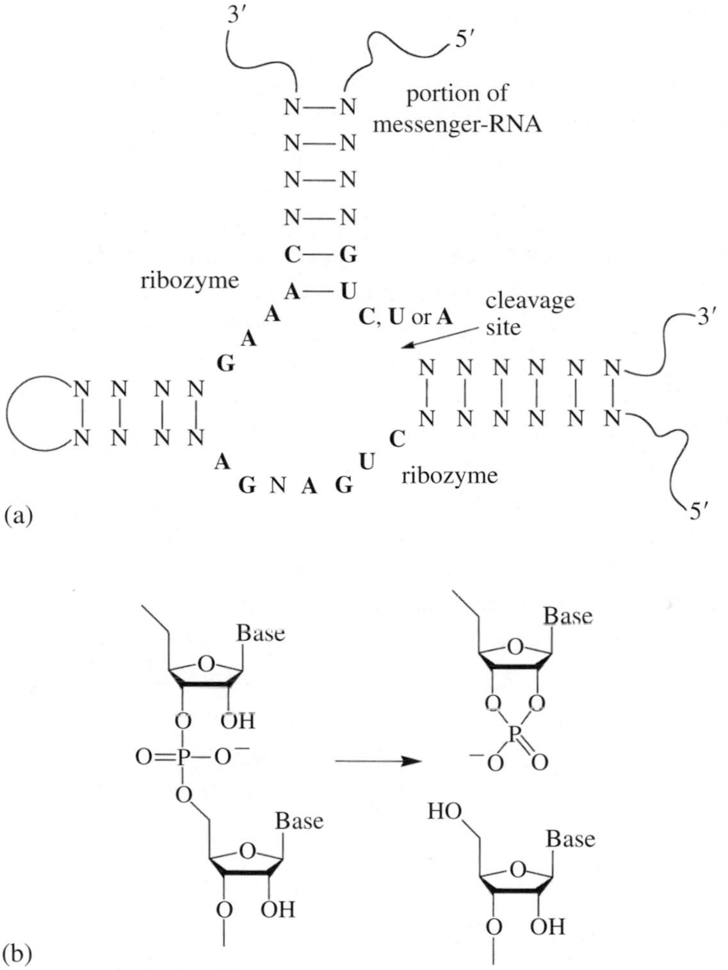

Figure A6 (a) Schematic diagram showing an RNA-cutting enzyme based on the naturally occurring 'hammerhead' RNA ribozyme from a plant viroid. An RNA molecule can be recognized by this ribozyme through complementary base-pairing, indicated by the lines connecting the bases; and it will then be cleaved at a specific point. The specific residues required for maintaining the appropriate structure for catalysis are in bold type, and all other bases are indicated by N. The RNA to be cut will co-fold with a ribozyme fragment in order to generate the cutting activity. (b) Chemical changes resulting from cutting by the ribozyme. The OH group on the 2′ position of the sugar is positioned by the hammerhead structure to initiate an attack on the phosphate. This breaks the linkage between the adjacent bases, leaving a cyclic 2′, 3′ product on the base with the reactive 2′ OH, and an OH group on the other fragment.

for the cleavage reaction. DNAzymes have been developed that catalyze other processes, such as incorporating metals into heme rings.
Most DNAzymes require cofactors, such as metals or amino acids,
that assist in the catalysis and may provide the catalytic groups.

si-RNA

In the cells of plants and animals, certain double-stranded RNA
molecules may repress the translation of specific genes in a controlled manner. There, an evolutionarily conserved process silences any
particular messenger-RNA which contains the same sequence as
that of the short double-stranded RNA. First, a ribonuclease called
'Dicer' (e.g. in plants, flies and worms) cuts long double-helical
RNA into many small pieces of size 21 to 23 base-pairs, with two-
base extended 3' ends. These are called 'small interfering RNA', or
si-RNA. Those small fragments then bind to a large ribonuclease
complex called 'RISC' (for 'RNA induced silencing complex'),
which is 'guided' by the small interfering RNA to destroy any
messenger-RNA that contains an identical sequence (Fig. A7).

Small interfering RNA may act also through a polymerase-
catalyzed amplification in worms or plants, where an RNA-directed
polymerase uses the starting RNA segment as a primer, to make
more interfering RNA along distant segments of the same message.
Thus, even a few RNA molecules can be amplified in the cell, and
can thereby have a very potent effect on gene expression.

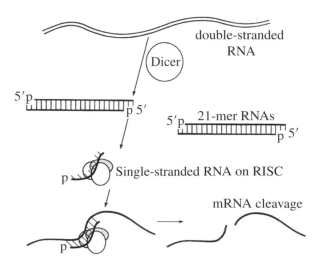

Figure A7 A probable mechanism of action for small interfering RNA. The double-
stranded complex that is formed between short si-RNA and long messenger-RNA
can be cleaved efficiently by a cellular ribonuclease assembly called 'RISC'.

But why should any cell have developed such a potentially hazardous system of RNA degradation? The si-RNA machinery probably evolved first as a mechanism of defence against RNA viruses, which may adopt a double-stranded form transiently during RNA-to-RNA replication. Later in evolution it may have became involved in regulating DNA methylation (see Chaper 11); and now it has proven very powerful at silencing gene expression in cells by the specific degradation of certain messenger-RNAs.

To activate the RNA interference system, one must first introduce long double-stranded RNA into a cell, one strand of which is complementary to the chosen messenger-RNA, for cells which contain Dicer; or else introduce small pieces of double-stranded RNA, 21 to 23 base-pairs in length with two-base 3′ overhangs, for cells which do not contain Dicer.

One important application of si-RNA so far has been to study gene usage in animal cellular development, for example in the worm *Caenorhabditis elegans*, where much progress has already been made. Scientists have found an easy method to inactivate any specific gene within that worm, by feeding it bacteria which make an RNA that targets the desired worm messenger-RNA. So, at least in the worm, gene inactivation by si-RNA works well. Soon, perhaps, anti-sense, ribozyme and si-RNA may become important as well for transgenic animals, or even for modern medicine, to silence defective single-copy genes in humans.

Further Reading and Bibliography

Antisense oligomers

Braasch, D.A. and Corey, D.R. (2002) Novel antisense and peptide nucleic acid strategies for controlling gene expression. *Biochemistry* **41**, 4504–10. A summary of chemically synthesized anti-sense oligomers which are currently available; their mechanisms of action which may reduce gene expression, as well as a summary of current clinical trials.

Sazani, P., Gemignani, F., Kang, S.-H., Maier, M.A. *et al.* (2002) Systemically delivered antisense oligomers upregulate gene expression in mouse tissues. *Nature Biotechnology* **20**, 1228–33. Cationic PNA-lysine or 2′-O-methoxyethyl-phosphorothioate anti-sense oligomers work well to influence messenger-RNA splicing in mouse.

Peptide nucleic acid

Borgatti, M., Lampronti, I., Romanelli, A., Pedone, C. *et al.* (2003) Transcription-factor decoy molecules based on a PNA-DNA chimera mimicking Sp1 sites. *Journal of Biological Chemistry* **278**, 7500–9. Double-helical mixtures of DNA and PNA can bind to transcription factors such as Sp1, and alter their *in vivo* activities.

Cartegni, L. and Krainer, A.R. (2003) Correction of disease-associaed exon skipping by synthetic exon-specific activators. *Nature Structural Biology* **10**, 120–5. Covalent coupling of PNA to a splicing activator can restore splicing.

Dias, N., Senamaud-Beaufort, C., le Forestier, E., Auvin, C. *et al.* (2002) RNA hairpin invasion and ribosome elongation arrest by mixed-base PNA oligomer. *Journal of Molecular Biology* **320**, 489–501. Certain PNA molecules which form triple helices can block translation of messenger-RNA at the ribosome.

Eriksson, M., Nielsen, P., and Good, L. (2002) Cell permeabilization and uptake of antisense peptide-PNA into *Escherichia coli*. *Journal of Biological Chemistry* **277**, 7144–7. The covalent coupling of PNA and a membrane-permeable peptide enables them to enter a bacterial cell.

Kaushik, N., Basu, A., and Pandey, V.N. (2002) Inhibition of HIV-1 replication by anti-trans-activation responsive polyamide nucleotide analog. *Antiviral Research* **56**, 13–27. Certain PNA molecules can block HIV replication in cell culture.

Ribozymes or DNAzymes and ribosomes

Doudna, J.A. and Cech, T.R. (2002) The chemical repertoire of natural ribozymes. *Nature* **418**, 222–8. A survey of self-cleaving RNA structures and their possible functions.

Feldman, A.R. and Sen, D. (2001) A new and efficient DNA enzyme for the sequence-specific cleavage of RNA. *Journal of Molecular Biology* **313**, 283–94. The use of DNAzymes as well as ribozymes.

Guerrier-Takada, C., Gardiner, K., Marsh, T., Pace, N., and Altman, S. (1983) The RNA moiety of ribonuclease P is the catalytic subunit of the enzyme. *Cell* **35**, 849–57. This work, and the study by Kruger *et al.*, showed for the first time that RNA itself can have a catalytic function.

Haseloff, J. and Gerlach, W.L. (1988) Simple RNA enzymes with new and highly specific endoribonuclease activities. *Nature* **334**, 585–91. Discovery of the hammerhead ribozyme from a plant viroid, used to cut a messenger-RNA.

Nissen, P., Hansen, J., Ban, N., Moore, P.B., and Steitz, T.A. (2000) The structural basis of ribosome activity in peptide bond synthesis. *Science* **289**, 920–30. The crystal structure of the large subunit of the ribosome, and the hypothesis that peptide-bond formation is catalyzed by the RNA itself.

Santoro, S.W. and Joyce, G.F. (1997) A general-purpose RNA-cleaving DNA enzyme. *Proceedings of the National Academy of Sciences, USA* **94**, 4262–6. Design of a general and efficient DNA enzyme that can cleave HIV-1 genomic RNA in the test tube.

Scott, W.G., Finch, J.T., and Klug, A. (1995) The crystal structure of an all-RNA hammerhead ribozyme – a proposed mechanism for RNA catalytic cleavage. *Cell* **81**, 991–1002.

Sullenger, B.A. and Gilboa, E. (2002) Emerging clinical applications of RNA. *Nature* **418**, 252–8. The use of ribozymes for clinical purposes such as treatment of HIV.

Takagi, Y., Suyama, E., Kawasaki, H., Miyagishi, M., and Taira, K. (2002) Group II introns and mRNA splicing. *Biochemical Society Transactions* **30**, 1145–9. The selection of variant ribozymes from randomized libraries.

si-RNA

Elbashir, S.M., Lendeckel, W., and Tuschl, T. (2001) RNA interference is mediated by 21- and 22-nucleotide RNAs. *Genes and Development* **15**, 188–200. How synthetic small RNAs activate the RNA interference process in the cell.

Gupta, B.P., Wang, M., and Sternberg, P.W. (2003) The *C. elegans* LIM homeobox gene lin-11 specifies multiple cell fates during vulval development. *Development* **130**, 2589–601. An example of how RNAi is used to study specific events in the development of the worm.

Hannon, G.J. (2002) RNA interference. *Nature* **418**, 244–51. An excellent summary of RNA-mediated gene silencing in plants, worms, flies and human cells, and its short-term therapeutic applications.

Hsu, A.L., Murphy, C.T., and Kenyon, C. (2003) Regulation of aging and age-related disease by DAF-16 and heat shock factor. *Science* **300**, 1142–5. Studies of specific genes which influence aging in a worm, often by use of si-RNA.

Miller, V.M., Xia, H., Marrs, G.L., Gouvion, C.M. *et al.* (2003) Allele-specific silencing of dominant disease genes. *Proceedings of the National Academy of Sciences, USA* **100**, 7195–200. Mutant protein which causes SCA-3 neural disease in humans may be destroyed, by using si-RNA targeted against messenger-RNA made from one chromosome but not the other.

Miyagishi, M., Hayashi, M., and Taira, K. (2003) Comparison of the suppressive effects of antisense oligonucleotides and si-RNAs directed against the same targets in mammalian cells. *Antisense and Nucleic Acid Drug Development* **13**, 1–7. Si-RNA works 100 times more efficiently per molecular copy than anti-sense, and far more than a ribozyme, at reducing gene expression in a model system.

Reinhart, B.J. and Bartel, D.P. (2002) Small RNAs correspond to centromere heterochromatic repeats. *Science* **297**, 1831. 'Silencing' of centromeric heterochromatin in yeast by small RNAs made by the centromeric repeat regions. Also describes possible links between methylation state and RNA interference, showing that the RNA interference mechanism can contribute to epigenetic effects as described in Chapter 11.

Tang, G., Reinhart, B.J., Bartel, D.P., and Zamore, P.D. (2003) A biochemical framework for RNA silencing in plants. *Genes and Development* **17**, 49–63. Si-RNA acts in plants as well as worms or mammals for gene silencing.

Tuschl, T. (2002) Expanding small RNA interference. *Nature Biotechnology* **20**, 446–8. Methods for expression of si-RNA from polymerase III promoters *in vivo*.

Yokota, T., Sakamoto, N., Enomoto, N., Tanabe, Y., Miyagishi, M., Maekawa, S., Yi, L., Kurosaki, M., Taira, K., Watanabe, M., and Mizusawa, H. (2003) Inhibition of intracellular hepatitis C virus replication by synthetic and vector-derived small interfering RNAs. *EMBO Reports* **4**, 602–8. Demonstrates the feasibility of using RNA interference to inhibit viral replication.

Answers to Selected Exercises

Note: most numerical values are given to more significant figures than are warranted by the data, in order to provide a better check on the arithmetic.

1.1 a Length of total DNA/diameter of cell = 200 000.
   b Volume of total DNA/volume of cell = 0.01.
   c Diameter of typical compact DNA ball = 6400 Å = 0.64 μm.
   Length of typical metaphase chromosome = 4.3 μm.

1.2 a Ala, Lys, Gln, Leu, Ile, Gln, Gly.
   b Pro, Ser, Asn, Ser, Phe, Lys.
   Gln, Ala, Thr, His, Ser, Arg.

1.3 a Ala, Lys, Gln, Arg, His, Ser, Arg.
   b Ala, Lys, Gln, Ser, Phe, Lys.

1.4 a Met, Ser, His, Gly, Thr, (Stop).
   b Met, Val, Ile, Arg, Asn, Ser, (Stop).

2.1 a 29°, 12.4 bp/turn.
   b 9.0 bp/turn.
   c 9.9 bp/turn.

2.2 a 330 Å, 10.0 turns.
   b 363 Å, 9.3 turns.
   c 495 Å, 6.4 turns.

2.5 a 2.
   b 21, 3.

3.1 Right-handed or clockwise.

3.6 'A': 4.7 Å, 22°, 2.5 Å
   'B': 0 Å, 0°, 3.3 Å
   'C': −2.2 Å, −9°, 3.1 Å.

4.2 b There are 3 AT-rich and 2 GC-rich regions.
   c There are 8 pyrimidine-purine steps.
   d GGCCC is the strongest region, and TATATA the weakest.

4.3 All bases except GCGC and CCGG at the two ends can make a cruciform about the central loop CTAG.

4.4 a 18, 10.8 m, 1.72 m.
  b 0.58 rad/m, 1.72 m.

4.5 a 30°, 0°, **b** 15°, −15°, **c** 9.3°, 0°, **d** 6°, −6°.

4.6 a 26.2°, **b** −26.2°, **c** 26.2°, **d** 30°, **e** 30°.

4.7 To adenine.

5.1 a −45°, 14 bp, −90 bp, 127 bp.
  b 45°, 14 bp, 90 bp, 127 bp.
  c 6°, 28 bp, 18 bp, 179 bp.
  d 84°, 3 bp, 178 bp, 179 bp.
Helix **d** has the smallest diameter, and helix **c** the largest. Helix **a** is left-handed, while helices **b**, **c**, **d** are right-handed.

5.2 a 'A': 30.7°, +11.2°
    'B': 36.0°, 0.0°
    'C': 39.4°, −7.0°.
  b 34.0°.

5.3 b first sum = 24.9°, second sum = 8.1°, k = 26.2°, phase = 72°, maximum roll R is at step 2.

5.4 a 0.86°/bp, step 6.
  b −1.7°/bp.
  c 45 Å, 189 bp.

5.5 a 1.07°/bp, step 7.
  b −1.5°/bp.
  c 60 Å, 195 bp.

5.6 a 1.05°/bp, step 7.5.
  b 1.7°/bp.
  c 50 Å, 180 bp.
A_4N_6 and A_6N_4 are left-handed, while A_6N_5 is right-handed.

6.1 a Once, approximately.
  b Once, approximately, depending on the kind of cord used.

6.4 a 2 turns, left-handed.
  b +1 turn, so that +1 −2 = −1. To reduce the supercoiling or tangling of DNA cell division.

6.5 Tw = 0, −0.50, −0.87, −1.00 turns
  Wr = −1.00, −0.50, −0.13, 0 turns.

7.1 a 214 ± 13, 429 ± 15, 644 ± 20, 855 ± 21.
  b 214.5 base-pairs.

7.2 a DNA occupies 53% of total volume.
  b 76 bp/hoop.

7.3 a a 291 Å. **b** 87 Å. **c** 100 Å.
b a 300 Å. **b** 100 Å. **c** 50 Å.

7.4 a No.
b No, they would join readily.

7.5 Nucleosome, 300 Å fiber, loops, metaphase coiling.

9.1 a Phosphorus, oxygen, nitrogen, carbon, hydrogen.
b The phosphate PO_4 at lower-center left.
c Because hydrogen scatters X-rays so weakly (36 times less strongly than carbon).

9.2 a 5. **b** 6. **c** 10.

9.3 a Four kinds of fragment, of size 100, 200, 250, 450 bp.
b 100 will be the fastest, and 450 the slowest.

9.4 a 125, **b** 7.5×10^6, i.e. 7.5 million.

9.5 For A_6N_4, pitch $p = 526$ Å, volume of circumscribing cylinder $= 8.01 \times 10^6$ Å^3, and ratio of volume for curved *versus* straight DNA $= 39.6$. For other sequences from Table 5.1, the volume ratios are $A_6N_2 = 1.4$, $A_6N_3 = 2.7$, $A_6N_5 = 29.5$, $A_6N_6 = 3.7$ and $A_6N_7 = 1.9$. Hence A_6N_4 and A_6N_5 go most slowly through a gel, while the speed of the other sequences is not much different from that of straight DNA.

9.6 Wr $= 0$ for slowest; and since Tw $= -12$, then Lk $= -12$ for slowest (check: this is within the given range). Fastest has the value of Wr which differs most from zero; and since Tw $= -12$, the range of Wr is from $+12$ to -8 for Lk $= 0$ to -20. Therefore Wr $= +12$, Lk $= 0$ is the fastest.

Index

(Page numbers in italics refer to illustrations and tables)